Position Sensorless Control Techniques for Permanent Magnet Synchronous Machine Drives

Gaolin Wang • Guoqiang Zhang • Dianguo Xu

Position Sensorless Control Techniques for Permanent Magnet Synchronous Machine Drives

Springer

Gaolin Wang
Harbin Institute of Technology
Harbin, Heilongjiang, China

Guoqiang Zhang
Harbin Institute of Technology
Harbin, Heilongjiang, China

Dianguo Xu
Harbin Institute of Technology
Harbin, Heilongjiang, China

ISBN 978-981-15-0049-7 ISBN 978-981-15-0050-3 (eBook)
https://doi.org/10.1007/978-981-15-0050-3

This Springer imprint is published by the registered company Springer Nature Singapore Pte Ltd.
The registered company address is: 152 Beach Road, #21-01/04 Gateway East, Singapore 189721, Singapore

Preface

Owing to the distinguished advantages such as high power density, fast dynamics and wide constant power speed range operations, permanent magnet synchronous machines (PMSM) have drawn increasing attentions from academia to industry applications. Since the merits such as cost reduction, system downsizing and also reliability enhancement can be achieved through position sensorless (also noted as self-sensing) control, many techniques for a wide range of rotor speeds from standstill to high speeds have been developed to estimate the rotor position in the last few decades.

The book focuses on position sensorless control for PMSM drives. Both basic principles and experimental evaluation have been addressed. This is achieved by providing in-depth study on a number of major topics such as model-based sensorless control, saliency-based sensorless control, position estimation error ripple elimination, acoustic noise reduction, etc. The comprehensive and systematic treatment of position sensorless control and practical issues are the major features of the book, which is particularly suited for readers who are interested to learn the sensorless control techniques for PMSM drives. The book can benefit researchers, engineers and graduate students in fields of AC motor drives, sensorless control, etc.

Harbin, China

Gaolin Wang
Guoqiang Zhang
Dianguo Xu

Contents

Chapter 1
Basic Structure and Mathematical Model

1.1 Application and Classification of PMSM

The motor could realize the energy conversion between the electrical type and the mechanical one, which is based on the interaction of the electricity and magnetism. According to the direction of the energy transformation, the motor could be divided into the electric generator and the electrical motor. The electric generator could transform the mechanical energy to the electrical one, and the electrical motor could transform the electrical energy to the mechanical one.

As for its convenient transformation, control, and application, the electric energy is the most widely applied energy in the modern society. The natural fossil energy (including coal, oil, and natural gas) and the renewable energy (bioenergy, solar energy, hydro energy, and wind energy) will be firstly transformed to the electric energy, which could be directly utilized by the industrial and economic society. More than 60% of the electric energy is consumed by the electrical motor, which plays an important role in the production practice. There are many types of electrical motors. According to the power supply, the electrical motor could be divided into the direct current (DC) motor and alternating current (AC) motor, and the AC motor consists of the synchronous motor and the induction motor (IM). According to the structure and the operation principle, the electrical motor could be divided as the IM, the electric excitation synchronous motor, the permanent magnet motor, and other specific motors. The electrical motor could also be classified by the volume, the phase number, the speed range, the control type, and the excitation type.

1.1.1 PMSM Applications

Generally, the electrical motor is divided into three types: the DC motor, the IM, and the synchronous motor. Among these types of motors, the permanent magnet

G. Wang et al., *Position Sensorless Control Techniques for Permanent Magnet Synchronous Machine Drives*, https://doi.org/10.1007/978-981-15-0050-3_1

synchronous motor (PMSM) has been widely developed in the industry and household applications for its high power density, high efficiency, high power factor, and excellent control performance. The PMSM has been applied in these areas [1]:

1. *The aeronautic and astronautic applications*. The requirements of the power density and the control performance are very strict in the aeronautic and astronautic applications. Hence the PMSM is a suitable choice for the aeronautic applications to reduce the launching cost of the whole system. Meanwhile, the high power density of the PMSM drive system could improve the performance of the astronautic applications.
2. *The electric vehicle (EV) applications.* The EV is the future traffic solution, which is supported by worldwide governments and companies. Due to the rapid development of the EV, the PMSM is widely applied as the power supply element due to the high power density and high efficiency.
3. *The industry applications.* For high-performance equipment, the excellent control performance of PMSM could satisfy the requirements, such as the computer numerical control (CNC), the robot drives, and the servo systems.
4. *The domestic appliance.* The high power density of the PMSM could reduce the power cost of the drive system effectively, which is an important concern in the household applications, such as the vacuum cleaner, the washing machine, the air-conditioner, the refrigerator, etc. Meanwhile, the direct drive traction system could reduce the volume and enhance the reliability, compared with the conventional traction drive system using the IM drive system and the gear box.
5. *The fan and pump applications.* The power consumption of the fan and pump applications could be effectively improved by applying the high-efficiency PMSM, which could reduce the whole power consumption and benefit the environment.

Hence the PMSM has a promising future in the energy conversion applications, which deserves to be studied further. However, there are also some disadvantages of PMSM influencing its application:

1. *The cost of the drive system.* The cost of the PMSM drive system is relatively higher than the IM drive system, which is suitable for the applications concerning for the high performance and the high efficiency, such as the aeronautics, astronautics, EV, CNC, and servo drive system applications. As for the general industry applications, the drive system cost is an important issue; hence it is necessary to optimize the design of the PMSM to reduce the cost.
2. *Irreversible demagnetization.* The PMSM will occur irreversible demagnetization when the motor operates in excessive high temperature and low temperature. If the PMSM suffers inrush current or severe mechanical vibration, the irreversible demagnetization could also be caused, which could reduce the performance of the motor or even disable the normal operation.

1.1.2 PMSM Classification

Despite these disadvantages, the PMSM is so far the solution of the future power energy conversion between the electric power and the mechanical power. The development of the PMSM focuses on the high performance, high power density, high speed, and high-level integration. There are many classification methods for the PMSM. According to the position of the armature winding, it can be divided into the inner rotor type and the outer rotor type. According to the direction of the working magnetic field, it can be divided into the radial magnetic field type and the axial magnetic field type. According to the current waveform flowing through the armature winding, it can be divided into the permanent magnet brushless DC motor (PMBLDCM) powered by square wave or trapezoidal wave and the PMSM powered by sine wave. According to the presence or absence of the starting winding on the rotor, it can be divided into the PMSM without the starting winding and the PMSM with the starting winding. The motor without the starting winding is powered by the inverter which starts and adjusts the speed along with the increase of the frequency. The motor with starting winding can not only be operated at a variable speed but also start at a certain frequency and voltage using the asynchronous torque generated by the starting winding [2, 3].

The most important difference between PMSM and other motors is the rotor magnetic circuit structure, which will be analyzed and discussed in detail. When the magnetic circuit is different, the performance, the control system, the manufacturing process, and the applicable occasions of the motor are also different. According to the installation position of the permanent magnet on the rotor, PMSM can be divided into two types: surface permanent magnet synchronous motor (SPMSM) and interior permanent magnet synchronous motor (IPMSM). The permanent magnet of SPMSM is located on the surface of the rotor core, whereas the permanent magnet of IPMSM is located inside the rotor. For each rotor structure, a sinusoidal distribution of air gap magnetic field should be produced as much as possible.

The rotor magnetic circuit structure of SPMSM can also be divided into the raised type and the plug-in type. For the application of the rare earth in the PMSM, the relative reversal permeability of permanent magnet materials is close to 1. Therefore, the surface raised rotor belongs to hidden pole rotor structure in terms of the electromagnetic performance, which leads to the symmetric orthogonal magnetic path, and the impedances of the d- and q-axes are the same ($X_d = X_q$). However, there is the ferromagnetic material between the two adjacent permanent magnetic poles of the surface plug-in rotor, so it belongs to the salient pole rotor structure in terms of electromagnetic performance. Hence the orthogonal magnetic path magnetic circuit is asymmetrical, and $X_d > X_q$.

Figure 1.1a shows the surface raised rotor magnetic circuit structure. As for the simple structure, the lower manufacturing cost, and the smaller moment of inertia, this kind of structure is widely used in PMBDCM and PMSM, which operate in a narrow constant power range. In addition, the permanent magnet poles in the surface

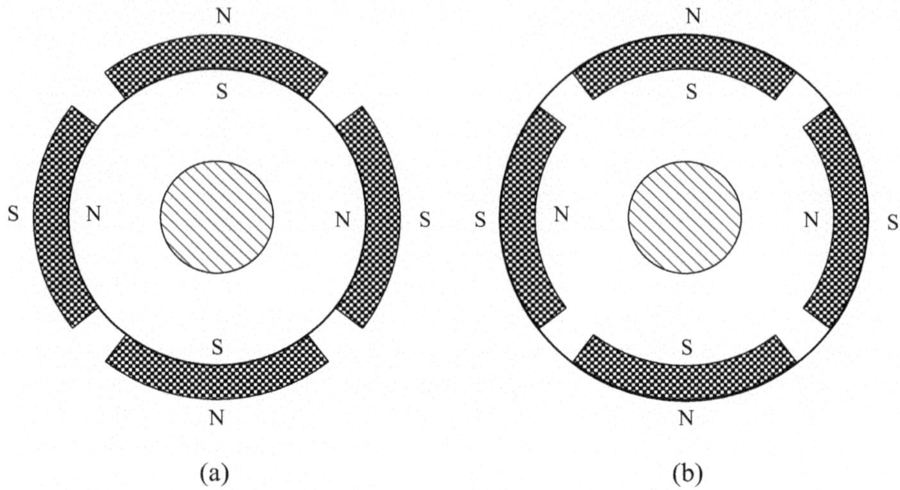

Fig. 1.1 Surface rotor magnetic circuit structure. (**a**) Raised type; (**b**) Plug-in type

raised rotor structure are easy to achieve the optimal design, which could make the air gap magnetic field close to the sinusoidal distribution. Hence the surface raised rotor magnetic circuit structure can significantly improve the performance of the motor and the entire drive system.

Figure 1.1b shows the surface plug-in rotor magnetic circuit structure. This kind of structure can make full use of the reluctance torque generated by the asymmetry of the rotor magnetic circuit. Therefore, the power density and the dynamic performance of the motor can be improved compared with the motor with surface plug-in rotor structure. In addition, its manufacturing process is relatively simple. Hence it is often used in some variable speed PMSMs. However, there are the larger magnetic flux leakage and the higher manufacturing costs of the motors with this kind of structure compared with the surface raised rotor structure.

As for the simpler manufacturing process and the lower cost of the rotor magnetic circuit structure, it is especially suitable to be applied in PMBDCM. However, as the starting winding cannot be installed on the rotor surface, there is no asynchronous starting capability, and it is only suitable for occasions when the starting requirements are not strict.

Figure 1.2 shows the interior rotor magnetic circuit structure. Permanent magnets in the interior rotor are protected by pole pieces. The reluctance torque generated by the asymmetry of the rotor magnetic circuit structure can improve the overload capacity and the power density of the motor. Meanwhile, the d-axis inductance of IPMSM is usually larger than that of SPMSM, so it is easy to weaken the magnetic field and expand the operation speed. It should be noted that the X_d, X_q, and X_d/X_q (salient ratio) of IPMSM are different when the rotor magnetic circuit structure is different. Larger salient ratio could improve the pull-in synchronization ability, the reluctance torque, and the overload capability of motor.

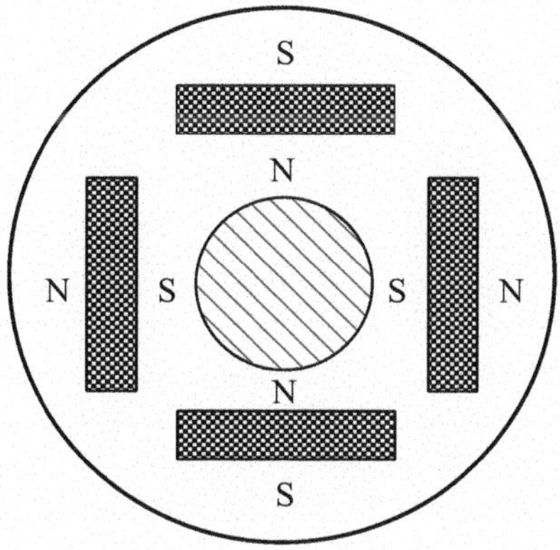

Fig. 1.2 Interior rotor magnetic circuit structure

1.2 Coordinate Transformation

1.2.1 Coordinate Representation of the Vector Space

Let $\boldsymbol{\alpha}_1$, $\boldsymbol{\alpha}_2$, ..., $\boldsymbol{\alpha}_n$ be the linear independent vectors of the vector space **F** [4, 5], where

$$\boldsymbol{\alpha}_1 = \begin{bmatrix} \alpha_{11} \\ \alpha_{12} \\ \vdots \\ \alpha_{1n} \end{bmatrix}, \boldsymbol{\alpha}_2 = \begin{bmatrix} \alpha_{21} \\ \alpha_{22} \\ \vdots \\ \alpha_{2n} \end{bmatrix}, \ldots, \boldsymbol{\alpha}_n = \begin{bmatrix} \alpha_{n1} \\ \alpha_{n2} \\ \vdots \\ \alpha_{nn} \end{bmatrix}. \tag{1.1}$$

Then any vector **x** can be represented as

$$\mathbf{x} = x_1\boldsymbol{\alpha}_1 + x_2\boldsymbol{\alpha}_2 + \ldots + x_n\boldsymbol{\alpha}_n. \tag{1.2}$$

The vectors $\boldsymbol{\alpha}_1, \boldsymbol{\alpha}_2, \ldots, \boldsymbol{\alpha}_n$ are a basis of **F** and $(x_1, x_2, \ldots, x_n)^{\mathrm{T}}$ are the coordinate of **x**, where **F** is called as the n-dimensional vector space. It can be seen that there are many different bases of **F**, whereas the coordinate representation of **x** is unique in a particular basis. Eq. (1.2) could be presented as

$$\mathbf{x} = (\boldsymbol{\alpha}_1, \boldsymbol{\alpha}_2, \ldots \boldsymbol{\alpha}_n) \begin{bmatrix} x_1 \\ x_2 \\ \vdots \\ x_n \end{bmatrix}. \tag{1.3}$$

Equation (1.3) is the general expression in the three-phase PMSM drive system, which could be generalized as

$$\begin{bmatrix} u_a \\ u_b \\ u_c \end{bmatrix} = \mathbf{R} \begin{bmatrix} i_a \\ i_b \\ i_c \end{bmatrix} + \mathbf{L}\frac{d}{dt} \begin{bmatrix} i_a \\ i_b \\ i_c \end{bmatrix} + \psi_f \mathbf{f}(\theta_e), \tag{1.4}$$

where u_a, u_b, u_c, i_a, i_b, i_c, $\mathbf{R}$, $\mathbf{L}$, ψ_f, and θ_e are the three-phase voltage vector, the three-phase current vector, the resistor matrix, the inductance matrix, the flux of the permanent magnet, and the rotor position, respectively. As for the three-phase SPMSM, the inductance matrix mainly contains the self-inductance and the mutual inductance of the three-phase windings. As the air gap of the SPMSM is uniform, hence the self-inductance and the mutual inductance of the three-phase windings are constant values and independent with the rotor position. As for the IPMSM, the air gap of the IPMSM is asymmetric; hence the self-inductance and the mutual inductance of the three-phase windings are closely related to the rotor position. The inductance matrices of both the SPMSM and the IPMSM are coupling and dependent with the rotor position.

1.2.2 Basis Transformation and Coordinate Transformation

Let vectors $\boldsymbol{\alpha}_1, \boldsymbol{\alpha}_2, \ldots, \boldsymbol{\alpha}_n$ and $\boldsymbol{\beta}_1, \boldsymbol{\beta}_2, \ldots, \boldsymbol{\beta}_n$ be two bases of the vector space. Any vector $\mathbf{x}$ could be presented as

$$\mathbf{x} = (\boldsymbol{\alpha}_1, \boldsymbol{\alpha}_2, \ldots \boldsymbol{\alpha}_n) \begin{bmatrix} x_{\alpha 1} \\ x_{\alpha 2} \\ \vdots \\ x_{\alpha n} \end{bmatrix} = (\boldsymbol{\beta}_1, \boldsymbol{\beta}_2, \ldots \boldsymbol{\beta}_n) \begin{bmatrix} x_{\beta 1} \\ x_{\beta 2} \\ \vdots \\ x_{\beta n} \end{bmatrix}, \tag{1.5}$$

where $\mathbf{x}_\alpha = (x_{\alpha 1}, x_{\alpha 2}, \ldots, x_{\alpha n})^{\mathrm{T}}$ and $\mathbf{x}_\beta = (x_{\beta 1}, x_{\beta 2}, \ldots, x_{\beta n})^{\mathrm{T}}$ are the coordinate representation at the two different bases. The relationship between the two bases could be presented as

$$(\boldsymbol{\beta}_1, \boldsymbol{\beta}_2, \ldots \boldsymbol{\beta}_n) = (\boldsymbol{\alpha}_1, \boldsymbol{\alpha}_2, \ldots \boldsymbol{\alpha}_n) \begin{bmatrix} p_{11} & p_{12} & \cdots & p_{1n} \\ p_{21} & p_{22} & \cdots & p_{2n} \\ \vdots & \vdots & & \vdots \\ p_{n1} & p_{n2} & \cdots & p_{nn} \end{bmatrix}. \tag{1.6}$$

Hence the $n \times n$ nonsingular matrix transition matrix **P** could be presented as

$$P = \begin{bmatrix} p_{11} & p_{12} & \cdots & p_{1n} \\ p_{21} & p_{22} & \cdots & p_{2n} \\ \vdots & \vdots & & \vdots \\ p_{n1} & p_{n2} & \cdots & p_{nn} \end{bmatrix}. \tag{1.7}$$

P could be applied to realize the transition between two bases $\boldsymbol{\alpha}_1, \boldsymbol{\alpha}_2, \ldots, \boldsymbol{\alpha}_n$ and $\boldsymbol{\beta}_1, \boldsymbol{\beta}_2, \ldots, \boldsymbol{\beta}_n$.

$$\begin{bmatrix} x_{\alpha 1} \\ x_{\alpha 2} \\ \vdots \\ x_{\alpha n} \end{bmatrix} = \begin{bmatrix} a_{11} & a_{12} & \cdots & a_{1n} \\ a_{21} & a_{22} & \cdots & a_{2n} \\ \vdots & \vdots & & \vdots \\ a_{n1} & a_{n2} & \cdots & a_{nn} \end{bmatrix} \begin{bmatrix} x_{\beta 1} \\ x_{\beta 2} \\ \vdots \\ x_{\beta n} \end{bmatrix} = \mathbf{P} \begin{bmatrix} x_{\beta 1} \\ x_{\beta 2} \\ \vdots \\ x_{\beta n} \end{bmatrix}. \tag{1.8}$$

1.2.3 Jordan Canonical Form Theorem and Diagonalization

The couple effect in the motor drive system mainly exists in the inductance matrix, which needs to be specially considered in the motor control strategy. Hence the decoupling control method is applied to eliminate the effect of the mutual inductance on the control performance. As for the existence of the mutual inductance, the inductance matrix is similar to the circulate matrix, which could be presented as

$$A = \begin{bmatrix} a_1 & a_2 & a_3 & \cdots & a_n \\ a_n & a_1 & a_2 & \cdots & a_{n-1} \\ a_{n-1} & a_n & a_1 & \cdots & a_{n-2} \\ \vdots & \vdots & \ddots & \ddots & \vdots \\ a_2 & a_3 & \cdots & a_n & a_1 \end{bmatrix}. \tag{1.9}$$

The matrix **A** is defined as the circulate matrix, and the previous row forward one step is the next row. Hence the decoupling method for **A** is essential to be concerned for the motor control strategy.

The diagonalization is an important tool to realize the decoupling control, which is closely related to the Jordan canonical form. The Jordan canonical form is formed of the Jordan block, and it is an upper triangular matrix that could be presented as

$$J_k(\lambda) = \begin{bmatrix} \lambda & 1 & & & \\ & \lambda & 1 & & \\ & & \ddots & \ddots & \\ & & & \lambda & 1 \\ & & & & \lambda \end{bmatrix}, \tag{1.10}$$

where λ is the unique Jordan block of a complex matrix. It could be transformed by the special matrix, which consists of the corresponding eigenvectors. There are k times of λ in the main diagonal and all other elements are zero. The Jordan matrix for consisting of the Jordan blocks could be presented as

$$J(\lambda) = \begin{bmatrix} J_{n1}(\lambda) & & & \\ & J_{n2}(\lambda) & & \\ & & \ddots & \\ & & & J_{nq}(\lambda) \end{bmatrix}. \tag{1.11}$$

The Jordan matrix $J(\lambda)$ plays an important role in the decoupling control strategy. The dynamic system could be totally decoupled by the transformation when the dimensional k of all the Jordan blocks of the coupled matrix is one. For the motor drive system, it is essential to study the diagonalization of the inductance matrix to realize the decoupling control.

The diagonalization transformation of a matrix is closely related to its eigenvalues and eigenvectors. The eigenvalue λ of the matrix $\mathbf{A}$ could be presented as

$$\mathbf{Ax} = \lambda\mathbf{x}, \mathbf{x} \in \mathbf{C}^n, \mathbf{x} \neq \mathbf{0}, \lambda \in \mathbf{C}, \tag{1.12}$$

where $\mathbf{x}$ is the eigenvector corresponding to eigenvalue λ. If the matrix $\mathbf{A}$ is regarded as a linear transformation, the effect of this transformation on the eigenvector $\mathbf{x}$ is to expand (or shorten) λ times compared with the original vector.

The transformation matrix $\mathbf{P}$ consists of all the eigenvectors, which could be presented as

$$\mathbf{P} = [\mathbf{P}_1 \;\; \mathbf{P}_2 \;\; \ldots \;\; \mathbf{P}_s]. \tag{1.13}$$

Hence the relationship between the matrix **A** and the transformation matrix **P** could be presented as

$$\mathbf{AP} = \mathbf{PJ}(\lambda) = \mathbf{P}\begin{bmatrix} \mathbf{J}_1 & & & \\ & \mathbf{J}_2 & & \\ & & \ddots & \\ & & & \mathbf{J}_s \end{bmatrix}, \tag{1.14}$$

where $\boldsymbol{J}(\lambda)$ is the Jordan matrix of matrix **A**. According to the division of the Jordan blocks, the transformation between the matrix **A** and the Jordan matrix $\boldsymbol{J}(\lambda)$ could be presented as

$$\begin{aligned} \mathbf{A}[\mathbf{P}_1\mathbf{P}_2\ldots\mathbf{P}_s] &= [\mathbf{P}_1\mathbf{P}_2\ldots\mathbf{P}_s]\mathbf{J}(\lambda) \\ &= [\mathbf{P}_1\mathbf{P}_2\ldots\mathbf{P}_s]\begin{bmatrix} \mathbf{J}_1 & & & \\ & \mathbf{J}_2 & & \\ & & \ddots & \\ & & & \mathbf{J}_s \end{bmatrix}, \end{aligned} \tag{1.15}$$

where

$$[\mathbf{AP}_1\mathbf{AP}_2\ldots\mathbf{AP}_s] = [\mathbf{P}_1\mathbf{J}_1\mathbf{P}_2\mathbf{J}_2\ldots\mathbf{P}_s\mathbf{J}_s]. \tag{1.16}$$

The relationship between every Jordan block and the matrix **A** could be presented as

$$\mathbf{AP}_i = \mathbf{P}_i\mathbf{J}_i(i = 1, 2, \ldots, s). \tag{1.17}$$

$\mathbf{P}_i$ could be presented as

$$\mathbf{P}_i = (\mathbf{X}_{i1}, \mathbf{X}_{i2}, \ldots, \mathbf{X}_{in_i}), \tag{1.18}$$

where $\mathbf{X}_{i1}$, $\mathbf{X}_{i2}$,…, $\mathbf{X}_{in_i}$ are the independent eigenvectors of the matrix **A**. Substituting (1.18) into (1.17),

$$\begin{cases} \mathrm{AX}_{i1} = \lambda_i\mathrm{X}_{i1} \\ \mathrm{AX}_{i2} = \mathrm{X}_{i1} + \lambda_i\mathrm{X}_{i1} \\ \vdots \\ \mathrm{AX}_{in_i} = \mathrm{X}_{in_i-1} + \lambda_i\mathrm{X}_{in_i}. \end{cases} \tag{1.19}$$

It can be seen that $\mathbf{X}_{i1}$ is the eigenvector of the matrix **A** corresponding to eigenvalue λ_i. All other eigenvectors $\mathbf{X}_{i1}$, $\mathbf{X}_{i1}$, …, $\mathbf{X}_{in_i}$ could be obtained by

(1.12). Hence all the column vectors of $\mathbf{P}$ could be calculated. Every Jordan block $\mathbf{J}_i$ is corresponded to the eigenvalue λ_i. As for the eigenvalue λ_i, the number of the Jordan block $\mathbf{J}_i$ is equal to its geometric multiplicity. The order of all the corresponding Jordan block $\mathbf{J}_i$ for the eigenvalue λ_i is equal to its algebraic multiplicity.

1.2.4 Coordinate Frame in Motor Drives

The three-phase currents i_a, i_b, and i_c of the three-phase PMSM could be presented as

$$\begin{cases} i_a = I_m \cos \omega_0 t \\ i_b = I_m \cos\left(\omega_0 t - \frac{2\pi}{3}\right) \\ i_c = I_m \cos\left(\omega_0 t - \frac{2\pi}{3}\right), \end{cases} \tag{1.20}$$

where I_m and ω_0 are the magnitude and the angular speed of the motor current, respectively. The symmetric three-phase currents could be presented in the space three-dimensional coordinate frame:

$$\begin{aligned} \mathbf{I} &= i_a\mathbf{i}_A + i_b\mathbf{i}_B + i_c\mathbf{i}_C \\ &= I_m \cos \omega_0 t\mathbf{i}_A + I_m \cos\left(\omega_0 t - \frac{2\pi}{3}\right)\mathbf{i}_B + I_m \cos\left(\omega_0 t - \frac{2\pi}{3}\right)\mathbf{i}_C, \end{aligned} \tag{1.21}$$

where $\boldsymbol{i}_A$, $\boldsymbol{i}_B$, and $\boldsymbol{i}_C$ are the unit vector of the three-dimensional orthogonal frame, respectively. The magnitude of the vector $\mathbf{I}$ could be presented as

$$|\mathbf{I}| = \sqrt{i_a^2 + i_b^2 + i_c^2} = I_m\sqrt{\cos^2\omega_0 t + \cos^2\omega_0 t + \cos^2\omega_0 t} = \sqrt{\frac{3}{2}}I_m. \tag{1.22}$$

It can be seen that the vector $\mathbf{I}$ rotates around the original point of the three-dimensional orthogonal frame. The magnitude and the angular speed of $\mathbf{I}$ are $\sqrt{3/2}I_m$ and ω_0, respectively. The trajectory of the rotation is a circle and the projection of the trajectory in the frame are the three-phase current, which is shown in Fig. 1.3.

The trajectory is located in a plane in the three-dimensional coordinate frame, which is defined as the *d*-*q* plane. All the transformation of the coordinate frame is carried out in this plane.

As analyzed in the previous section, the stationary *a*-*b*-*c* frame in the *d*-*q* plane is the projection of the three three-dimensional orthogonal coordinate frames, which is shown in Fig. 1.4.

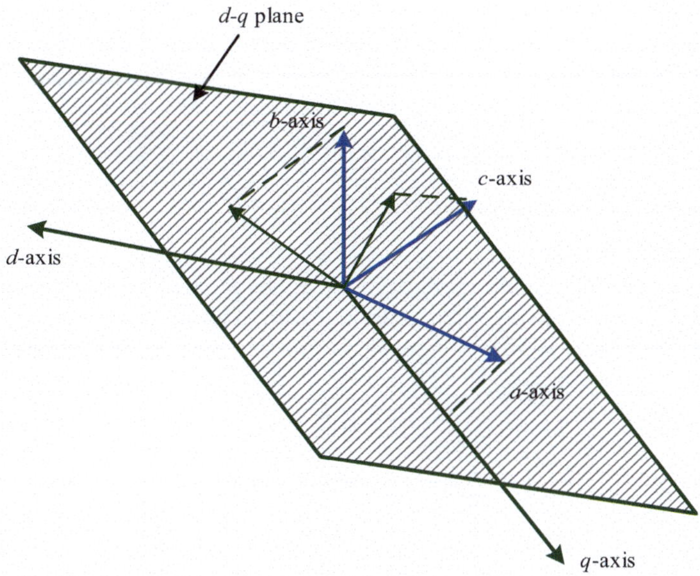

Fig. 1.3 The current vector **I** in the three-dimensional coordinate frame

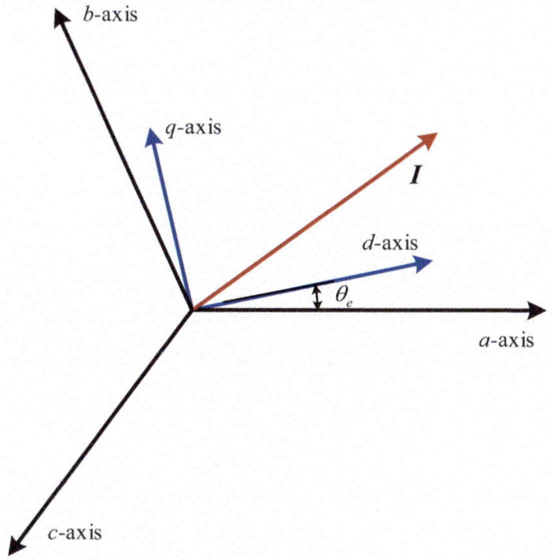

Fig. 1.4 The current vector **I** in the d-q frame

It can be seen that the α-β orthogonal coordinate frame is defined in Fig. 1.4. The coordinate transformation is to transform the vector in the *a-b-c* frame to the *d-q* orthogonal coordinate frame. According to the previous analysis, the vector **I** rotates around the original point with the angular speed ω_0.

The coordinate transformation could be derived from the geometric view. The relationship between the α-β orthogonal coordinate frame and the a-b-c coordinate frame could be presented as

$$\begin{aligned} N_2 i_\alpha &= N_1 i_a + N_1 i_b \cos\left(\frac{2\pi}{3}\right) + N_1 i_c \cos\left(\frac{4\pi}{3}\right), \\ N_2 i_\beta &= N_1 i_b \sin\left(\frac{2\pi}{3}\right) + N_1 i_c \sin\left(\frac{4\pi}{3}\right), \end{aligned} \tag{1.23}$$

where i_α, i_β, N_1, and N_2 are the currents in the α- and β-axes and the number of turns in the windings, respectively. Hence the transformation matrix between the a-b-c coordinate frame and the α-β orthogonal coordinate frame could be presented as

$$\begin{bmatrix} i_\alpha \\ i_\beta \\ i_0 \end{bmatrix} = \frac{N_1}{N_2} \begin{bmatrix} 1 & -\frac{1}{2} & -\frac{1}{2} \\ 0 & \frac{\sqrt{3}}{2} & -\frac{\sqrt{3}}{2} \\ \frac{1}{\sqrt{2}} & \frac{1}{\sqrt{2}} & \frac{1}{\sqrt{2}} \end{bmatrix} \begin{bmatrix} i_a \\ i_b \\ i_c \end{bmatrix}, \tag{1.24}$$

where i_0 is the zero-sequence component of the three-phase current. As for the three-phase star connection PMSM, the sum of the three-phase currents is zero and the corresponding zero-sequence component i_0 is also zero. Hence i_0 will be ignored in this section to make the derivation process clearer. Hence the simplified Clarke transformation could be presented as

$$\begin{bmatrix} i_\alpha \\ i_\beta \end{bmatrix} = \frac{N_1}{N_2} \begin{bmatrix} 1 & -\frac{1}{2} & -\frac{1}{2} \\ 0 & \frac{\sqrt{3}}{2} & -\frac{\sqrt{3}}{2} \end{bmatrix} \begin{bmatrix} i_a \\ i_b \\ i_c \end{bmatrix}. \tag{1.25}$$

The Clarke transformation could also be evaluated from the matrix transformation. The inductance matrix in the SPMSM could be presented as

$$L_{SPM} = \begin{bmatrix} L_{s\sigma} + L_m & -\frac{1}{2} L_m & -\frac{1}{2} L_m \\ -\frac{1}{2} L_m & L_{s\sigma} + L_m & -\frac{1}{2} L_m \\ -\frac{1}{2} L_m & -\frac{1}{2} L_m & L_{s\sigma} + L_m \end{bmatrix}, \tag{1.26}$$

where $L_{s\sigma}$ and L_m are the leakage inductance and the excitation inductance of the three-phase windings, respectively. The coordinate transformation aims to realize the decoupling control of the coupled inductance matrix. According to the basis of the diagonalization transformation, the transformation matrix of the eigenvectors could be applied to realize the Jordan canonical form transformation. The eigenvalues of the inductance matrix are the single root λ_1 and the double root $\lambda_{2,3}$, which could be presented as

$$\begin{aligned} \lambda_1 &= L_{s\sigma}, \\ \lambda_{2,3} &= L_{s\sigma} + \frac{3}{2}L_m. \end{aligned} \tag{1.27}$$

Hence the eigenvectors can be presented as

$$\mathbf{X}_1 = \begin{bmatrix} x_{11} \\ x_{11} \\ x_{11} \end{bmatrix}, \mathbf{X}_2 = \begin{bmatrix} x_{12} \\ x_{22} \\ -x_{12} - x_{22} \end{bmatrix}, \mathbf{X}_3 = \begin{bmatrix} x_{13} \\ x_{23} \\ -x_{13} - x_{23} \end{bmatrix}. \tag{1.28}$$

The transformation matrix can be presented as

$$\mathbf{P} = \begin{bmatrix} x_{12} & x_{13} & x_{11} \\ x_{22} & x_{23} & x_{11} \\ -x_{12} - x_{22} & -x_{13} - x_{23} & x_{11} \end{bmatrix}. \tag{1.29}$$

The coordinate transformation between the *a*-*b*-*c* coordinate frame and the α-β orthogonal coordinate frame could be presented as

$$\begin{bmatrix} i_a \\ i_b \\ i_c \end{bmatrix} = \begin{bmatrix} x_{12} & x_{13} & x_{11} \\ x_{22} & x_{23} & x_{11} \\ -x_{12} - x_{22} & -x_{13} - x_{23} & x_{11} \end{bmatrix} \begin{bmatrix} i_\alpha \\ i_\beta \\ i_0 \end{bmatrix}. \tag{1.30}$$

The Jordan matrix for the inductance matrix of the three-phase SPMSM could be presented as

$$J(\lambda) = \begin{bmatrix} L_{s\sigma} + \frac{3}{2}L_m & & \\ & L_{s\sigma} + \frac{3}{2}L_m & \\ & & L_{s\sigma} + \frac{3}{2}L_m \end{bmatrix}. \tag{1.31}$$

Let the transformation matrix **P** be a unit orthogonal matrix; the transformation matrix **P** could be presented as

$$\mathbf{P}=\begin{bmatrix}\frac{\sqrt{6}}{3} & 0 & \frac{1}{\sqrt{3}}\\ -\frac{\sqrt{6}}{6} & \frac{\sqrt{2}}{2} & \frac{1}{\sqrt{3}}\\ -\frac{\sqrt{6}}{6} & -\frac{\sqrt{2}}{2} & \frac{1}{\sqrt{3}}\end{bmatrix}. \tag{1.32}$$

The equivalent transformation could be presented as

$$\begin{bmatrix} i_\alpha \\ i_\beta \\ i_0 \end{bmatrix} = \mathrm{P}^{-1}\begin{bmatrix} i_a \\ i_b \\ i_c \end{bmatrix} = \sqrt{\frac{2}{3}}\begin{bmatrix} 1 & -\frac{1}{2} & -\frac{1}{2} \\ 0 & \frac{\sqrt{3}}{2} & -\frac{\sqrt{3}}{2} \\ \frac{1}{\sqrt{2}} & \frac{1}{\sqrt{2}} & \frac{1}{\sqrt{2}} \end{bmatrix}\begin{bmatrix} i_a \\ i_b \\ i_c \end{bmatrix}. \tag{1.33}$$

Actually, the coordinate transformation with the orthogonal matrix is the power invariance method, which means that the drive system power in the *a-b-c* coordinate frame and the *α-β* coordinate frame remains constant. As for another phase magnitude invariance method, the magnitude of the voltage and current in the *a-b-c* coordinate frame and the *α-β* coordinate frame remains constant, whereas the torque and the power should be multiplied by 3/2 in the *α-β* frame to obtain the same values in the *a-b-c* coordinate frame. The transformation matrix **P** in the *α-β* frame could be presented as

$$\begin{bmatrix} i_\alpha \\ i_\beta \\ i_0 \end{bmatrix} = \frac{2}{3}\begin{bmatrix} 1 & -\frac{1}{2} & -\frac{1}{2} \\ 0 & \frac{\sqrt{3}}{2} & -\frac{\sqrt{3}}{2} \\ \frac{1}{\sqrt{2}} & \frac{1}{\sqrt{2}} & \frac{1}{\sqrt{2}} \end{bmatrix}\begin{bmatrix} i_a \\ i_b \\ i_c \end{bmatrix}. \tag{1.34}$$

The *α-β* coordinate frame is the stationary frame, and the elements of the inductance matrix are closely related to the rotor angle; hence it is essential to eliminate the change caused by the different value of the rotor angle. The Park transformation is applied to transform the variables in the stationary *α-β* coordinate frame to the rotating *d-q* coordinate frame. The transformation could be presented as

$$\begin{bmatrix} i_d \\ i_q \end{bmatrix} = \begin{bmatrix} \cos\theta_e & -\sin\theta_e \\ \sin\theta_e & \cos\theta_e \end{bmatrix} \begin{bmatrix} i_\alpha \\ i_\beta \end{bmatrix}$$
$$\begin{bmatrix} i_\alpha \\ i_\beta \end{bmatrix} = \begin{bmatrix} \cos\theta_e & \sin\theta_e \\ -\sin\theta_e & \cos\theta_e \end{bmatrix} \begin{bmatrix} i_d \\ i_q \end{bmatrix}. \quad (1.35)$$

1.3 Mathematical Model of Three-Phase SPMSM

The mathematical modeling of PMSM is the key to realize the vector control. To simplify the analysis, it is necessary to make the following assumptions: (1) The three-phase stator winding is distributed symmetrically in space and the discrete properties of its structure are ignored. Therefore, the armature reactive magnetomotive force generated by stator windings in the air gap is sinusoidal, and the induced inverse electromotive force (EMF) is also the sinusoidal wave. (2) The internal permeability of a permanent magnet is consistent with that of air, and the excitation EMF produced in the air gap is also a sinusoidal distribution. (3) The iron loss, the terminal effect, and the magnetic saturation effect are ignored. And the constant rotor permeability is infinite. (4) Regardless of the influence of temperature and load effect on the motor parameters, there is no damper winding on the rotor.

Figure 1.5 shows the surface PMSM physical model. The voltage equation of its three-phase winding can be presented as

$$\begin{bmatrix} u_a \\ u_b \\ u_c \end{bmatrix} = \begin{bmatrix} R_s & 0 & 0 \\ 0 & R_s & 0 \\ 0 & 0 & R_s \end{bmatrix} \begin{bmatrix} i_a \\ i_b \\ i_c \end{bmatrix} + p \begin{bmatrix} \psi_a \\ \psi_b \\ \psi_c \end{bmatrix}, \quad (1.36)$$

where ψ_a, ψ_b, and ψ_c are flux leakage of three-phase *a-b-c*, R_s is the stator winding resistance, and p is differential operator d/dt, respectively.

The three-phase flux leakage of the SPMSM could be presented as

$$\begin{bmatrix} \psi_a \\ \psi_b \\ \psi_c \end{bmatrix} = L_{abc} \begin{bmatrix} i_a \\ i_b \\ i_c \end{bmatrix} + \psi_f \begin{bmatrix} \cos(\theta_e) \\ \cos\left(\theta_e - \frac{2\pi}{3}\right) \\ \cos\left(\theta_e + \frac{2\pi}{3}\right) \end{bmatrix}$$
$$= \begin{bmatrix} L_a & M_{ab} & M_{ac} \\ M_{ba} & L_b & M_{bc} \\ M_{ca} & M_{cb} & L_c \end{bmatrix} \begin{bmatrix} i_a \\ i_b \\ i_c \end{bmatrix} + \begin{bmatrix} \psi_{fa} \\ \psi_{fb} \\ \psi_{fc} \end{bmatrix}, \quad (1.37)$$

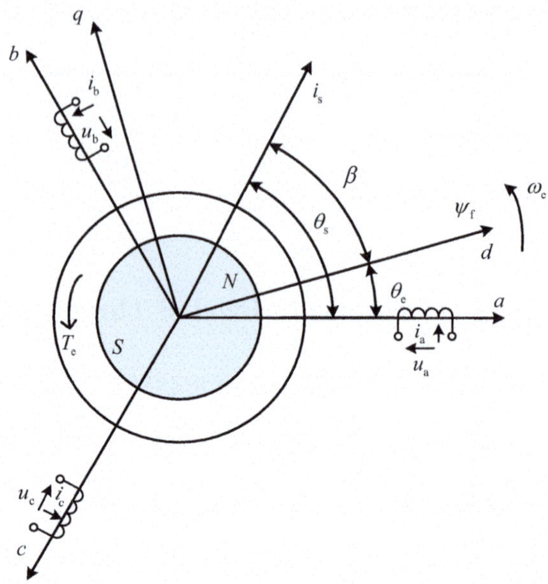

Fig. 1.5 Physical model of SPMSM

where ψ_{fa}, ψ_{fb}, and ψ_{fc} are the flux of the permanent magnet passing through *a-b-c* winding and ψ_f is the flux of the permanent magnet, respectively.

The SPMSM is a three-phase hidden pole synchronous motor, and the air gap is uniform. The self-inductance and the mutual inductances of the *a-b-c* winding are constant, which are independent of the rotor position. So the self-inductance and mutual inductance of the three-phase windings in SPMSM could be presented as

$$L_a = L_b = L_c = L_{s\sigma} + L_m, \tag{1.38}$$

$$M_{ab} = M_{ba} = M_{ac} = M_{ca} = M_{bc} = M_{cb} = L_m \cos 120^\circ = -\frac{1}{2}L_m. \tag{1.39}$$

Considering $i_a + i_b + i_c = 0$, (1.37) could be presented as

$$\psi_a = \left(L_{s\sigma} + \frac{3}{2}L_{m1}\right)i_a + \psi_{fa} = (L_{s\sigma} + L_m)i_a + \psi_{fa} = L_s i_a + \psi_{fa}, \tag{1.40}$$

where L_m is equivalent excitation inductance and L_s is synchronous inductance [1, 6]. ψ_b and ψ_c can be similarly expressed in the form of (1.40). Therefore (1.37) could be presented as

$$\begin{bmatrix} \psi_a \\ \psi_b \\ \psi_c \end{bmatrix} = (L_{s\sigma} + L_m)\begin{bmatrix} i_a \\ i_b \\ i_c \end{bmatrix} + \begin{bmatrix} \psi_{fa} \\ \psi_{fb} \\ \psi_{fc} \end{bmatrix}. \tag{1.41}$$

The stator current vector $\boldsymbol{i}_s$ composed of the current i_a, i_b, and i_c in the three-phase winding. The stator flux vector $\boldsymbol{\psi}_s$ can be presented by the full flux of the three-phase winding, and the rotor flux vector $\boldsymbol{\psi}_f$ can be presented by ψ_{fa}, ψ_{fb}, and ψ_{fc}. According to (1.40), the stator flux vector $\boldsymbol{\psi}_s$ can be presented as

$$\boldsymbol{\psi}_s = L_{s\sigma}\mathbf{i}_s + L_m\mathbf{i}_s + \psi_f = L_s\mathbf{i}_s + \psi_f, \tag{1.42}$$

where $L_{s\sigma}\boldsymbol{i}_s$ is the flux leakage produced by $\boldsymbol{i}_s$ corresponding to leakage magnetic field of the stator winding, $L_m\boldsymbol{i}_s$ is produced by $\boldsymbol{i}_s$ corresponding to armature field, and $\boldsymbol{\psi}_f$ is produced by the permanent magnet.

Hence (1.36) could be transformed into the vector form, which could be presented as

$$\mathbf{u}_s = R_s\mathbf{i}_s + \frac{d\boldsymbol{\psi}_s}{dt}. \tag{1.43}$$

Substituting (1.42) into (1.43),

$$\mathbf{u}_s = R_s\mathbf{i}_s + L_s\frac{d\mathbf{i}_s}{dt} + \frac{d\boldsymbol{\psi}_f}{dt}. \tag{1.44}$$

Considering $\boldsymbol{\psi}_f = \psi_f e^{j\theta_e}$, (1.44) could be presented as

$$\mathbf{u}_s = R_s\mathbf{i}_s + L_s\frac{d\mathbf{i}_s}{dt} + j\omega_e\boldsymbol{\psi}_f, \tag{1.45}$$

where $j\omega_e\boldsymbol{\psi}_f$ is the induced EMF caused by the rotation of the rotor magnetic field. In the steady state, the amplitude of $\boldsymbol{i}_s$ constant. Therefore (1.45) could be presented as

$$\dot{U}_s = R_s\dot{I}_s + j\omega_e L_s\dot{I}_s + j\omega_e\dot{\psi}_f = R_s\dot{I}_s + j\omega_s L_s\dot{I}_s + \dot{E}_0. \tag{1.46}$$

The phasor diagram of SPMSM is shown in Fig. 1.6, which can be derived from (1.46).

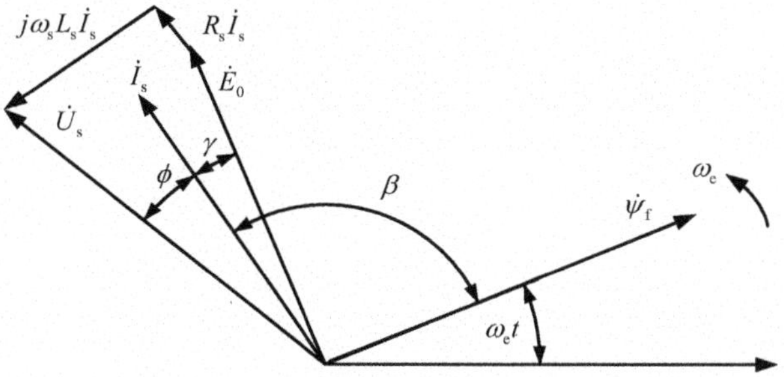

Fig. 1.6 The phasor diagram of SPMSM

The electromagnetic torque of the SPMSM could be presented as

$$T_e = P_n \psi_f i_s \sin\beta = P_n \boldsymbol{\psi}_\mathbf{f} \times \mathbf{i}_\mathbf{s}, \tag{1.47}$$

where P_n is the pole pairs number.

Equation (1.47) indicates that in the vector control of torque, the amplitude of the stator current vector i_s and the space phase angle β relative to ψ_f are used for control. In the sinusoidal steady state, it is equivalent to control the amplitude of the stator current phasor and the phase angle β.

1.4 Mathematical Model of Three-Phase IPMSM

Figure 1.7 shows the IPMSM physical model. The air gap of IPMSM is not uniform. Hence the armature reaction magnetic field will not be the same due to the difference of space phase angle β [7, 8].

As shown in (1.36) and (1.37), the voltage equation and full flux linkage equation of the *a*-*b*-*c* three-phase winding of IPMSM have the same form as SPMSM. However, since the air gap of IPMSM is not uniform, the self-inductances and mutual inductances of *a*-*b*-*c* three-phase winding are periodic functions with the rotor position.

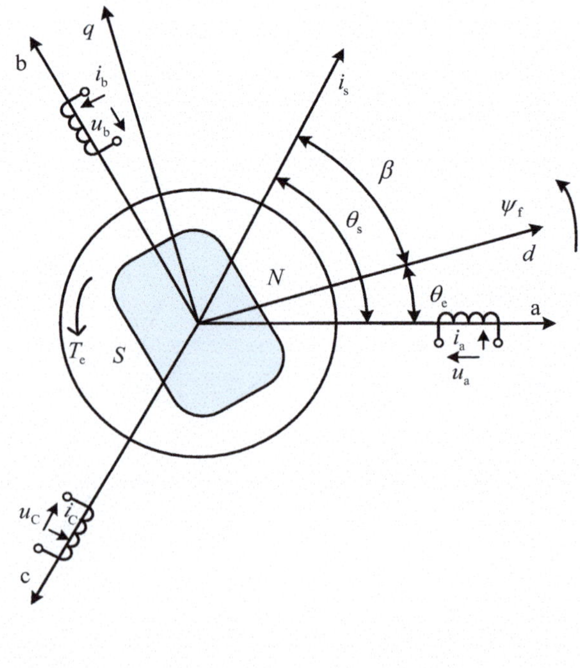

Fig. 1.7 Physical model of three-phase IPMSM

$$
\begin{cases}
L_a = L_{s0} + L_{s2}\cos(2\theta_e) \\
L_b = L_{s0} + L_{s2}\cos 2\left(\theta_e - \frac{2\pi}{3}\right) \\
L_c = L_{s0} + L_{s2}\cos 2\left(\theta_e + \frac{2\pi}{3}\right)
\end{cases}
$$
$$
\begin{cases}
M_{ab} = M_{ba} = -M_{s0} + M_{s2}\cos 2\left(\theta_e + \frac{2\pi}{3}\right) \\
M_{bc} = M_{cb} = -M_{s0} + M_{s2}\cos(2\theta_e) \\
M_{ca} = M_{ac} = -M_{s0} + M_{s2}\cos 2\left(\theta_e - \frac{2\pi}{3}\right),
\end{cases} \quad (1.48)
$$

where L_{s0}, L_{s2}, M_{s0}, and M_{s2} are the average magnitude value and the second harmonic of the self-inductance and the average magnitude value and the second harmonic of the mutual inductance, respectively.

As can be seen from the above equation, since the inductances of the *a-b-c* three-phase winding are the periodic functions of the rotor position, the voltage equations are the time-varying differential forms. It is essential to apply the decoupling control method to transform the voltage equations.

By applying the Clarke transformation, (1.36) and (1.37) are transformed into the stationary coordinate system of α-β. The voltage equation, the flux leakage, and the inductance matrix of α-β axes could be presented as

$$
\begin{bmatrix} u_\alpha \\ u_\beta \end{bmatrix} = R_s \begin{bmatrix} i_\alpha \\ i_\beta \end{bmatrix} + p \begin{bmatrix} \psi_\alpha \\ \psi_\beta \end{bmatrix} \quad (1.49)
$$

$$
\begin{bmatrix} \psi_\alpha \\ \psi_\beta \end{bmatrix} = L_{\alpha\beta} \begin{bmatrix} i_\alpha \\ i_\beta \end{bmatrix} + \psi_f \begin{bmatrix} \cos\theta_e \\ \sin\theta_e \end{bmatrix} \quad (1.50)
$$

$$
L_{\alpha\beta} = \begin{bmatrix} L_\alpha & M_{\alpha\beta} \\ M_{\beta\alpha} & L_\beta \end{bmatrix} = \begin{bmatrix} \Sigma L + \Delta L\cos(2\theta_e) & \Delta L\sin(2\theta_e) \\ \Delta L\sin(2\theta_e) & \Sigma L\text{-}\Delta L\cos(2\theta_e) \end{bmatrix}, \quad (1.51)
$$

where u_α and u_β are the stator voltage components of α-β axes, i_α and i_β are the stator current components of α-β axes, ψ_α and ψ_β are the stator flux components of α-β axes, $L_{\alpha\beta}$ is the inductance matrices stator of α-β axes, $\Sigma L = (L_d + L_q)$ /2 and $\Delta L = (L_d + L_q)$ /2 is average inductance, and L_d and L_q are the equivalent inductances of the *d-q* axes.

As can be seen from (1.51), the inductances at α-β axes are still periodic functions of the rotor position. By applying the Park transformation, (1.49) and (1.50) are

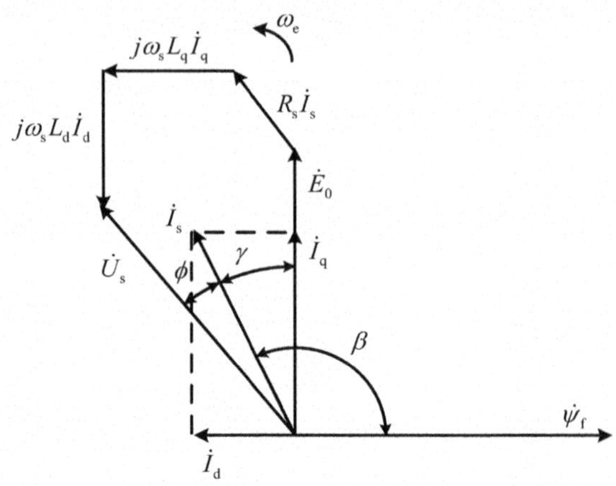

Fig. 1.8 Phasor diagram of three-phase IPMSM

transformed into the *d*-*q* coordinate frame. The voltage equation, the flux leakage, and the inductance matrix of the *d*-*q* coordinate frame could be presented as

$$\begin{bmatrix} u_d \\ u_q \end{bmatrix} = R_s \begin{bmatrix} i_d \\ i_q \end{bmatrix} + p \begin{bmatrix} \psi_\alpha \\ \psi_\beta \end{bmatrix} + \omega_e \begin{bmatrix} -\psi_q \\ \psi_d \end{bmatrix} \tag{1.52}$$

$$\begin{bmatrix} \psi_d \\ \psi_q \end{bmatrix} = L_{dq} \begin{bmatrix} i_d \\ i_q \end{bmatrix} + \begin{bmatrix} \psi_f \\ 0 \end{bmatrix} \tag{1.53}$$

$$L_{dq} = \begin{bmatrix} L_d & 0 \\ 0 & L_q \end{bmatrix}, \tag{1.54}$$

where u_d and u_q are the stator voltage components of *d*-*q* axes, i_d and i_q are the stator current components of *d*-*q* axes, ψ_d and ψ_q are the stator flux components of *d*-*q* axes, and $\boldsymbol{L}_{dq}$ is the inductance matrices stator of *d*-*q* axes. As can be seen from (1.54), the inductance matrix under the *d*-*q* axes becomes the constant matrix and is no longer the periodic function of the rotor position; thus the decoupling of the *d*-*q* axes could be achieved. The phasor diagram of three-phase IPMSM is shown in Fig. 1.8; the voltage equation could be presented as

$$\dot{U}_s = R_s \dot{I}_s + j\omega_s L_d \dot{I}_d + j\omega_s L_q \dot{I}_q + \dot{E}_0. \tag{1.55}$$

The electromagnetic torque expression under d-q axes of IPMSM can be expressed as

$$T_e = \frac{3}{2} P_n \left[\psi_f i_q + \left(L_d - L_q \right) i_d i_q \right], \tag{1.56}$$

and the electromechanical dynamic characteristics could be presented as

$$T_{\mathrm{e}} - T_{\mathrm{L}} = J\frac{d\omega_{\mathrm{r}}}{dt} + B\omega_{\mathrm{r}}, \tag{1.57}$$

where T_{L} is the torque load, J is the inertia of rotor and load, B is the coefficient of viscous friction, and $\omega_{\mathrm{r}} = \omega_{\mathrm{e}}/P_n$ is rotor mechanical angular velocity.

1.5 Per-Unit System of PMSM

There are many power levels of the PMSM, which leads to the complexity of the control system design, because the rated power, the rated voltage, the rated current, and the motor parameters are varied in a wide range [9]. Hence the normalization method is applied in the motor drive system to make the per-unit value of the system and motor parameters in a relatively concentrated range. The per-unit value is obtained by the ration between the actual parameters of the drive system and the base values.

The base power is the rated power of the drive system. The base voltage U_b could be presented as

$$U_b = \frac{U_{dc}}{\sqrt{3}}, \tag{1.58}$$

where U_{dc} is the DC-link voltage of the drive system. Hence the base current I_b could be presented as

$$I_b = \frac{2P_b}{3U_b}. \tag{1.59}$$

The base impedance Z_b of the drive system could be obtained by the ratio between the base voltage and the base current, which could be presented as

$$Z_b = \frac{U_b}{I_b}. \tag{1.60}$$

As for the base frequency, the rated frequency of the motor is preferred. The base angular frequency ω_b is presented as

$$\omega_b = 2\pi f_b, \tag{1.61}$$

and the base inductance L_b could be presented as

$$L_b = \frac{Z_b}{\omega_b}. \tag{1.62}$$

Hence the base capacitance could be presented as

$$C_b = \frac{1}{\omega_b Z_b}. \tag{1.63}$$

The base flux linkage could be presented as

$$\psi_b = L_b i_b. \tag{1.64}$$

As for the mechanical part of the drive system, the base value of the mechanical angular speed Ω_b could be presented as

$$\Omega_b = \frac{\omega_b}{P_n}. \tag{1.65}$$

The base torque is presented as

$$T_b = \frac{P_b}{\Omega_b}. \tag{1.66}$$

Hence the base moment of inertia J_b could be presented as

$$J_b = \frac{P_b}{\Omega_b^2}. \tag{1.67}$$

Taking the three-phase IPMSM as an example, the *d*-*q* axis voltages are presented as

$$\begin{aligned} u_d &= R_s i_d + L_d \frac{di_d}{dt} - \omega_e L_q i_q \\ u_q &= R_s i_q + L_q \frac{di_q}{dt} + \omega_e L_d i_d + \omega_e \psi_f. \end{aligned} \tag{1.68}$$

Hence the *d*-*q* axis voltages divided by base voltage U_b could be presented as

$$\begin{aligned} \frac{u_d}{U_b} &= \frac{R_s}{Z_b}\frac{i_d}{I_b} + \frac{L_d}{L_b}\frac{1}{\omega_b}\frac{di_d}{dt} - \frac{\omega_e}{\omega_b}\frac{L_q}{L_b}\frac{i_q}{I_b} \\ \frac{u_q}{U_b} &= \frac{R_s}{Z_b}\frac{i_q}{I_b} + \frac{L_q}{L_b}\frac{1}{\omega_b}\frac{di_q}{dt} + \frac{\omega_e}{\omega_b}\frac{L_d}{L_b}\frac{i_d}{I_b} + \frac{\omega_e}{\omega_b}\frac{\psi_f}{\psi_b}. \end{aligned} \tag{1.69}$$

The normalized value of the parameters is defined as

$$
\begin{aligned}
u_{d,pu} &= \frac{u_d}{U_b}, u_{q,pu} = \frac{u_q}{U_b}, L_{d,pu} = \frac{L_d}{L_b}, L_{q,pu} = \frac{L_q}{L_b}, \\
i_{d,pu} &= \frac{i_d}{I_b}, i_{q,pu} = \frac{i_q}{I_b}, R_{s,pu} = \frac{R_s}{Z_b}, \omega_{r,pu} = \frac{\omega_e}{\omega_b}, \psi_{f,pu} = \frac{\psi_f}{\psi_b}.
\end{aligned} \tag{1.70}
$$

Hence the normalized voltages could be presented as

$$
\begin{aligned}
u_{d,pu} &= R_{s,pu} i_{d,pu} + \frac{L_{d,pu}}{\omega_b} \frac{di_{d,pu}}{dt} - \omega_{e,pu} L_{q,pu} i_{q,pu}, \\
u_{q,pu} &= R_{s,pu} i_{q,pu} + \frac{L_{q,pu}}{\omega_b} \frac{di_{q,pu}}{dt} + \omega_{e,pu} L_{d,pu} i_{d,pu} + \omega_{e,pu} \psi_{f,pu}.
\end{aligned} \tag{1.71}
$$

The electromagnetic torque of the IPMSM could be presented as

$$
T_e = \frac{3P_n}{2} i_q \left(\psi_f + (L_d - L_q) i_d\right). \tag{1.72}
$$

According to the relationship among the base values, the base torque could be presented as

$$
T_b = \frac{3P_n}{2} \psi_b I_b. \tag{1.73}
$$

Hence the normalized torque is presented as

$$
T_{e,pu} = i_{q,pu} \left(\psi_{f,pu} + (L_{d,pu} - L_{q,pu}) i_{d,\mathrm{pu}}\right). \tag{1.74}
$$

The electromechanical dynamic characteristics of the IPMSM could be presented as

$$
T_e = J \frac{d\omega_r}{dt} + T_L + B\omega_r. \tag{1.75}
$$

According to the relationship among the base values, the base torque could also be presented as

$$
T_b = J_b \Omega_b. \tag{1.76}
$$

Therefore, the normalized electromechanical dynamic characteristics can be presented as

$$
T_{e,pu} = J_{pu} \frac{d\omega_{r,pu}}{dt} + T_{L,pu} + \frac{B}{J_b} \omega_{r,pu}. \tag{1.77}
$$

1.6 Summary

The basic characteristics of the PMSM have been introduced in this section. The voltage equations of the PMSM in the *a-b-c* coordinate frame are severely coupled, which is mainly caused by the mutual inductances. In order to realize the decoupling control, the α-β and *d-q* coordinate frames are introduced, which aim to eliminate the effect of the mutual inductances. The PMSM models in the α-β and *d-q* coordinate frames are obtained by the coordinate transformation, which is an important tool for PMSM motor drives. In order to make the parameters of the different power levels PMSM in a concentrated range, the normalization method can be applied in the drives, which benefits the design of the control system.

References

1. S.K. Sul, *Control of Electric Machine Drive Systems* (Wiley, Hoboken, 2011)
2. T. Wildi, *Electrical Machines, Drives, and Power Systems*, 2nd edn. (Prentice-Hall, Englewood Cliffs, 1991)
3. T. Kenjo, S. Nagamori, *Permanent-Magnet and Brushless DC Motors* (Clarendon Press, Oxford, 1986)
4. P.C. Krause, *Analysis of Electric Machinery* (McGraw-Hill, New York, 1987)
5. A.E. Fitzgerald, C. Kingslay, S.D. Umans, *Electric Machinery* (McGraw-Hill, New York, 1983)
6. R. H. Park, Two-reaction theory of synchronous machines-generalized method of analysis-part 1. AIEE Trans. **48**, 716–727 (July 1929)
7. I. Bolea, S.A. Nasar, *Electric Drives* (CRC Press, Boca Raton, 1998)
8. R. Krishnan, *Electric Motor Drives* (Prentice Hall, Upper Saddle River, 2001)
9. R. Krishnan, *Permanent Magnet Synchronous and Brushless DC Motor Drives* (CRC Press, Boca Raton, 2009)

Chapter 2
Introduction of PMSM Control Methods

2.1 Vector Control

Vector control is based on the rotor flux orientation. Torque control of the PMSM can be achieved by controlling the stator current vector. Using vector control method, the PMSM can obtain the same satisfactory control performance as the DC motor. Nowadays, vector control method has been widely used in the high-performance motor drives.

In order to control the electromagnetic torque precisely, the allocation of the excitation current component i_d and the torque current component i_q should be considered according to the electromagnetic torque as shown in (1.56). At present, $i_d = 0$ control and maximum torque per ampere (MTPA) control are the most popular vector control modes.

2.1.1 $\mathbf{i_d} = 0$ Vector Control Mode

$i_d = 0$ vector control mode means controlling the excitation current i_d to zero. Consequently, the electromagnetic torque only related to the torque current i_q. As shown in Fig. 2.1, the permanent flux linkage ψ_f is on the *d*-axis and rotates with the rotor electrical position θ_e to achieve the rotor flux orientation. The excitation current is zero, and the torque current equals to the amplitude of the stator current vector. In the diagram, the stator voltage vector is $\boldsymbol{u}$, and the stator current vector is $\boldsymbol{i}$.

In this control mode, the torque generation is realized by controlling i_q. Therefore, the control system is easy to implement, and high-performance torque control can be achieved. This control mode is especially suitable for the surface-mounted SPMSM drives, since the stator current can be minimized with the same output torque. However, for interior PMSM, the sacrifice of the reluctance torque component will reduce the efficiency of the motor. With the increase of the load, the stator current

G. Wang et al., *Position Sensorless Control Techniques for Permanent Magnet Synchronous Machine Drives*, https://doi.org/10.1007/978-981-15-0050-3_2

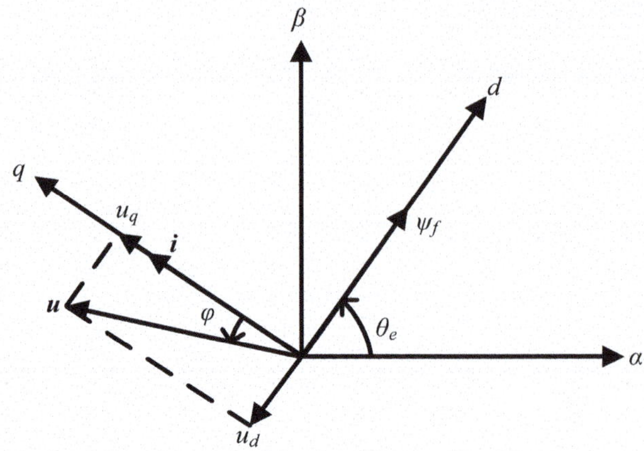

Fig. 2.1 Space vector diagram of $i_d = 0$ control

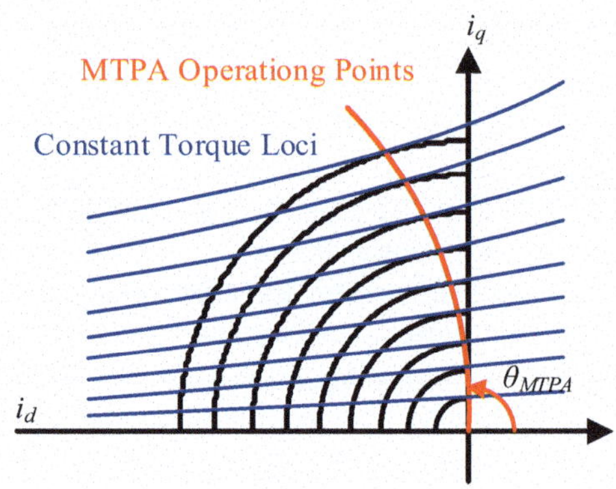

Fig. 2.2 MTPA operating points and the constant torque loci in the current plane

increases gradually, and the cross-axis flux increases accordingly, which increases the angle between the stator voltage vector and the stator current vector, resulting in the decrease of the power factor.

2.1.2 Maximum Torque per Ampere Control

MTPA is a vector control method to minimize the stator current under the same output torque for interior PMSM [1, 2]. The MTPA point is the nearest point to the origin among the points on the constant torque curve in the current plane. And the MTPA point is also the point with a minimum magnitude in the constant torque curve as shown in Fig. 2.2.

Fig. 2.3 Locus of the constant torque in the current vector plane

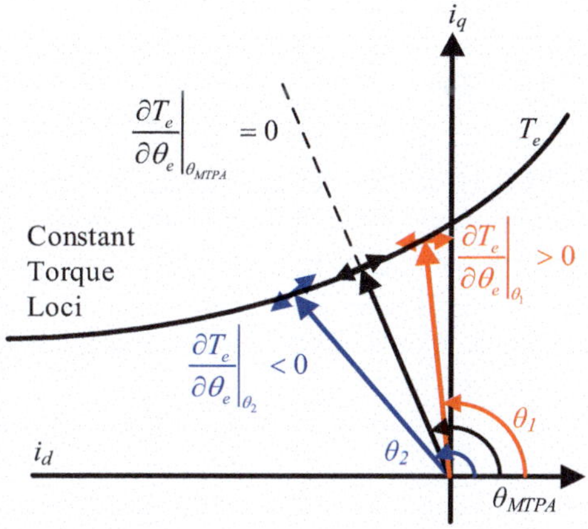

In Fig. 2.3, the constant torque locus is shown in the current plane at the synchronous reference frame. As can be seen, no matter when $\partial T_e/\partial \theta_e > 0$ or $\partial T_e/\partial \theta_e < 0$, the magnitude of the vector is larger than the middle vector. In order to achieve MTPA vector control, the stator current should satisfy the following constraint:

$$\frac{\partial T_e}{\partial \theta_e} = \frac{3P_n}{4} I_S \{\psi_f \cos\theta_e + (L_d - L_q) I_S \cos 2\theta_e\} = 0, \tag{2.1}$$

where I_S the magnitude of the stator current vector.

Substituting (1.56) into (2.1), the current angle of the MTPA point θ_{MTPA} can be derived as follows:

$$\theta_{\text{MTPA}} = \cos^{-1}\left(\frac{-\psi_f + \sqrt{\psi_f^2 + 8(L_d - L_q)^2 I_S^2}}{4(L_d - L_q) I_S}\right). \tag{2.2}$$

By using (2.2), the exciting current and the torque current components in MTPA control mode can be obtained as

$$\begin{cases} i_d^*\big|_{\text{MTPA}} = I_s \cos\theta_{\text{MTPA}} \\ i_q^*\big|_{\text{MTPA}} = I_s \sin\theta_{\text{MTPA}} \end{cases}, \tag{2.3}$$

where $i_d^*|_{\text{MTPA}}$ is the d-axis current reference in MTPA control mode and $i_q^*|_{\text{MTPA}}$ is the q-axis current reference in MTPA control mode.

Thus, the optimal exciting current and torque current components can be obtained for a desired torque by using the MTPA control strategy.

2.2 Flux Weakening Control

2.2.1 Principle of the Flux Weakening Control

In PMSM drives, the amplitudes of the stator current and the voltage vector are limited by the rated current of the insulated gate bipolar transistor (IGBT) and the DC-link voltage, respectively. The maximum voltage and current can be denoted as the voltage limit U_{smax} and the current limit I_{smax}, respectively. Generally, the amplitude of the stator current and the voltage vector should meet the following constraints:

$$|\boldsymbol{i}| \leq I_{smax}, \tag{2.4}$$

$$|\boldsymbol{u}| \leq U_{smax}, \tag{2.5}$$

where $|\boldsymbol{i}|$ and $|\boldsymbol{u}|$ are the amplitude of the stator current vector and stator voltage vector, respectively.

As can be seen from (1.52), (1.53), and (1.54), if the PMSM operates in steady state, the differential terms in the voltage equation are equal to zero. Meanwhile, when the motor operates beyond the based speed, the voltage drop on the stator resistance can be neglected. Ultimately, the motor voltage equations can be represented as

$$u_d = -\omega_e L_q i_q, \tag{2.6}$$

$$u_q = \omega_e L_d i_d + \omega_e \psi_f. \tag{2.7}$$

Substituting (2.6) and (2.7) into (2.5), the voltage boundary can be obtained as

$$\left(L_d i_d + \psi_f\right)^2 + \left(L_q i_q\right)^2 \leq \left(\frac{U_{smax}}{\omega_e}\right)^2. \tag{2.8}$$

It can be seen from (2.8) that the voltage boundary of the interior PMSM is an elliptical cluster with the point $(-\psi_f/L_d, 0)$ as the center, and the radius is reduced proportionally with the increase of the operation speed. The current boundary is shown as (2.9), which is a circle centered on the origin. The radius of the current limit circle is I_{smax}, as shown in Fig. 2.4.

$$i_d^2 + i_q^2 \leq I_{smax}^2. \tag{2.9}$$

As the speed increases, the voltage limit ellipse shrinks toward the center point. When the current trajectory runs along the arrow line to the intersection of the voltage boundary and the current boundary, the *d*-axis current can only move to the left, which is the current trajectory in the weak magnet operation.

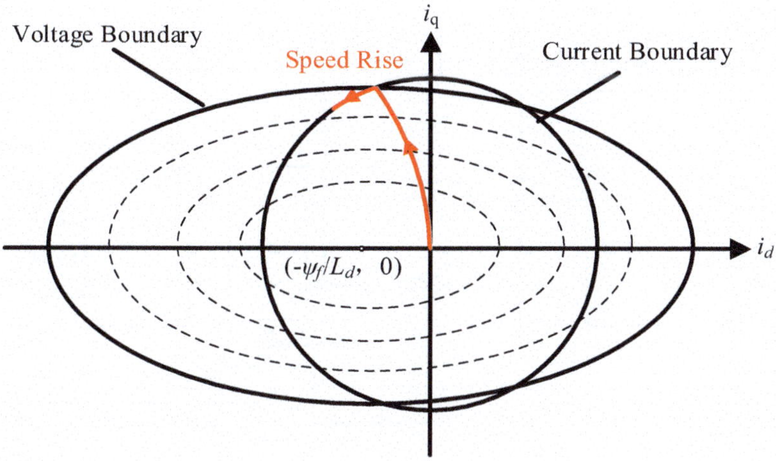

Fig. 2.4 Schematic diagram of voltage and current boundary

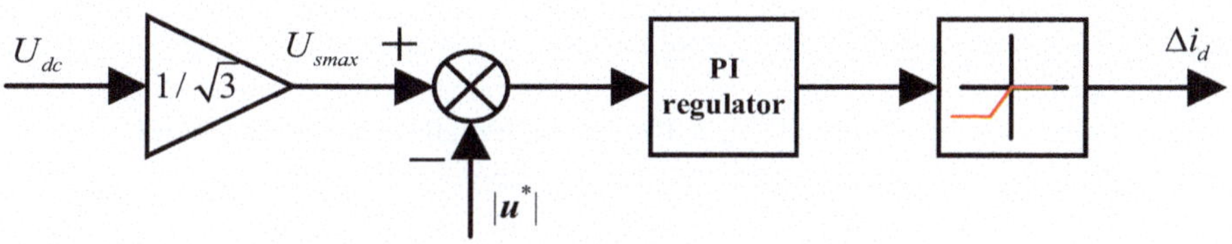

Fig. 2.5 Block diagram of the voltage closed-loop flux-weakening scheme

2.2.2 *Flux Weakening Control Scheme*

The voltage closed-loop flux-weakening method is the most commonly used flux-weakening scheme [3], as shown in Fig. 2.5. The amplitude of the stator voltage vector $|\boldsymbol{u}^*|$ is compared with the limit voltage U_{smax}. When $|\boldsymbol{u}^*| < U_{\text{smax}}$, the PI regulator is positively saturated and therefore $\Delta i_d = 0$. Otherwise, when $|\boldsymbol{u}^*| > U_{\text{smax}}$, the PI regulator will work to produce a negative Δi_d to reduce the reference of the d-axis current.

The voltage closed-loop flux-weakening method integrated with the vector control strategy is shown in Fig. 2.6. The vector control scheme consists of the current loop and the speed loop. The dq-axis current references are allocated by MTPA control strategy to maximize the electromagnetic torque when the motor operates below the based speed. The voltage references u_d^* and u_q^* are used to calculate the amplitude of the stator voltage vector. When $|\boldsymbol{u}^*| > U_{\text{smax}}$, the negative Δi_d is added to the d-axis current reference directly.

Actually, many flux-weakening control strategies have been designed during past decades, such as the single current regulation flux-weakening method [4] and the

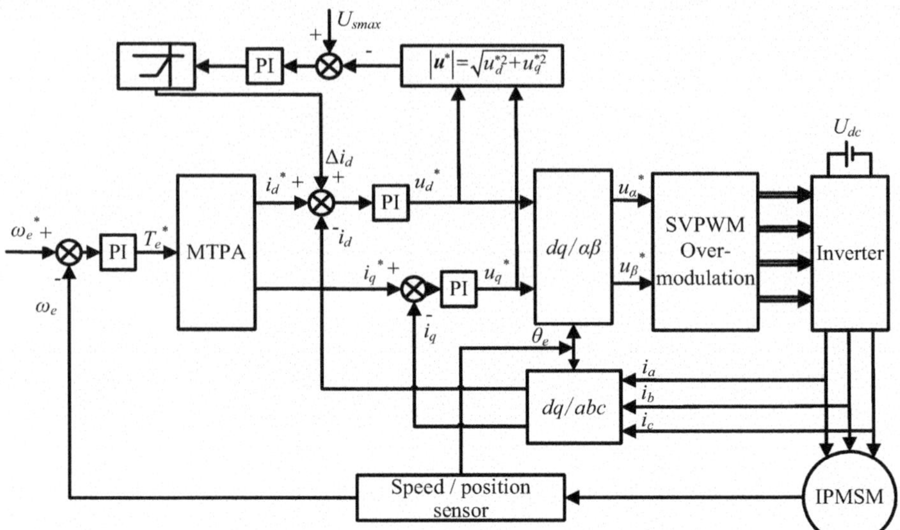

Fig. 2.6 Block diagram of the voltage closed-loop flux weakening control

voltage angle adjusting method [5]. Although each method has their special features, they should obey the same design principles as described above.

2.3 Space Vector Pulse Width Modulation (SVPWM)

2.3.1 Principle of SVPWM

SVPWM is often used in AC motor drive system. The main purpose of classical SVPWM control is to make the fundamental voltage output by the inverter close to the sine wave. SVPWM control technology uses the different combinations of control signals of each bridge arm in the inverter to make the fundamental voltage vector trajectory as close to circular as possible. The equivalent switch model is shown in Fig. 2.7.

Fig. 2.7 PWM equivalent switch model

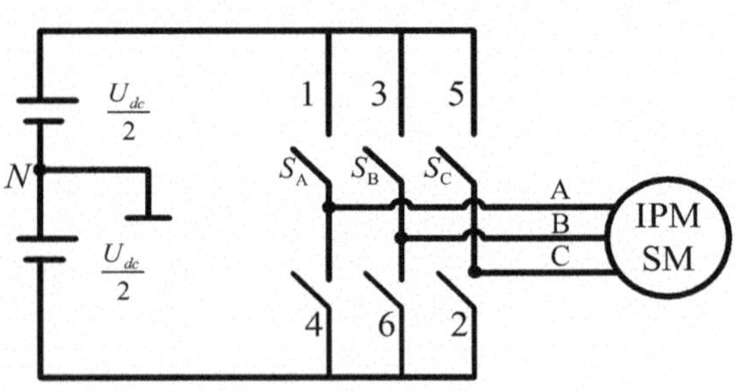

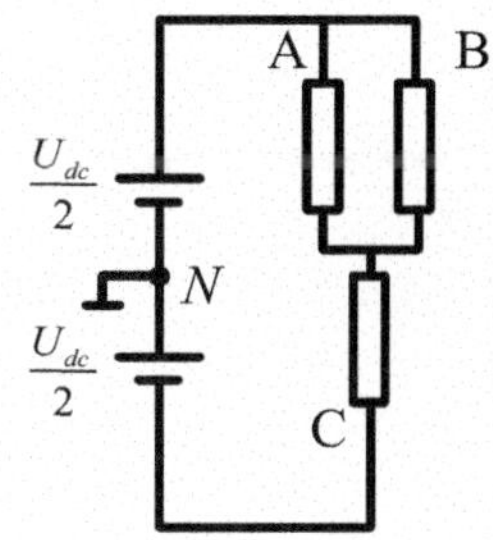

Fig. 2.8 Equivalent circuit diagram of inverter when $S_A = S_B = 1$, and $S_C = 0$

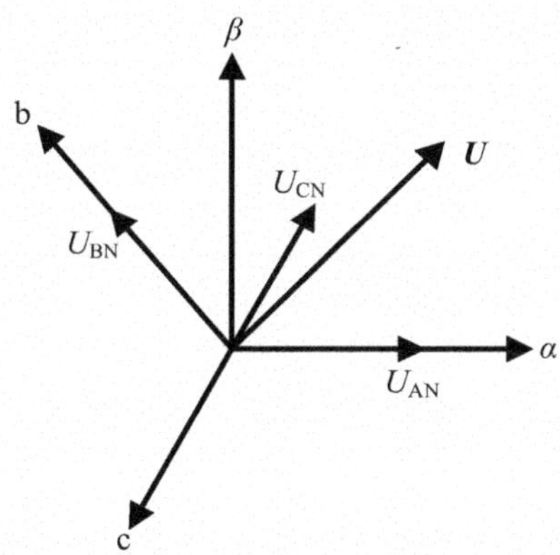

Fig. 2.9 Diagram of voltage vector when $S_A = S_B = 1$, $S_C = 0$

Usually, voltage source inverters (VSIs) use 180° conduction mode. S_A, S_B, and S_C are used to mark the states of the three arms respectively. State 1 indicates that the upper bridge arm is turned on, and state 0 indicates that the lower bridge arm is turned on. Eight switching modes can be obtained as the changes of the three arms and each switching mode corresponds to a basic voltage vector. There are totally eight basic voltage vectors that can be formed, including six non-zero vectors and two zero vectors.

In order to understand the principle of SVPWM, one switching mode is selected to describe in detail. When transistors 1, 2, 3 are in the open state, which means $S_A = S_B = 1$, and $S_C = 0$. Therefore, the equivalent circuit of the inverter can be represented as in Fig. 2.8.

Under this circumstance, the voltage drop on each phase can be denoted as

$$U_{AN} = \frac{1}{3} U_{dc}, U_{BN} = \frac{1}{3} U_{dc}, U_{CN} = -\frac{2}{3} U_{dc}. \tag{2.10}$$

The voltage vector $\boldsymbol{U}$ generated by U_{AN}, U_{BN}, and U_{CN} is shown as in Fig. 2.9. In the $\alpha\beta$-axes, the voltage vector $\boldsymbol{U}$ can be expressed as

$$\boldsymbol{U} = k\left(U_{AN} + U_{BN} e^{j\frac{2}{3}\pi} + U_{CN} e^{j\frac{4}{3}\pi}\right), \tag{2.11}$$

Table 2.1 Basic voltage vectors of SVPWM modulation strategy

$S_A S_B S_C$	A phase voltage	B phase voltage	C phase voltage	Vector expression	Vector symbol
000	0	0	0	0	u_0
001	$-\frac{1}{3}U_{dc}$	$-\frac{1}{3}U_{dc}$	$\frac{2}{3}U_{dc}$	$\frac{2}{3}U_{dc}e^{j\frac{4}{3}\pi}$	u_1
010	$-\frac{1}{3}U_{dc}$	$\frac{2}{3}U_{dc}$	$-\frac{1}{3}U_{dc}$	$\frac{2}{3}U_{dc}e^{j\frac{2}{3}\pi}$	u_2
011	$-\frac{2}{3}U_{dc}$	$\frac{1}{3}U_{dc}$	$\frac{1}{3}U_{dc}$	$\frac{2}{3}U_{dc}e^{j\pi}$	u_3
100	$\frac{2}{3}U_{dc}$	$-\frac{1}{3}U_{dc}$	$-\frac{1}{3}U_{dc}$	$\frac{2}{3}U_{dc}e^{j0}$	u_4
101	$\frac{1}{3}U_{dc}$	$-\frac{2}{3}U_{dc}$	$\frac{1}{3}U_{dc}$	$\frac{2}{3}U_{dc}e^{j\frac{5}{3}\pi}$	u_5
110	$\frac{1}{3}U_{dc}$	$\frac{1}{3}U_{dc}$	$-\frac{2}{3}U_{dc}$	$\frac{2}{3}U_{dc}e^{j\frac{1}{3}\pi}$	u_6
111	0	0	0	0	u_7

where k is the transformation coefficient to transform the vector from three-phase stationary coordinate system to two-phase stationary coordinate system. The transformation can be divided into equal power transformation and equal quantity transformation. Here, the equal quantity transformation is used, and $k = 2/3$.

$$\boldsymbol{U} = \frac{2}{3}\left(U_{\mathrm{AN}} + U_{\mathrm{BN}}e^{j\frac{2}{3}\pi} + U_{\mathrm{CN}}e^{j\frac{4}{3}\pi}\right). \tag{2.12}$$

Substituting (2.10) into (2.12), the voltage vector can be expressed as

$$\boldsymbol{U} = \frac{2}{3}U_{dc}e^{j\frac{1}{3}\pi}. \tag{2.13}$$

The corresponding voltage vectors of each switching state can be obtained through the same method as shown in Table 2.1. As can be seen in Fig. 2.10, the six non-zero voltage vectors and the two zero voltage vectors form a voltage hexagon, and the amplitude of each voltage vector is $2U_{dc}/3$. Meanwhile, the voltage hexagon is divided into six sectors by the voltage vectors.

If only eight basic voltage vectors are used individually, a hexagonal rotating voltage vector can be obtained. This is far away from the circular rotating fundamental voltage vector as expected. The additional intermediate voltage vectors should be formed to achieve the circular fundamental voltage vector trajectory. Actually, the intermediate vectors can be synthesized by six non-zero voltage vectors and two zero voltage vectors.

It is impossible to realize two switching states at the same time. However, if the switching frequency of the inverter is much higher than the frequency of its output voltage, the desired voltage vector can be synthesized by using the basic voltage vector based on the state-space averaging method. As shown in Fig. 2.11, where $\boldsymbol{u}_a$ and $\boldsymbol{u}_b$ are two adjacent basic voltage vectors, T_s is the switching time and T_a and T_b are corresponding opening times. In one switching period, the basic voltage vectors

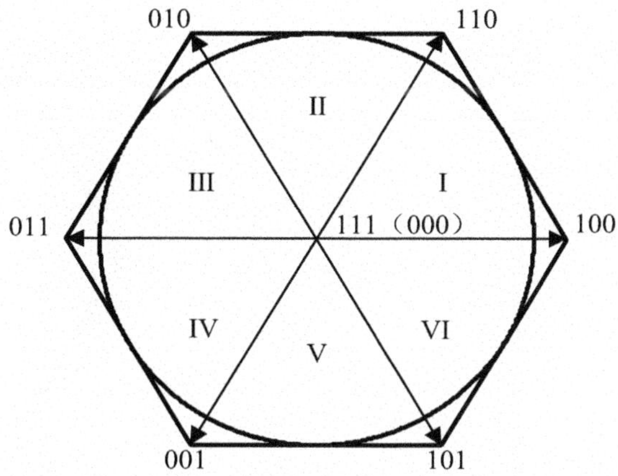

Fig. 2.10 Voltage hexagon and sectors

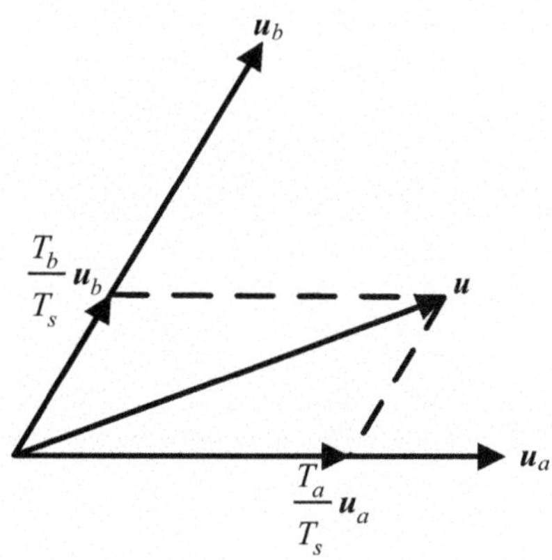

Fig. 2.11 Diagram of space vector constitute by adjacent basic voltage vectors

$\boldsymbol{u}_a$ and $\boldsymbol{u}_b$ act for T_a and T_b, respectively, and the desired voltage vector can be synthesized by these two vectors. Ultimately, the circular voltage vector trajectories can be obtained.

2.3.2 SVPWM Overmodulation

In the previous section, the circular fundamental voltage vector trajectory is aimed, which means the voltage vector instruction cannot exceed the inscribed circle of the hexagon. However, in some condition, the utilization rate of the DC-link voltage is desired to be increased. As shown in Fig. 2.12, the voltage hexagon is divided into the linear modulation region and the overmodulation region. In order to expand the output power of the inverter, some overmodulation strategies were designed in past decades.

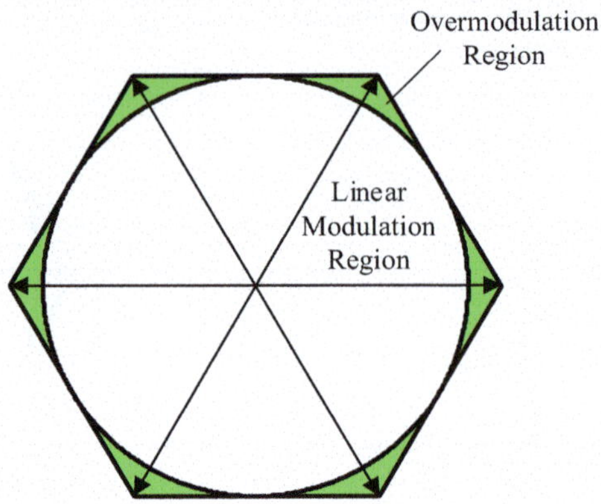

Fig. 2.12 SVPWM modulation region

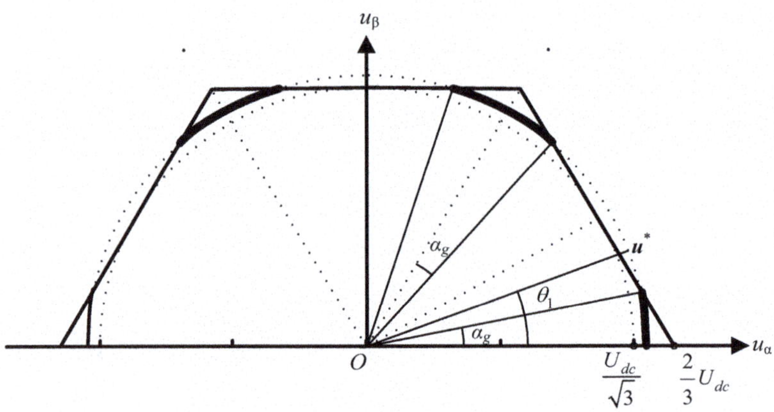

Fig. 2.13 Diagram of Bolognani overmodulation strategy

The modulation index M for Pulse-Width Modulation (PWM) inverter is defined here as

$$M = \frac{|\boldsymbol{U}|}{2U_{dc}/\pi}, \tag{2.14}$$

where $|\boldsymbol{U}|$ is the amplitude of the stator voltage vector.

When the modulation index exceeds 0.9069, the inverter enters into the overmodulation region, and the maximum modulation index is 1.

Bolognani proposed an overmodulation method. The basic idea is keeping the output voltage vector as the fixed vector when the reference voltage vector exceeds the hexagonal boundary [6]. When the reference voltage vector trajectory is within the hexagon, the linear modulation method is used. Since the modulation methods in each sector are similar, sector I is described here. As shown in Fig. 2.13, θ_1 is the phase of the reference voltage vector $\boldsymbol{u}^*$. The linear modulation method is used before $\boldsymbol{u}^*$ runs to the holding angle α_g. When θ_1 runs from α_g to π/6, the phase of $\boldsymbol{u}^*$ is limited to $\theta = \alpha_g$. And when θ_1 runs from π/6 to π/3-α_g, the phase of $\boldsymbol{u}^*$ is limited to $\theta = \pi/3\text{-}\alpha_g$. When θ_1 runs from π/3-α_g to π/3, a linear modulation is used again.

The phase θ linked to the phase θ_1 can be expressed as

$$\theta = \begin{cases} \theta_1 & 0 \le \theta_1 \le \alpha_g \\ \alpha_g & \alpha_g \le \theta_1 \le \pi/6 \\ \pi/3 - \alpha_g & \pi/6 \le \theta_1 \le \pi/3 - \alpha_g \\ \theta_1 & \pi/3 - \alpha_g \le \theta_1 \le \pi/3. \end{cases} \tag{2.15}$$

The holding angle α_g can be derived as

$$\alpha_g = \frac{\pi}{6} - \arccos\left(\frac{U_{dc}}{\sqrt{3}|\boldsymbol{u}^*|}\right). \tag{2.16}$$

The idea of this control method is clear. The output voltage vector is limited to hexagon and can achieve six-step operation when $|\boldsymbol{u}^*|$ reaches $2/3U_{dc}$. The DC-link voltage of the inverter can be effectively utilized. However, the phase error is bigger than the minimum phase error overmodulation method and the minimum amplitude error overmodulation method [7].

2.4 Summary

In this section, the control methods of PMSM based on vector control are introduced, including the MTPA control method, the flux-weakening control method, the SVPWM modulation strategy, and the corresponding overmodulation strategy. By using these methods, high-performance motor drives can be realized. Nowadays, as the expanded usage of PMSM in industry and home appliance, higher requirements of motor drives are proposed, such as lower cost and higher reliability. Therefore, position sensorless control method becomes an attractive research area. In the following sections, position sensorless control methods in full speed region will be introduced.

References

1. S. Kim, Y. Yoon, S. Sul, K. Ide, Maximum torque per ampere (MTPA) control of an IPM machine based on signal injection considering inductance saturation. IEEE Trans. Power Electron., **28**(1), 488–497, (January 2013)
2. G. Wang, R. Yang, D. Xu, DSP-based control of sensorless IPMSM drives for wide-speed-range operation. IEEE Trans. Ind. Electron. **60**(2), 720–727 (February 2013)

3. D. S. Maric, S. Hiti, C. C. Stancu, J. M. Nagashima, D. B. Rutledge, Two flux weakening schemes for surface-mounted permanent-magnet synchronous drives. Design and transient response considerations, *ISIE '99. Proceedings of the IEEE International Symposium on Industrial Electronics*, vol. 2, pp. 673–678, 1999
4. X. Fang, T. Hu, F. Lin, Z. Yang, A novel flux-weakening control method based on single current regulator for permanent magnet synchronous motor, *2014 International Power Electronics Conference (IPEC-Hiroshima 2014 – ECCE ASIA)*, Hiroshima, 2014, pp. 335–340
5. Liu Wei, Liang Hui, Wang Chao, Study on flux-weakening control based on Single Current Regulator for PMSM, *2014 IEEE Conference and Expo Transportation Electrification Asia-Pacific (ITEC Asia-Pacific)*, Beijing, 2014, pp. 1–3
6. S. Bolognani, M. Zigliotto, Novel digital continuous control of SVM inverters in the overmodulation range. IEEE Trans. Ind. Appl. **33**(2), 525–530 (March–April 1997)
7. F. Briz, A. Diez, M. W. Degner, R. D. Lorenz, Current and flux regulation in field-weakening operation [of induction motors]. IEEE Trans. Ind. Appl. **37**(1), 42–50 (January–February 2001)

Chapter 3
Saliency-Tracking-Based Sensorless Control for PMSM Drives

3.1 HF Model and Basic Principle of Saliency-Tracking-Based Methods

PMSMs have been widely applied in industry applications, due to their superiorities in efficiency, power density, and dynamic performance. The elimination of the position sensor helps for not only a decrease in cost but also an enhancement in reliability and robustness. Sensorless control has attracted increasing attentions and obtained extended investigations in the last few decades.

According to the operation speed range of PMSMs, the sensorless control methods can be classified into two categories: saliency tracking-based methods and EMF model-based methods. For medium- to high-speed operation, EMF model-based methods have been well founded and presented satisfying performance, which will be explained further in Chap. 5. However, the capability to obtain the rotor position deteriorates dramatically in low-speed operation, since the amplitude of back electromotive force (EMF) signal becomes too weak to be detected precisely.

Saliency-tracking-based methods, by contrast, perform well at zero and low speeds [1–8]. High-frequency (HF) signal injection-based method and fundamental PWM excitation-based method both have been proved to be effective.

Since the winding inductance can be expressed equations related to rotor position when the HF signal is injected, the rotor position information can be extracted effectively. In general, the HF voltage signal is preferred. The impedance increases with the injection frequency, and it results in the increased voltage. If the HF current is injected, the bandwidth of current regulator must be high enough, which is more difficult to be achieved. Besides, the stationary reference frame (SRF) or the estimated rotary reference frame (RRF) can be selected as the signal injection axes generally. The injection mode can be superimposed injection or separate injection, that is, the HF signal is superimposed onto the fundamental-frequency signal, or

G. Wang et al., *Position Sensorless Control Techniques for Permanent Magnet Synchronous Machine Drives*, https://doi.org/10.1007/978-981-15-0050-3_3

injected directly into the stator windings when the field-oriented control (FOC) is interrupted.

It is worth noting that the injection amplitude and frequency are crucial for sensorless control performance. The determination of injection amplitude is a compromise between position estimation accuracy and control performance. The larger the injection amplitude is, the better position estimation accuracy will be. However, the current and torque ripples will increase, resulting in the increased copper loss and the deteriorated control performance. In addition, the determination of the injection frequency is also worth considering. The higher injection frequency helps to increase the system bandwidth and dynamic performance, but the additional motor loss will increase.

In order to facilitate subsequent theoretical analysis of HF signal injection-based method, the HF mathematical model of PMSM should be first established. According to the PMSM model as expressed in (1.4), the voltage equation in the RRF can be obtained as in (1.49).

When PMSM operates at low speed and the injection frequency is set high enough, the voltage drops on stator resistance and the terms associated with the rotor speed can be neglected. Hence, the PMSM can be regarded as a purely inductive load, and the HF mathematical model in the RRF can be expressed as follows:

$$\begin{bmatrix} u_{dh} \\ u_{qh} \end{bmatrix} = \begin{bmatrix} L_d & 0 \\ 0 & L_q \end{bmatrix} \frac{d}{dt} \begin{bmatrix} i_{dh} \\ i_{qh} \end{bmatrix}, \tag{3.1}$$

where subscript h represents HF components.

By utilizing the coordinate transformation, the above equation can be transformed as follows:

$$\boldsymbol{R}(\theta_e) \begin{bmatrix} u_{dh} \\ u_{qh} \end{bmatrix} = \boldsymbol{R}(\theta_e) \begin{bmatrix} L_d & 0 \\ 0 & L_q \end{bmatrix} \frac{d}{dt} \boldsymbol{R}^{-1}(\theta_e) \begin{bmatrix} i_{\alpha h} \\ i_{\beta h} \end{bmatrix}. \tag{3.2}$$

Then the HF mathematical model of PMSM in the SRF is deduced as follows:

$$\begin{bmatrix} u_{\alpha h} \\ u_{\beta h} \end{bmatrix} = \begin{bmatrix} \Sigma L + \Delta L \cos(2\theta_e) & \Delta L \sin(2\theta_e) \\ \Delta L \sin(2\theta_e) & \Sigma L - \Delta L \cos(2\theta_e) \end{bmatrix} \frac{d}{dt} \begin{bmatrix} i_{\alpha h} \\ i_{\beta h} \end{bmatrix}, \tag{3.3}$$

$$\frac{d}{dt} \begin{bmatrix} i_{\alpha h} \\ i_{\beta h} \end{bmatrix} = \frac{1}{\Sigma L^2 - \Delta L^2} \begin{bmatrix} \Sigma L - \Delta L \cos(2\theta_e) & -\Delta L \sin(2\theta_e) \\ -\Delta L \sin(2\theta_e) & \Sigma L + \Delta L \cos(2\theta_e) \end{bmatrix} \begin{bmatrix} u_{\alpha h} \\ u_{\beta h} \end{bmatrix}, \tag{3.4}$$

where $u_{\alpha h}$, $u_{\beta h}$, $i_{\alpha h}$, $i_{\beta h}$ are HF voltages and currents in the SRF, respectively. ΣL and ΔL represent the average inductance and the difference inductance, respectively. Specifically, they are defined as $\Sigma L = (L_d + L_q)/2$, $\Delta L = (L_d - L_q)/2$.

As for HF signal injection-based method, there are three methods widely used: HF sinusoidal signal injection-based method [1–3], HF square-wave signal injection-based method [4–7], and HF pulse signal injection-based method [8], respectively. The above three methods will be explained further in Sect. 3.2, 3.3, and 3.4. Fundamental PWM excitation-based method combines derivative calculations of current and zero voltage vector (ZVV) injection, which avoids the acoustic noises and torque ripples caused by the HF injection effectively. The position estimation method will be explained in Sect. 3.5.

3.2 HF Sinusoidal Signal Injection-Based Method

HF sinusoidal signal injection-based methods were proposed relatively early, which laid the solid foundation for HF signal injection-based method. According to the injection frame, two methods of the HF sinusoidal signal are widely used: HF rotating sinusoidal signal injection-based method and HF pulsating sinusoidal signal injection-based method. HF rotating sinusoidal signal injection-based method is the earliest proposed HF signal injection-based method, where a rotating HF voltage vector was injected into the SRF, and the position information can be implied in the phase of the induced HF currents. The HF pulsating sinusoidal signal injection-based method usually injects the pulsating voltage signal into the estimated RRF, and the amplitude of the induced HF currents contains the position information. The above two methods are further explained in the following.

3.2.1 HF Rotating Sinusoidal Signal Injection-Based Method

HF rotating sinusoidal signal injection-based method was first proposed by Robert D. Lorenz [1]. The block diagram of HF rotating sinusoidal signal injection-based sensorless PMSM drive is shown in Fig. 3.1. Two orthogonal HF voltage signals are injected into the SRF as follows:

$$\begin{bmatrix} u_{\alpha h} \\ u_{\beta h} \end{bmatrix} = U_h \begin{bmatrix} \cos \omega_h t \\ \sin \omega_h t \end{bmatrix}, \tag{3.5}$$

where U_h and ω_h denote the amplitude and frequency of the injected HF rotating voltage, respectively. When PMSM operates at zero or low speed, the injection frequency is much higher than the fundamental frequency; thus the HF mathematical model of PMSM in the SRF can be obtained. Then the induced HF currents can be deduced as follows:

$$\begin{bmatrix} i_{\alpha h} \\ i_{\beta h} \end{bmatrix} = \frac{U_h}{\omega_h \left(\Sigma L^2 - \Delta L^2 \right)} \begin{bmatrix} \Sigma L \sin \omega_h t + \Delta L \sin \left(2\theta_e - \omega_h t \right) \\ -\Sigma L \cos \omega_h t - \Delta L \cos \left(2\theta_e - \omega_h t \right) \end{bmatrix}. \tag{3.6}$$

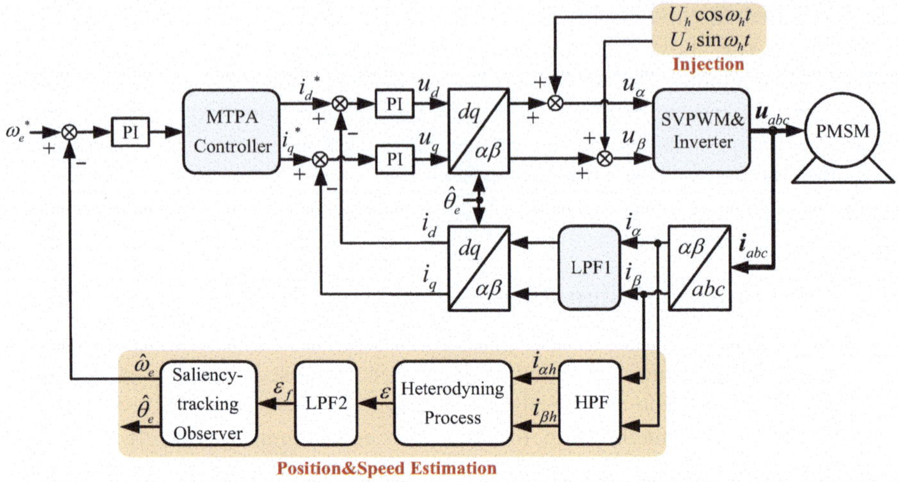

Fig. 3.1 Block diagram of HF rotating sinusoidal signal injection-based sensorless PMSM drive

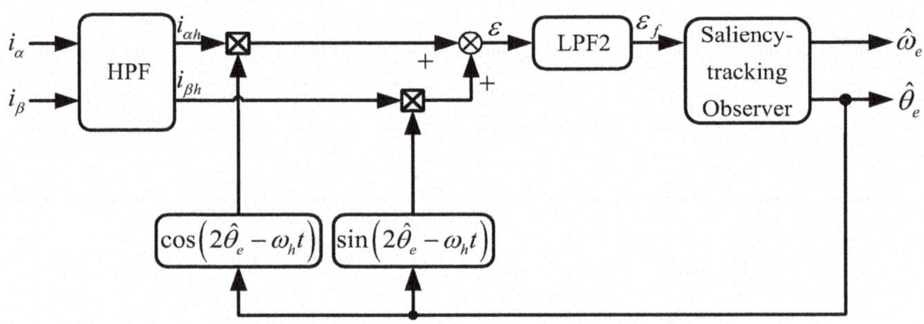

Fig. 3.2 Position and speed estimation

As can be seen, the position information θ_e is contained in the phase of the negative sequence current component. In order to extract rotor position from the induced HF currents, the heterodyning process and the saliency-tracking observer can be utilized, as shown in Fig. 3.2. Firstly, the high-pass filter (HPF) serves to separate the induced HF currents. Then HF currents are multiplied in SRF by $\cos\left(2\hat{\theta}_e - \omega_h t\right)$ and $\sin\left(2\hat{\theta}_e - \omega_h t\right)$,

$$\begin{aligned}\varepsilon =& i_{\alpha h}\cos\left(2\hat{\theta}_e - \omega_h t\right) + i_{\beta h}\sin\left(2\hat{\theta}_e - \omega_h t\right) \\ =& \frac{U_h}{\omega_h\left(\Sigma L^2 - \Delta L^2\right)}\left(\Delta L\sin\left(2\tilde{\theta}_e\right) + \Sigma L\sin\left(2\omega_h t - 2\hat{\theta}_e\right)\right),\end{aligned} \tag{3.7}$$

where $\hat{\theta}_e$ is the estimated position via the saliency-tracking observer, and $\tilde{\theta}_e$ represents the position estimation error, i.e., $\tilde{\theta}_e = \theta_e - \hat{\theta}_e$. Coordinate distribution of each axis is shown in Fig. 3.3. Furthermore, the low-pass filter (LPF) is used to suppress

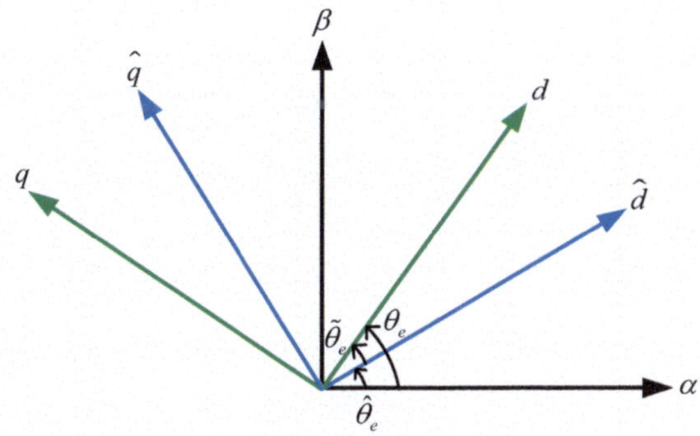

Fig. 3.3 Coordinate distribution of each axis

the HF component and obtain the term related to position estimation error. In the assumption that $\tilde{\theta}_e$ is small enough, $\varepsilon_f$is proportional to $\tilde{\theta}_e$, i.e.,

$$\begin{aligned}\varepsilon_f =& \mathrm{LPF}\left(i_\alpha \cos\left(2\hat{\theta}_e - \omega_h t\right) + i_\beta \sin\left(2\hat{\theta}_e - \omega_h t\right)\right)\\ \approx& \frac{2U_h \Delta L}{\omega_h\left(\Sigma L^2 - \Delta L^2\right)}\tilde{\theta}_e.\end{aligned} \tag{3.8}$$

Then ε_f acts as the input of the saliency-tracking observer to estimate the motor speed and position. By driving ε_f to approach to zero, saliency-tracking observer forces the convergence of the estimated position to the actual one, i.e., $\hat{\theta}_e \to \theta_e$. The specific design process of saliency-tracking observer will be introduced in Sect. 3.5.

HF rotating sinusoidal signal injection-based method has potential stability since it injects symmetrical HF voltages into the SRF. However, this method inevitably brings about current fluctuations and torque ripples, which have an adverse effect on control performance. In addition, the injection frequency is limited in order to ensure the sinusoid of the injected signal. HPFs and LPFs are needed to extract the signals with different frequencies, while the existence of LPFs would restrict the bandwidth of controllers.

3.2.2 HF Pulsating Sinusoidal Signal Injection-Based Method

In order to alleviate current fluctuations and torque ripples, the HF pulsating sinusoidal signal injection-based method has been proposed by injecting a pulsating voltage expressed as (3.9) along the estimated RRF [2–3]. The block diagram of HF pulsating sinusoidal signal injection-based sensorless PMSM drive is shown in Fig. 3.4.

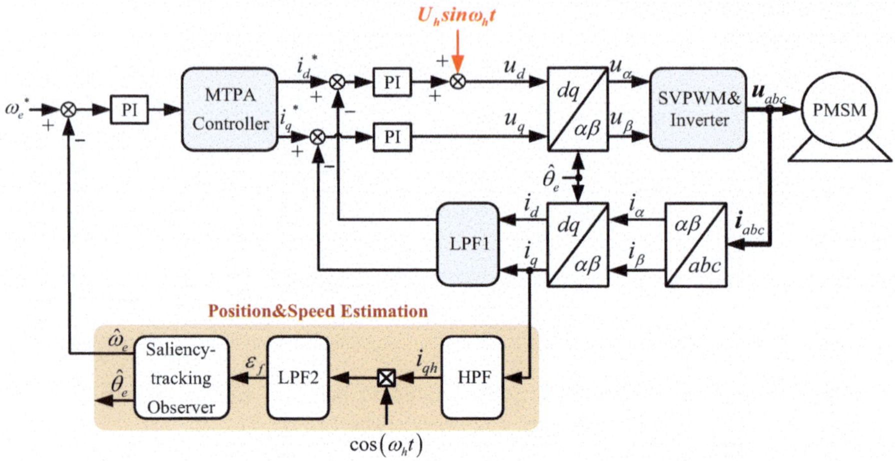

Fig. 3.4 Block diagram of HF pulsating sinusoidal signal injection-based sensorless PMSM drive

$$\begin{bmatrix} u_{\hat{d}h} \\ u_{\hat{q}h} \end{bmatrix} = \begin{bmatrix} U_h \sin(\omega_h t) \\ 0 \end{bmatrix}. \tag{3.9}$$

According to the HF mathematical model in the RRF, (3.10) can be obtained via coordinate transformation. Accordingly, the HF mathematical model in the estimated RRF is expressed as (3.11), where $u_{\hat{d}h}$, $u_{\hat{q}h}$, $i_{\hat{d}h}$, $i_{\hat{q}h}$ are HF voltages and currents, respectively.

$$\boldsymbol{R}(\tilde{\theta}_e)\begin{bmatrix} u_{dh} \\ u_{qh} \end{bmatrix} = \boldsymbol{R}(\tilde{\theta}_e)\begin{bmatrix} L_d & 0 \\ 0 & L_q \end{bmatrix}\frac{d}{dt}\boldsymbol{R}^{-1}(\tilde{\theta}_e)\begin{bmatrix} i_{\hat{d}h} \\ i_{\hat{q}h} \end{bmatrix}, \tag{3.10}$$

$$\frac{d}{dt}\begin{bmatrix} i_{\hat{d}h} \\ i_{\hat{q}h} \end{bmatrix} = \frac{1}{\Sigma L^2 - \Delta L^2}\begin{bmatrix} \Sigma L - \Delta L\cos(2\tilde{\theta}_e) & -\Delta L\sin(2\tilde{\theta}_e) \\ -\Delta L\sin(2\tilde{\theta}_e) & \Sigma L + \Delta L\cos(2\tilde{\theta}_e) \end{bmatrix} \times \begin{bmatrix} u_{\hat{d}h} \\ u_{\hat{q}h} \end{bmatrix}. \tag{3.11}$$

Further, substitute (3.9) into (3.11); the induced HF currents in the estimated RRF can be deduced as follows:

$$\begin{bmatrix} i_{\hat{d}h} \\ i_{\hat{q}h} \end{bmatrix} = \frac{U_h}{\omega_h(\Sigma L^2 - \Delta L^2)}\begin{bmatrix} -\Sigma L + \Delta L\cos(2\tilde{\theta}_e) \\ \Delta L\sin(2\tilde{\theta}_e) \end{bmatrix}\cos(\omega_h t). \tag{3.12}$$

As can be seen, $i_{\hat{q}h}$ is approximately zero on the assumption that the position estimation error is small enough. In view of this, only $i_{\hat{q}h}$ is processed by multiplying it by $\cos(\omega_h t)$; thus ε_f associated with position estimation error can be obtained as

(3.13) via LPF. Further, ε_f serves as the input of saliency-tracking observer; the position information can be obtained by driving ε_f to zero.

$$\begin{aligned}\varepsilon_f =& \mathrm{LPF}\left(i_{\hat{q}h} \times \cos(\omega_h t)\right) \\ =& \frac{U_h \Delta L}{2\omega_h\left(\Sigma L^2 - \Delta L^2\right)} \sin\left(2\tilde{\theta}_e\right) \\ \approx& \frac{U_h \Delta L}{\omega_h\left(\Sigma L^2 - \Delta L^2\right)} \tilde{\theta}_e.\end{aligned} \tag{3.13}$$

It is worth noting that the retrieve of rotor position depends on the difference inductance ΔL, i.e., on the incremental saliency of PMSM. If ΔL becomes zero, ε_f associated with position estimation error will disappear.

In addition, there is another way to estimate rotor position. Different from the above method, the following one defines the rotational measurement axes ($d^m q^m$) that lag behind the estimated RRF $\pi/4$. The coordinate distribution of each axis, the block diagram of HF pulsating sinusoidal signal injection-based sensorless PMSM drive, and the position and speed estimation process are shown in Figs. 3.5, 3.6, and 3.7, respectively.

By coordinate transformation, the HF mathematical model in the RRF can be transformed into rotational measurement axes, as explained in (3.14). Accordingly, the HF mathematical model in $d^m q^m$ is deduced as (3.15), where u_{dh}^m, u_{qh}^m, i_{dh}^m, i_{qh}^m are HF voltages and currents, respectively.

$$\boldsymbol{R}\left(\tilde{\theta}_e + \frac{\pi}{4}\right)\begin{bmatrix} u_{dh} \\ u_{qh} \end{bmatrix} = \boldsymbol{R}\left(\tilde{\theta}_e + \frac{\pi}{4}\right)\begin{bmatrix} L_d & 0 \\ 0 & L_q \end{bmatrix} \frac{d}{dt}\boldsymbol{R}^{-1}\left(\tilde{\theta}_e + \frac{\pi}{4}\right)\begin{bmatrix} i_{dh}^m \\ i_{qh}^m \end{bmatrix}, \tag{3.14}$$

$$\frac{d}{dt}\begin{bmatrix} i_{dh}^m \\ i_{qh}^m \end{bmatrix} = \frac{1}{\Sigma L^2 - \Delta L^2}\begin{bmatrix} \Sigma L + \Delta L \sin\left(2\tilde{\theta}_e\right) & -\Delta L \cos\left(2\tilde{\theta}_e\right) \\ -\Delta L \cos\left(2\tilde{\theta}_e\right) & \Sigma L - \Delta L \sin\left(2\tilde{\theta}_e\right) \end{bmatrix}\begin{bmatrix} u_{dh}^m \\ u_{qh}^m \end{bmatrix}. \tag{3.15}$$

In Fig. 3.5, the injected voltage in estimated RRF can be transformed into rotational measurement axes, as explained in (3.16); thus the induced HF currents in $d^m q^m$ can be deduced as (3.17). Multiply the HF currents by 2 cos $(\omega_h t)$, and the amplitude containing the position estimation error can be obtained via LPF, as shown in (3.18).

$$\begin{bmatrix} u_{dh}^m \\ u_{qh}^m \end{bmatrix} = \frac{U_h}{\sqrt{2}}\begin{bmatrix} \sin(\omega_h t) \\ \sin(\omega_h t) \end{bmatrix}, \tag{3.16}$$

$$\begin{bmatrix} i_{dh}^m \\ i_{qh}^m \end{bmatrix} = \frac{-U_h \cos(\omega_h t)}{\sqrt{2}\omega_h\left(\Sigma L^2 - \Delta L^2\right)}\begin{bmatrix} \Sigma L - \Delta L \cos\left(2\tilde{\theta}_e\right) + \Delta L \sin\left(2\tilde{\theta}_e\right) \\ \Sigma L - \Delta L \cos\left(2\tilde{\theta}_e\right) - \Delta L \sin\left(2\tilde{\theta}_e\right) \end{bmatrix}, \tag{3.17}$$

Fig. 3.5 Coordinate distribution of each axis

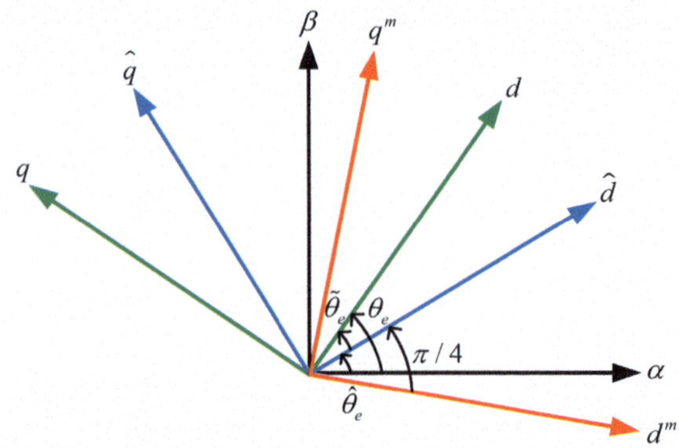

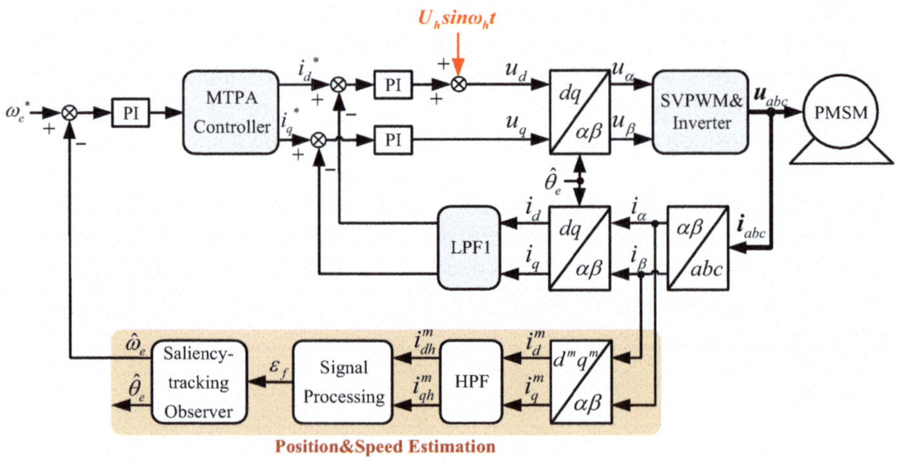

Fig. 3.6 Block diagram of HF pulsating sinusoidal signal injection-based sensorless PMSM drive

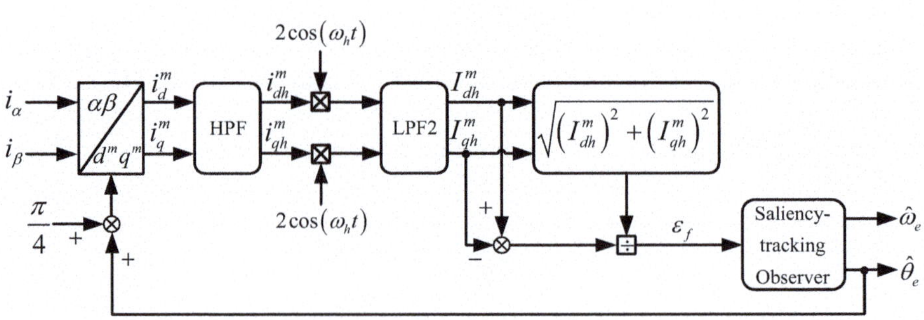

Fig. 3.7 Position and speed estimation process

$$\begin{aligned}\begin{bmatrix} I_{dh}^{m} \\ I_{qh}^{m} \end{bmatrix} &= \mathrm{LPF}\left(\begin{bmatrix} i_{dh}^{m} \\ i_{qh}^{m} \end{bmatrix} \times 2\cos(\omega_h t)\right) \\ &= \frac{-U_h}{\sqrt{2}\omega_h(\Sigma L^2 - \Delta L^2)}\begin{bmatrix} \Sigma L - \Delta L\cos(2\tilde{\theta}_e) + \Delta L\sin(2\tilde{\theta}_e) \\ \Sigma L - \Delta L\cos(2\tilde{\theta}_e) - \Delta L\sin(2\tilde{\theta}_e) \end{bmatrix}\end{aligned}. \tag{3.18}$$

After that, the difference is processed and followed by the normalization; ε_f associated with position estimation error can be expressed as (3.19). The saliency-tracking observer is used to further extract the rotor position and speed information.

$$\varepsilon_f = \frac{I_{dh}^{m} - I_{qh}^{m}}{\sqrt{I_{dh}^{m\,2} + I_{qh}^{m\,2}}} \approx \sqrt{2}\left(1 - \frac{L_d}{L_q}\right)\tilde{\theta}_e. \tag{3.19}$$

Compared with the HF sinusoidal signal injection-based method, the HF pulsating sinusoidal signal injection-based method usually injects the pulsating voltage into the $\hat{d}$ axis, resulting in the smaller current and torque ripples. Besides, the signal processing is relatively easy to implement. However, in order to ensure the sinusoid of the injected voltage, the injection frequency is limited, which is not conducive to the improvement of dynamic performance.

3.3 HF Square-Wave Signal Injection-Based Method

HF square-wave signal injection-based method has attracted sustaining attention for higher injection frequency and better dynamic performance. Similar to the HF pulsating sinusoidal signal injection-based method, it also injects the HF voltage into the estimated RRF [4–7]. The difference is that the injected voltage is a square-wave, not a sinusoidal one. Block diagram of HF square-wave signal injection-based sensorless PMSM drive is shown in Fig. 3.8. The process for rotor position and speed estimation is shown in Fig. 3.9a.

The injected voltage is expressed as follows:

$$\begin{bmatrix} u_{\hat{d}h} \\ u_{\hat{q}h} \end{bmatrix} = \begin{bmatrix} U_h(-1)^k \\ 0 \end{bmatrix}, \tag{3.20}$$

where U_h and k indicate the amplitude and sequence of the injected voltage, respectively. Substitute (3.20) into (3.11) and replace the current differentiation *di*/*dt* with $\Delta i/\Delta T$; thus the HF current variation in one sampling period ΔT can be deduced as follows:

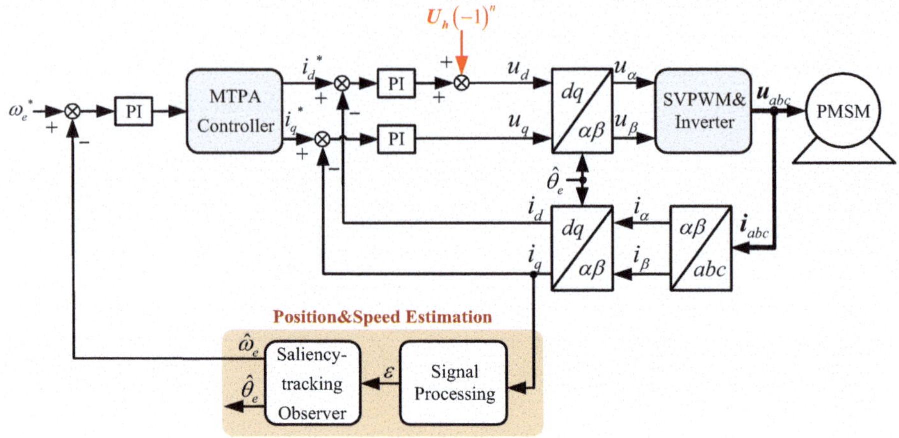

Fig. 3.8 Block diagram of HF square-wave signal injection-based sensorless PMSM drive

$$\begin{bmatrix} \Delta i_{\hat{d}h} \\ \Delta i_{\hat{q}h} \end{bmatrix} = \frac{\Delta T u_{\hat{d}h}}{\Sigma L^2 - \Delta L^2} \begin{bmatrix} \Sigma L - \Delta L \cos(2\tilde{\theta}_e) \\ -\Delta L \sin(2\tilde{\theta}_e) \end{bmatrix}. \tag{3.21}$$

As can be seen from (3.21), $\Delta i_{\hat{q}h}$ contains position estimation error. Since the injection frequency is relatively high, it can be assumed that the fundamental-frequency component is almost constant in one sampling period. In view of this, the variation of the sampling current approximates to that of HF current, i.e.,

$$\Delta i_{\hat{q}h} = i_{\hat{q}h}[k] - i_{\hat{q}h}[k-1] \approx i_{\hat{q}}[k] - i_{\hat{q}}[k-1]. \tag{3.22}$$

By utilizing the variation of the sampling current in $\hat{q}$ axis and the injected voltage, τ_f associated with position estimation error can be obtained as (3.23). The saliency-tracking observer is further used to obtain the rotor position and speed information. In this way, no digital filter is needed in the process of rotor position and speed estimation. Besides, the LPFs in the current loop can be neglected, due to the higher injection frequency.

$$\begin{aligned} \tau &= \frac{\Delta i_{\hat{q}h}}{u_{\hat{d}h}[k]} \approx \frac{i_{\hat{q}}[k] - i_{\hat{q}}[k-1]}{u_{\hat{d}h}[k]} \\ &\approx \frac{-2\Delta L \Delta T}{\Sigma L^2 - \Delta L^2} \tilde{\theta}_e. \end{aligned} \tag{3.23}$$

It is worth noting that the effects of voltage errors caused by inverter nonlinearities and voltage drops on the position estimation accuracy are not considered in the scheme shown in Fig. 3.9a. In order to solve the problem better, an enhanced scheme for position and speed estimation is introduced and shown in Fig. 3.9b. The current

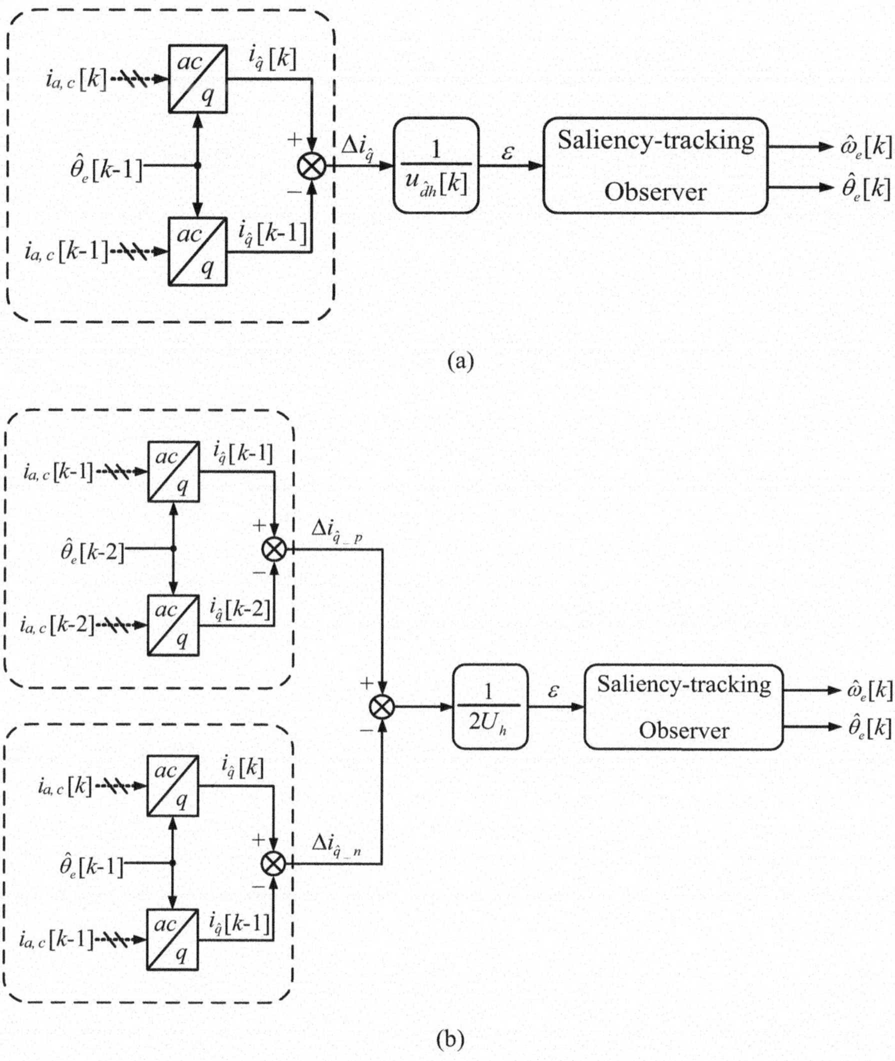

Fig. 3.9 Rotor position and speed estimation process. (**a**) Conventional scheme without taking voltage errors into account. (**b**) Enhanced scheme taking voltage errors into account

variations $\Delta i_{\hat{q}h\_p}$ and $\Delta i_{\hat{q}h\_n}$ can be obtained when the positive and negative voltage components are injected. Subsequently, two current variations are needed for subtraction to offset the effects of voltage errors on the position estimation accuracy. Considering that there is a delay of one sampling period from the given voltage reference to the execution of the digital control system, an additional FOC period is required after the injection of the positive and negative HF voltages. The sequence of the enhanced HF square-wave signal injection-based method is shown in Fig. 3.10. It is worth noting that the enhanced method introduces a delay of one third of the

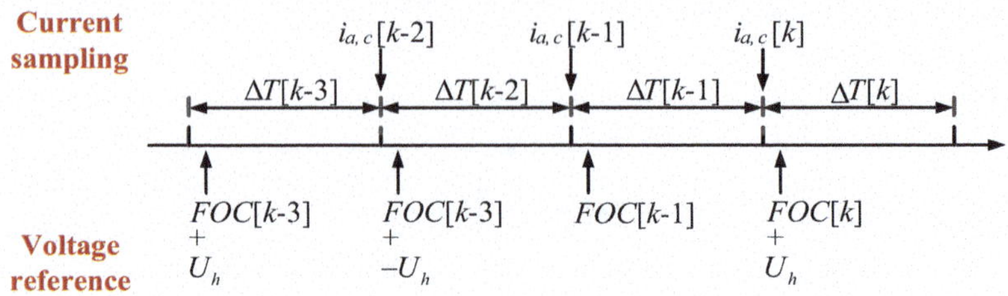

Fig. 3.10 Sequence of the enhanced HF square-wave signal injection-based method

control period, so that the bandwidth of the current loop is reduced compared to the conventional HF square-wave signal injection-based method. However, due to the higher injection frequency, the LPF can be omitted in the feedback of the current loop. Therefore, the current-loop bandwidth of the enhanced HF square-wave signal injection-based motor drive system is still higher than that of the HF sinusoidal signal injection-based one. The specific theoretical analysis is introduced as follows.

When the voltage errors are taken into account, the injected voltage will be distorted from U_{INJ} to $\overline{\mathrm{U}}_{INJ}$ as shown in Fig. 3.11. The voltage in the estimated RRF can be modified as follows:

$$\begin{bmatrix} u'_{\hat{d}h} \\ u'_{\hat{q}h} \end{bmatrix} = \begin{bmatrix} U_h(-1)^k - U_{\hat{d},err} \\ -U_{\hat{q},err} \end{bmatrix}, \tag{3.24}$$

where $U_{\hat{d},err}$ and $U_{\hat{q},err}$ are the voltage errors in the estimated RRF, and assume they are approximately invariant in one injection period. Substitute (3.24) into HF mathematical model of PMSM as shown in (3.11), and replace the current differentiation di/dt with $\Delta i/\Delta T$. Thus, the HF current variation in one sampling period ΔT can be deduced as follows:

$$\begin{bmatrix} \Delta i_{\hat{d}h} \\ \Delta i_{\hat{q}h} \end{bmatrix} = \frac{\Delta T}{\Sigma L^2 - \Delta L^2} \begin{bmatrix} [\Sigma L - \Delta L\cos(2\tilde{\theta}_e)]\left[U_h(-1)^k\right] \\ -\Delta L\sin(2\tilde{\theta}_e)\left[U_h(-1)^k\right] \end{bmatrix} \tag{3.25}$$
$$\begin{bmatrix} -[\Sigma L - \Delta L\cos(2\tilde{\theta}_e)]U_{\hat{d},err} + \Delta L\sin(2\tilde{\theta}_e)U_{\hat{q},err} \\ \Delta L\sin(2\tilde{\theta}_e)U_{\hat{d},err} - [\Sigma L + \Delta L\cos(2\tilde{\theta}_e)]U_{\hat{q},err} \end{bmatrix}.$$

As can be seen from (3.25), the current variation in $\hat{q}$ axis $\Delta i_{\hat{q}h}$ is affected by the voltage errors; thus the retrieved rotor position and speed information would be also adversely affected. In order to improve the robustness to voltage errors, two current variations $\Delta i_{\hat{q}h\_p}$ and $\Delta i_{\hat{q}h\_n}$ are all calculated when the positive and negative voltage components are injected, respectively. After that, $\Delta i_{\hat{q}h\_p}$ and $\Delta i_{\hat{q}h\_n}$ are needed for

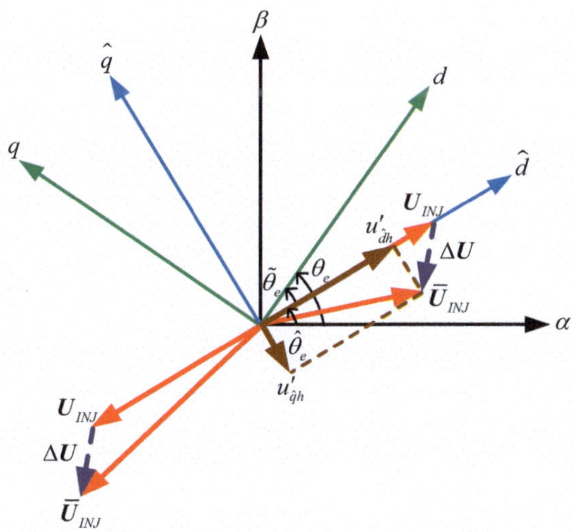

Fig. 3.11 Coordinate distribution of each axis

the subtraction to offset the effects of voltage errors and obtaining τ' associated with the position estimation error, and τ' can be expressed as follows:

$$\begin{aligned}\tau' &= \frac{\Delta i_{\hat{q}h\_p} - \Delta i_{\hat{q}h\_n}}{2U_h} \\ &\approx \frac{\left(i_{\hat{q}}[k-1] - i_{\hat{q}}[k-2]\right) - \left(i_{\hat{q}}[k] - i_{\hat{q}}[k-1]\right)}{2U_h} \\ &\approx \frac{-2\Delta L \Delta T}{\Sigma L^2 - \Delta L^2}\tilde{\theta}_e.\end{aligned} \tag{3.26}$$

Similarly, τ_f' associated with position estimation error is used as the input of the saliency-tracking observer to further obtain rotor position and speed information.

The conventional scheme and the enhanced one are compared via an experimental platform with a 400 W PMSM, the parameters of which are listed in Table 3.1. The dead time of the inverter is set to 2 μs, 2.5 μs, 3 μs, and 4 μs. The actual rotor position of the PMSM is obtained by the incremental encoder with a resolution of 2048 lines. The DC motor mechanically couples with PMSM as a load. In the experiment, the PWM frequency is 5 kHz, and the sampling and control frequency is 10 kHz; that is, the digital signal processor (DSP) is configured as a double-sampling, double-update mode.

The experimental comparison between the conventional and the enhanced HF square-wave signal injection-based methods at different injection voltages and dead time values is performed, as shown in Fig. 3.12. From top to bottom, the actual operation frequency f_r of PMSM, the q-axis current i_q, the position estimation error $\tilde{\theta}_e$, and the phase current i_a are shown, respectively. In the experiment, the rotor position used in the Park and inverse Park transformation is the estimated one, and the actual speed obtained by the encoder serves as the feedback of the speed loop,

Table 3.1 Prototype parameters for position sensorless control

Rated power (W)	400
Rated frequency (Hz)	97.9
Stall current (A)	2.9
Number of poles	4
d-axis inductance (mH)	10
q-axis inductance (mH)	13
Stator resistance (Ω)	2.3
Flux linkage (Wb)	0.12

which reduces the effects of the position estimation result. The PMSM operates at 0.5 Hz with no load, then the rated step load is added, and the motor runs in the motoring state. After about 12 s, the load direction is reversed to make the motor work in the regenerating state. Then after 12 s, the PMSM is operating again with no load.

It can be seen from Fig. 3.12a, c, e that the fluctuation of the position estimation error decreases with the increase of the injection amplitude. The similar conclusion can be drawn from Fig. 3.12b, d, f. As the amplitude of the injection voltage increases, its ratio with respect to the voltage errors is increased, so that the effects of injection voltage on the induced HF currents are increased accordingly.

Besides, as can be seen from Fig. 3.12e, g, when the conventional HF square-wave signal injection-based method is applied and the dead time is increased from 2 to 4 μs at the injection voltage of 30 V, the fluctuation of the position estimation error ascends significantly. Further, the fluctuation of torque and speed fluctuation will be increased severely. In Fig. 3.12f, h, when the enhanced HF square-wave signal injection-based method is applied, the fluctuation of the position estimation error is almost insensitive to the dead time. The enhanced HF square-wave signal injection-based method achieves the stronger robustness to the voltage errors, compared with the conventional one. More importantly, the fluctuation of the position estimation error is much smaller than that of the conventional HF square-wave signal injection-based method, as can be seen in Fig. 3.12.

Under the same conditions that the injection voltage is 30 V and the dead time is 2 μs, the harmonic components comparison of the HF pulsating sinusoidal signal injection-based method, the conventional HF square-wave signal injection-based method, and the enhanced HF square-wave signal injection-based method is performed, as shown in Fig. 3.13a, b, c, respectively. Since the digital control system is configured as the double-sampling, double-update mode, the sampling frequency is 10 kHz. In order to ensure a sufficient carrier ratio, the injection frequency of the HF pulsating sinusoidal signal injection-based method is determined as one tenth of the sampling frequency, i.e., 1 kHz. Besides, the injection frequency of the conventional and the enhanced HF square-wave signal injection-based method can reach one half and one third of the sampling frequency, respectively, that is, 5 kHz and 3.33 kHz.

As can be seen from Fig. 3.13, the induced HF current of the conventional and the enhanced HF square-wave signal injection-based method is significantly smaller, compared with that of the HF pulsating sinusoidal signal injection-based method. Considering that the enhanced HF square-wave signal injection-based method can use a smaller injection voltage, the induced HF current can be further reduced, resulting in the decreased additional loss and noise.

Compared with the HF sinusoidal signal injection-based method, the HF square-wave signal injection-based method can achieve a higher injection frequency and usually can reach half of the switching frequency. By using the double-sampling, double-update mode, the injection frequency can even reach switching frequency. Accordingly, the speed feedback is rather timely, which helps for the increased bandwidth and the improved dynamic performance. However, the losses of HF square-wave signal injection-based method are relatively high.

3.4 HF Pulse Signal Injection-Based Method

The aforementioned three HF signal injection-based methods superimpose HF signals onto the fundamental-frequency components. In view of this, the HPF or band-pass filter (BPF) may be used to extract the HF components for further retrieving the rotor position and speed information. The LPF may be also used in the process for position and speed estimation, resulting in the decreased bandwidth and deteriorated dynamic performance. Besides, the estimated rotor position and speed may be distorted by voltage errors due to the inverter nonlinearities, the motor resistance, and the rotational voltage drops. To overcome this drawback, HF pulse signal injection-based method was proposed by researchers in recent years [8]. Block diagram of HF pulse signal injection-based sensorless PMSM drive is shown in Fig. 3.14. Digital filters are not used and the robustness to the voltage errors is improved greatly without any compensation.

The sequence of HF pulse signal injection-based method is shown in Fig. 3.15. By separating the FOC period and the injection period, the extraction of the fundamental and the induced currents becomes easy without using any digital filters. Moreover, the opposite voltage components are required to offset the effects of voltage errors on the position estimation accuracy. When the FOC period is enabled, no HF signal is injected. Besides, when the injection period is enabled, the FOC is interrupted. Thus, the sampling current is induced HF current, and no HPF or BPF is needed to extract HF components.

When the positive injection period is enabled, the positive voltage component U_h is injected into the $\hat{d}$ axis, and the corresponding voltage component injected into the $\hat{q}$ axis is zero. Taking the voltage errors into consideration, the injected voltage vector will be distorted from U_{INJ} to $\overline{\mathrm{U}}_{INJ}$ as shown in Fig. 3.16. The voltage in the estimated RRF is modified from (3.27)–(3.28),

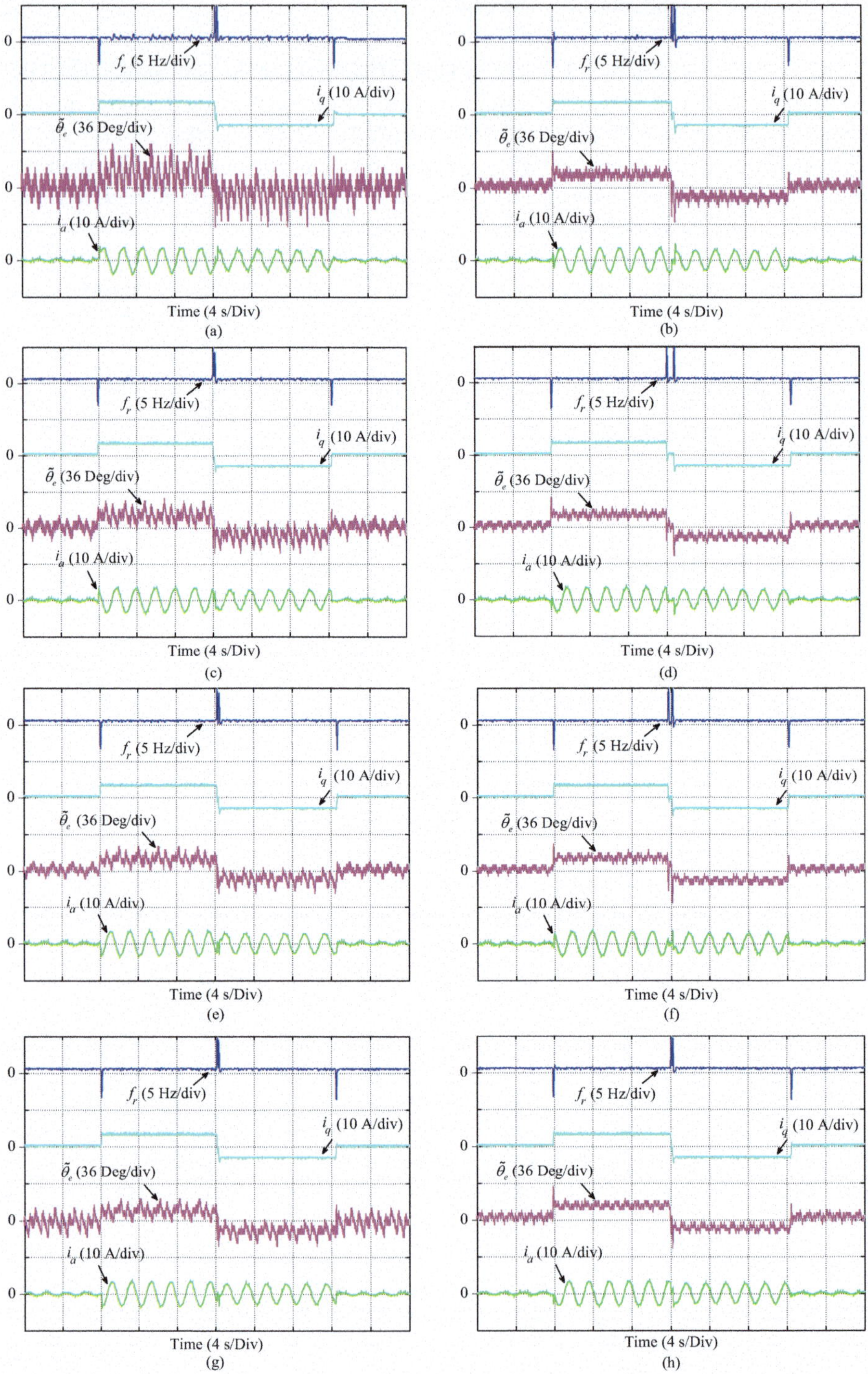

Fig. 3.12 Experimental waveform comparison between conventional and enhanced HF square-wave signal injection-based methods. (**a**) Conventional scheme with injected voltage of 10 V and

$$\begin{bmatrix} u_{\hat{d}h\_p} \\ u_{\hat{q}h\_p} \end{bmatrix} = \begin{bmatrix} U_h \\ 0 \end{bmatrix}, \tag{3.27}$$

$$\begin{bmatrix} u'_{\hat{d}h\_p} \\ u'_{\hat{q}h\_P} \end{bmatrix} = \begin{bmatrix} U_h - U_{\hat{d},err} \\ -U_{\hat{q},err} \end{bmatrix}, \tag{3.28}$$

where $U_{\hat{d},err}$ and $U_{\hat{q},err}$ are the voltage errors in the estimated RRF. Subscripts p and n indicate the positive and the negative injection period, respectively. Besides, $u'_{\hat{d},\hat{q}h}$ and $u_{\hat{d},\hat{q}h}$ are the voltages in the estimated RRF with and without voltage errors taken into account, respectively. By replacing the current differentiation di/dt with $\Delta i/\Delta t$, the HF current variation in the SRF can be deduced as follows:

$$\begin{bmatrix} \Delta i_{\alpha h\_p} \\ \Delta i_{\beta h\_P} \end{bmatrix} = \frac{\Delta T}{L_d L_q} \begin{bmatrix} \left(U_h - U_{\hat{d},err}\right) \cdot \left(L_d \sin\theta_e \sin\tilde{\theta}_e + L_q \cos\theta_e \cos\tilde{\theta}_e\right) + U_{\hat{q},err} \cdot \left(L_d \sin\theta_e \cos\tilde{\theta}_e - L_q \cos\theta_e \sin\tilde{\theta}_e\right) \\ \left(U_h - U_{\hat{d},err}\right) \cdot \left(-L_d \cos\theta_e \sin\tilde{\theta}_e + L_q \sin\theta_e \cos\tilde{\theta}_e\right) - U_{\hat{q},err} \cdot \left(L_d \cos\theta_e \cos\tilde{\theta}_e + L_q \sin\theta_e \sin\tilde{\theta}_e\right) \end{bmatrix}, \tag{3.29}$$

where ΔT denotes sampling period. Similarly, when the negative injection period is enabled, -U_h is injected into the $\hat{d}$ axis and the corresponding voltage component injected into the $\hat{q}$ axis is zero. Assuming that the voltage errors are approximately constant in one injection period, the HF current variation in the SRF can be deduced as follows:

$$\begin{bmatrix} \Delta i_{\alpha h\_n} \\ \Delta i_{\beta h\_n} \end{bmatrix} = \frac{\Delta T}{L_d L_q} \begin{bmatrix} \left(-U_h - U_{\hat{d},err}\right) \cdot \left(L_d \sin\theta_e \sin\tilde{\theta}_e + L_q \cos\theta_e \cos\tilde{\theta}_e\right) + U_{\hat{q},err} \cdot \left(L_d \sin\theta_e \cos\tilde{\theta}_e - L_q \cos\theta_e \sin\tilde{\theta}_e\right) \\ \left(-U_h - U_{\hat{d},err}\right) \cdot \left(-L_d \cos\theta_e \sin\tilde{\theta}_e + L_q \sin\theta_e \cos\tilde{\theta}_e\right) - U_{\hat{q},err} \cdot \left(L_d \cos\theta_e \cos\tilde{\theta}_e + L_q \sin\theta_e \sin\tilde{\theta}_e\right) \end{bmatrix}. \tag{3.30}$$

Fig. 3.12 (continued) dead time of 2 μs. (**b**) Enhanced scheme with injected voltage of 10 V and dead time of 2 μs. (**c**) Conventional scheme with injected voltage of 20 V and dead time of 2 μs. (**d**) Enhanced scheme with injected voltage of 20 V and dead time of 2 μs. (**e**) Conventional scheme with injected voltage of 30 V and dead time of 2 μs. (**f**) Enhanced scheme with injected voltage of 30 V and dead time of 2 μs. (**g**) Conventional scheme with injected voltage of 30 V and dead time of 4 μs. (**h**) Enhanced scheme with injected voltage of 30 V and dead time of 4 μs

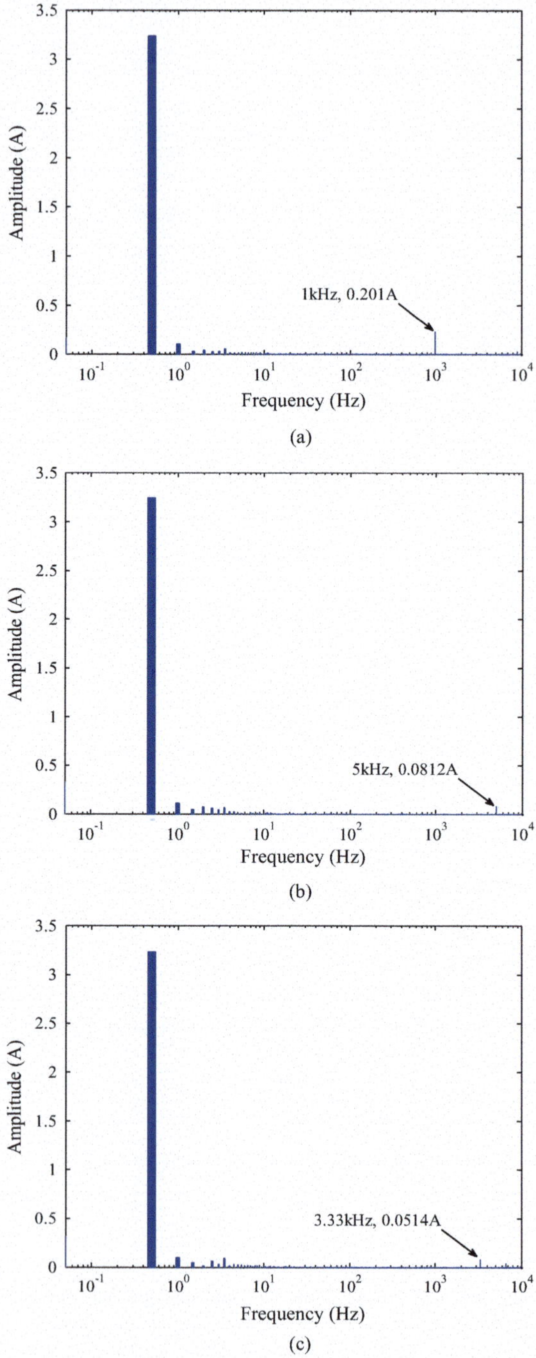

Fig. 3.13 Phase current harmonic comparison among three HF signal injection-based methods in the RRF at the same load with injected voltage of 30 V and dead time of 2 μs. (**a**) HF pulsating sinusoidal signal injection-based method. (**b**) HF square-wave signal injection-based method with conventional signal processing. (**c**) HF square-wave signal injection-based method with enhanced signal processing

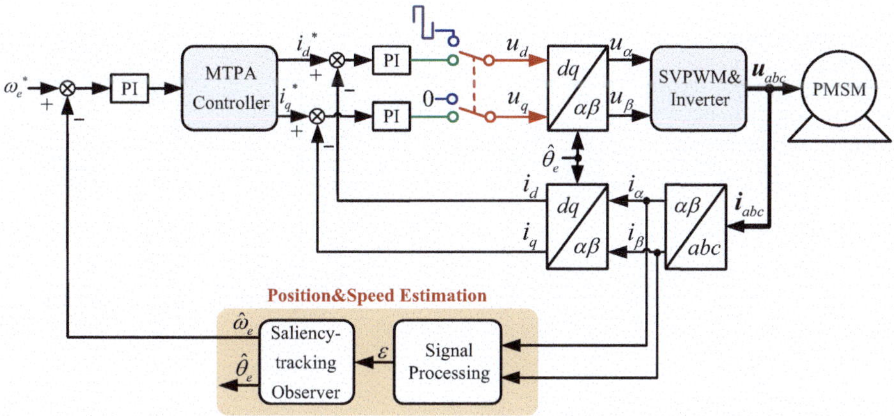

Fig. 3.14 Block diagram of HF pulse signal injection-based sensorless PMSM drive

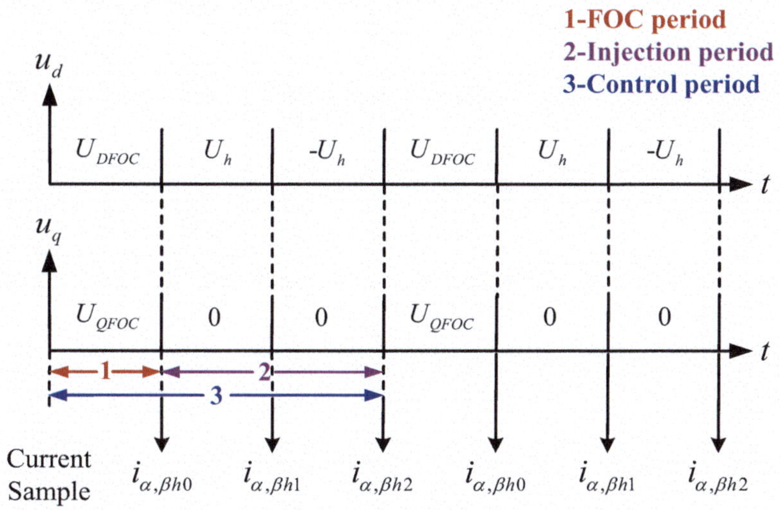

Fig. 3.15 Sequence of HF pulse signal injection-based method

As can be seen from (3.28) to (3.30), the injected voltage and the induced HF current are severely distorted, which eventually causes a deteriorated position estimation result. Hence the opposite voltage components are indispensable to cancel the voltage errors. Subsequently, two current variations in the positive and negative injection periods are needed for subtraction to offset the impacts of voltage errors on the position estimation accuracy, yielding

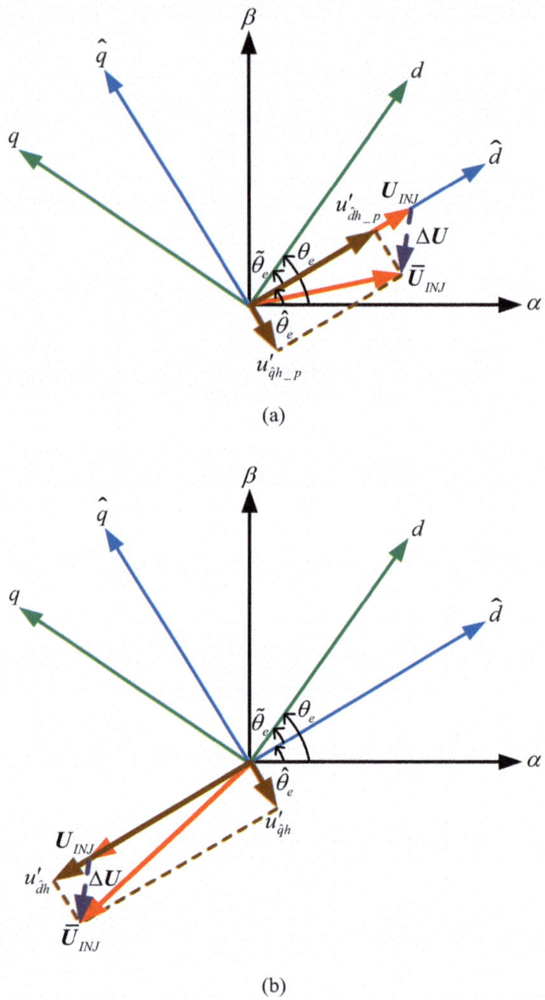

Fig. 3.16 Coordinate distribution of each axis. (**a**) Distorted voltage in the positive injection period. (**b**) Distorted voltage in the negative injection period

$$\begin{bmatrix} \Delta i_{\alpha h\_p} - \Delta i_{\alpha h\_n} \\ \Delta i_{\beta h\_p} - \Delta i_{\beta h\_n} \end{bmatrix} = \begin{bmatrix} 2i_{\alpha h1} - i_{\alpha h0} - i_{\alpha h2} \\ 2i_{\beta h1} - i_{\beta h0} - i_{\beta h2} \end{bmatrix} = 2K \begin{bmatrix} \Delta i_{\alpha h\_pu} \\ \Delta i_{\beta h\_pu} \end{bmatrix}, \tag{3.31}$$

where

$$K = \frac{\Delta T U_h}{L_d}, \tag{3.32}$$

$$\begin{bmatrix} \Delta i_{\alpha h\_pu} \\ \Delta i_{\beta h\_pu} \end{bmatrix} = \begin{bmatrix} \cos(\theta_e) \\ \sin(\theta_e) \end{bmatrix}. \tag{3.33}$$

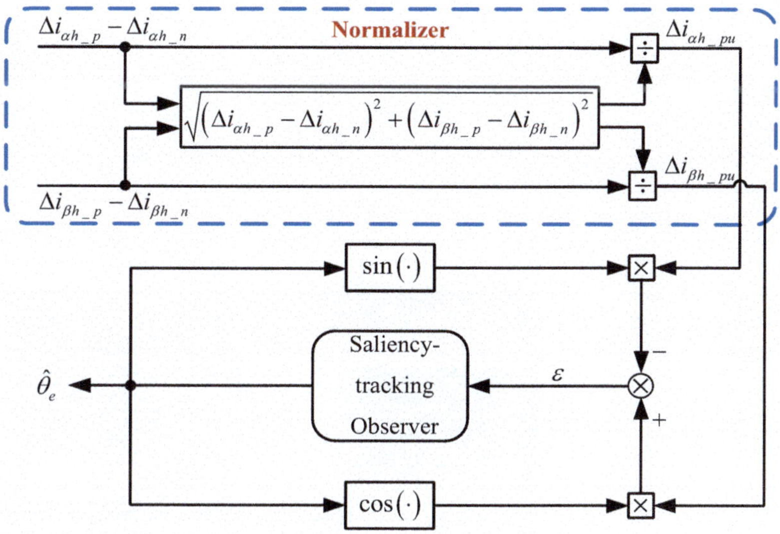

Fig. 3.17 Process for position and speed estimation

In order to reduce the effects of the inductance variety and the injected amplitude fluctuation on 2 K, normalization is essential as shown in Fig. 3.17. Then, ε associated with the position estimation error can be obtained as (3.34) through the heterodyne processing, which acts as the input of the saliency-tracking observer to further estimate the rotor position and speed.

$$\begin{aligned} \varepsilon &= \Delta i_{\beta h\_pu} \cos\left(\hat{\theta}_e\right) - \Delta i_{\alpha h\_pu} \sin\left(\hat{\theta}_e\right) \\ &= \sin\left(\tilde{\theta}_e\right) \approx \tilde{\theta}_e. \end{aligned} \tag{3.34}$$

HF pulse signal injection-based method is verified via an experimental platform with a 2.2-kW PMSM, the parameters of which are listed in Table 3.2. As shown in Fig. 3.18, an induction motor mechanically couples with PMSM as the load. A low-cost ARM chip STM32F103VCT6 is applied to execute the entire algorithm. In the experiments, the switching frequency of the inverter and the sampling frequency are both 6 kHz. The amplitude and frequency of the injected voltage are 120 V and 1.5 kHz, respectively.

In the following experiment, the observed rotor speed and position from the saliency-tracking observer are applied as the feedback, and the experimental results are obtained based on the sensorless control method.

Figure 3.19a, b show the waveforms of $\Delta i_{\alpha h\_pu}$, $\Delta i_{\beta h\_pu}$, and their Lissajous figure at the operation of 100 r/min with no load, respectively. It can be seen from Fig. 3.19a that $\Delta i_{\alpha h\_pu}$ and $\Delta i_{\beta h\_pu}$ achieve the good sinusoidal signal, and the HF noise is smaller. Besides, as can be seen in Fig. 3.19b, the Lissajous figure is near a circle, indicating that $\Delta i_{\alpha h\_pu}$ and $\Delta i_{\beta h\_pu}$ have good orthogonality.

Table 3.2 PMSM parameters

Parameter	Value	Parameter	Value
Rated power (kW)	2.2	Rated torque (N·m)	21
Rated voltage (V)	380	Stator resistance (Ω)	2.75
Rated current (A)	5.6	*d*-axis inductance (mH)	31.6
Rated speed (r/min)	1000	*q*-axis inductance (mH)	62.8
Rated frequency (Hz)	50	Flux linkage (Wb)	0.56

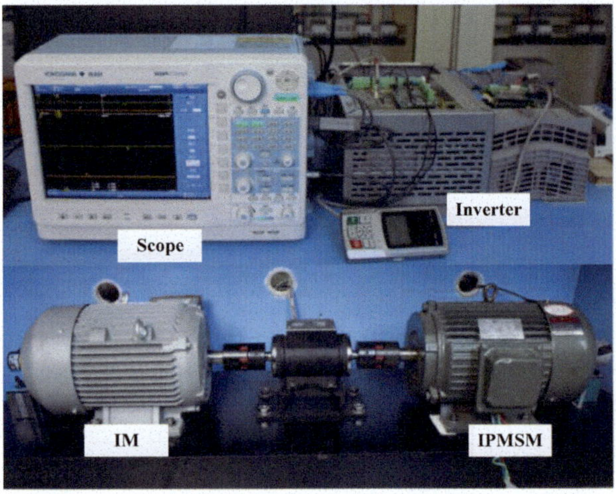

Fig. 3.18 Experimental platform with a 2.2-kW PMSM drive

Figure 3.20a, b show the waveforms of the actual rotor position, the position estimation error, and the phase current, when the IPMSM operates at 100 r/min and 20 r/min with step rated-load torque change, respectively. According to the experimental results, it can be noticed that the transient position estimation error is within 0.20 rad when the IPMSM operates at 100 r/min with rated load, and the maximum position error is 0.19 rad in the dynamic response at 20 r/min. Hence, the position estimation errors can fulfill the observation accuracy requirements.

In Fig. 3.21, the speed reversal at ±100 r/min with half- and full-rated load torque is depicted. As can be seen, the HF pulse voltage injection-based sensorless IPMSM drive can well survive in the reversal operation.

3.5 Fundamental PWM Excitation-Based Method

High frequency signal injection methods have been widely used in sensorless control of PMSM drives. However, the acoustic noise and torque ripples caused by the injected high frequency signal limit the application of the methods. To avoid these drawbacks, this section proposes a position estimation method combining derivative

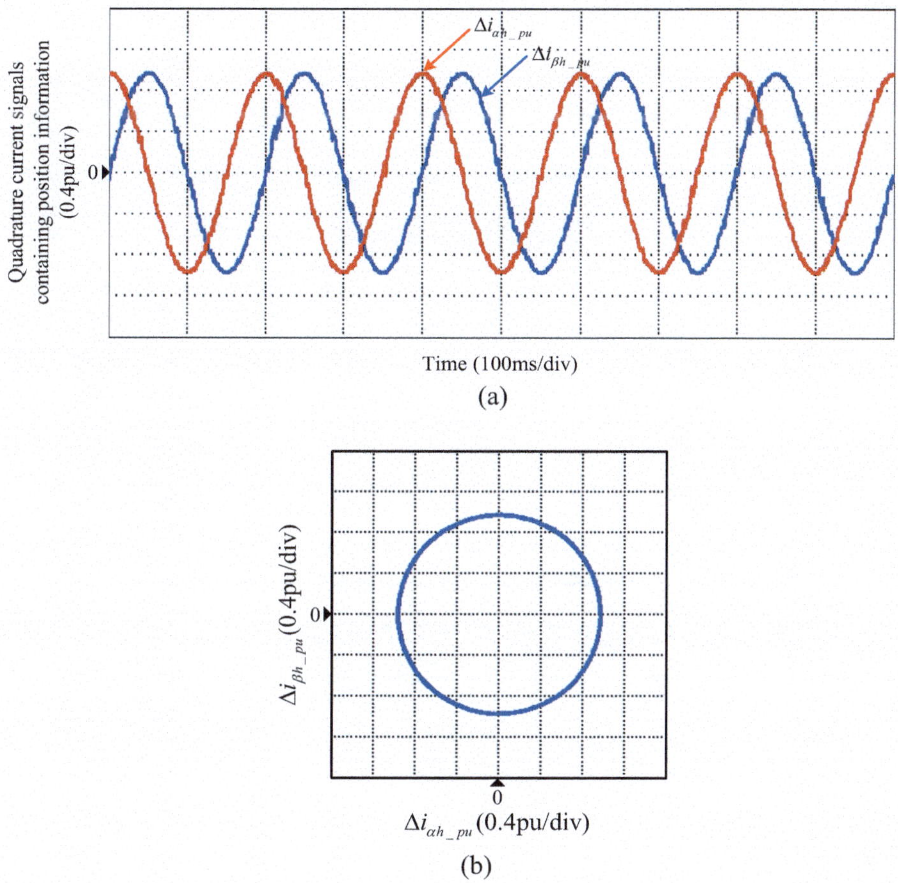

Fig. 3.19 $\Delta i_{\alpha h\_pu}$, $\Delta i_{\beta h\_pu}$, and their Lissajous figure. (**a**) $\Delta i_{\alpha h\_pu}$ and $\Delta i_{\beta h\_pu}$. (**b**) the Lissajous figure of $\Delta i_{\alpha h\_pu}$ and $\Delta i_{\beta h\_pu}$

calculations of current and zero voltage vector (ZVV) injection, which is especially effective for zero- and low-speed operation of sensorless PMSM drives [9]. This investigation mainly focuses on the extraction of rotor position from ZVV period and the improvements of the accuracy of estimated position. The PMSM equation is developed in the stationary frame when ZVV is applied. The proposed d-axis current allocation scheme ensures the application of the proposed method from no load to rated load. A modified SVPWM control strategy and a ZVV extension scheme are employed to improve the accuracy of estimated position, which also makes the method more realizable. Further, a novel current sampling method selecting two suitable phase currents for derivative calculation according to the voltage sector location is proposed, which reduces current derivative ripples and decreases position estimation errors. Finally, the effectiveness of the proposed position estimation strategy is verified on a 1.0 kW interior PMSM (IPMSM) drive platform.

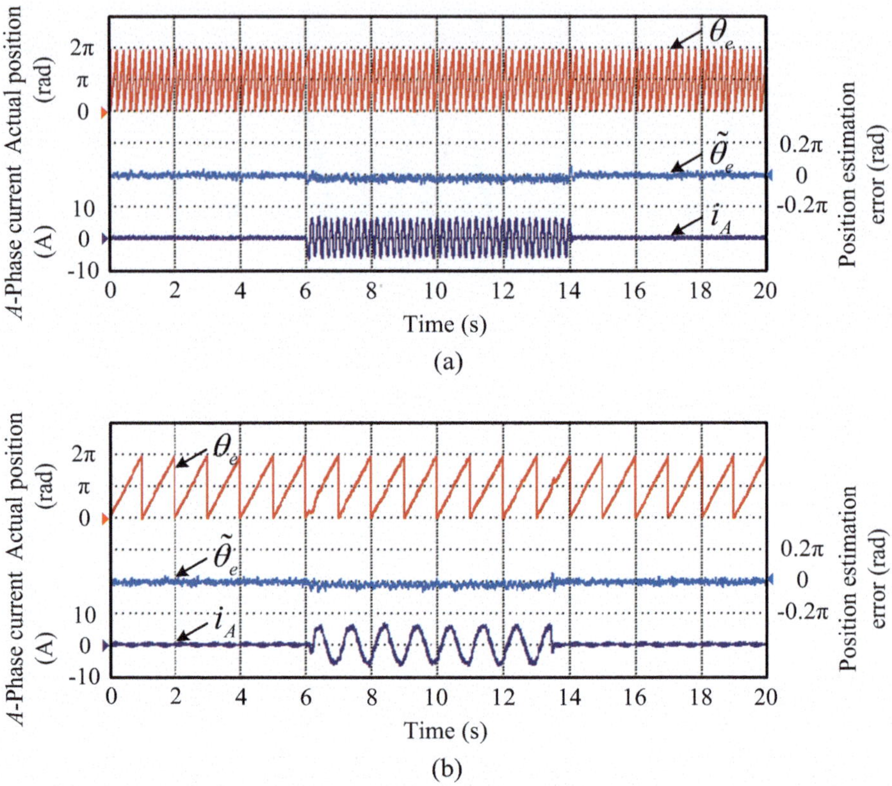

Fig. 3.20 Sensorless operation with step rated-load torque change. (**a**) Operation at 100 r/min. (**b**) Operation at 20 r/min

3.5.1 Mathematical Analysis of Rotor Position Estimation with Zero Voltage Vector (ZVV)

The voltage equation of PMSMs in the rotor reference frame can be expressed as (1.52) where ψ_d and ψ_q are the flux linkages and can be expressed as

$$\begin{bmatrix} \psi_d \\ \psi_q \end{bmatrix} = \begin{bmatrix} L_d & 0 \\ 0 & L_q \end{bmatrix} \begin{bmatrix} i_d \\ i_q \end{bmatrix} + \begin{bmatrix} \psi_f \\ 0 \end{bmatrix}, \tag{3.35}$$

where L_d, L_q, and ψ_f are the d-q synchronous frame inductances and the permanent magnet flux linkage, respectively.

Substituting ψ_d and ψ_q in (3.34) with (3.35), yielding

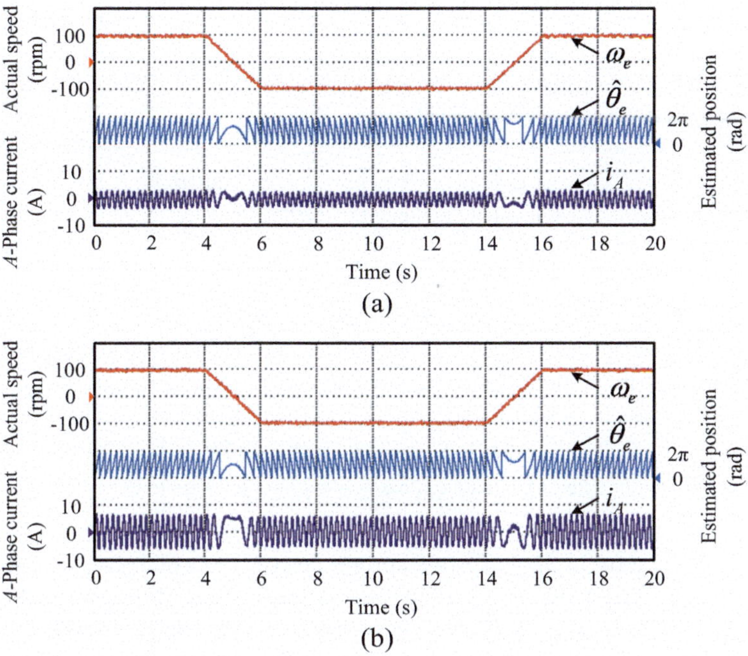

Fig. 3.21 Speed reversal at ±100 r/min with half- and full-rated load torque. (**a**) Half-rated load torque. (**b**) Full-rated load torque

$$\begin{bmatrix} u_d \\ u_q \end{bmatrix} = R\begin{bmatrix} i_d \\ i_q \end{bmatrix} + \begin{bmatrix} pL_d & -L_q\omega_e \\ L_d\omega_e & pL_q \end{bmatrix}\begin{bmatrix} i_d \\ i_q \end{bmatrix} + \omega_e\begin{bmatrix} 0 \\ \psi_f \end{bmatrix}. \tag{3.36}$$

Transforming (3.35) from the rotating frame to the stationary frame, the voltage equation can be expressed as

$$\begin{aligned} \begin{bmatrix} u_\alpha \\ u_\beta \end{bmatrix} &= R\begin{bmatrix} i_\alpha \\ i_\beta \end{bmatrix} + \begin{bmatrix} \Sigma L + \Delta L\cos 2\theta_e & \Delta L\sin 2\theta_e \\ \Delta L\sin 2\theta_e & \Sigma L\text{-}\Delta L\cos 2\theta_e \end{bmatrix} p\begin{bmatrix} i_\alpha \\ i_\beta \end{bmatrix} \\ &+\omega_e\begin{bmatrix} -2\Delta L\sin 2\theta_e & 2\Delta L\cos 2\theta_e \\ 2\Delta L\cos 2\theta_e & 2\Delta L\sin 2\theta_e \end{bmatrix}\begin{bmatrix} i_\alpha \\ i_\beta \end{bmatrix} + \psi_f\omega_e\begin{bmatrix} -\sin\theta_e \\ \cos\theta_e \end{bmatrix}, \end{aligned} \tag{3.37}$$

where u_α, u_β, i_α, and i_β are the voltages and currents components in the SRF and θ_e is the electrical rotor position.

When the ZVV is applied, the superimposed voltage to the PMSM is null, and the PMSM equation is expressed as

$$\begin{aligned}\mathbf{Y}_d \cdot p\begin{bmatrix} i_\alpha \\ i_\beta \end{bmatrix} &= -R\begin{bmatrix} i_\alpha \\ i_\beta \end{bmatrix} - \psi_f \omega_e \begin{bmatrix} -\sin\theta_e \\ \cos\theta_e \end{bmatrix} \\ &\quad -\omega_e \begin{bmatrix} -2\Delta L \sin 2\theta_e & 2\Delta L \sin 2\theta_e \\ 2\Delta L \cos 2\theta_e & 2\Delta L \cos 2\theta_e \end{bmatrix} \begin{bmatrix} i_\alpha \\ i_\beta \end{bmatrix},\end{aligned} \tag{3.38}$$

where $\boldsymbol{Y}_d$ represents the inductance matrix and can be expressed as

$$\mathbf{Y}_d = \begin{bmatrix} \Sigma L + \Delta L \cos 2\theta_e & \Delta L \sin 2\theta_e \\ \Delta L \sin 2\theta_e & \Sigma L\text{-}\Delta L \cos 2\theta_e \end{bmatrix}. \tag{3.39}$$

In (3.38), the term in the left is the current derivatives which can be detected or calculated. The first term in the right side is the voltage drop across the stator resistance. Generally, it is neglected because of the very low amplitude compared to the rest terms. However, the voltage drop in the stator resistance contributes a fairly large proportion since the other two terms which contain the speed parameter are all small variables when the motor operates in low-speed region and standstill. The second term is induced by permanent magnet flux, while the third term exists due to saliency effect.

To simplify (3.38), the equation can be expressed as functions of the stationary frame currents i_α and i_β by multiplying the inverse matrix of $\boldsymbol{Y}_d$ in both sides, as shown in (3.40).

Therefore, the PMSM equations in the stationary frame are obtained when the ZVV is applied. From (3.40), it can be observed that the speed and the position information exists both in the coefficients of the stationary frame current terms and the back-EMF terms. The amplitude of the third term in the right dies down with the decrease of the operating speed. Hence, separating the coupled speed and position is necessary in the position estimation process. There are two solutions considered to solve this problem. One is to eliminate the speed in the two equations and calculate the rotor position directly. The other is to eliminate the θ_e by manipulating the two equations in (3.40) and then calculating the rotor position by integrating the remaining ω_e. Actually, it is not practical for the first solution since the speed dies down to low amplitude signals and even can be regarded as noises when the operating speed approaches standstill. Thus, an advantageous method eliminating terms pertaining to rotor position will be introduced in the next part.

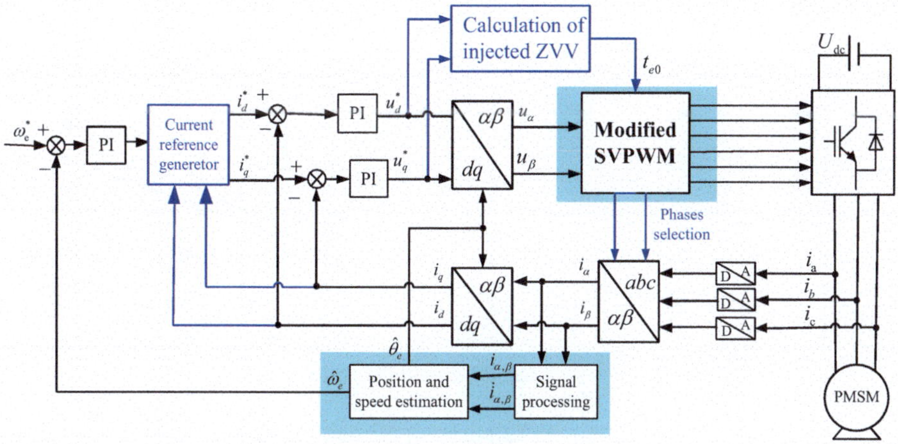

Fig. 3.22 Proposed sensorless FOC scheme for the PMSM

$$
\begin{cases}
\frac{di_\alpha}{dt} = \frac{1}{L_d L_q}\{2\Sigma L \Delta L \omega_e \sin 2\theta_e - R(\Sigma L - \Delta L \cos 2\theta_e) \cdot i_\alpha \\
\qquad -2\Delta L(\Sigma L \omega_e \cos 2\theta_e - 2\Delta L \omega_e - R \sin 2\theta_e) \cdot i_\beta\} \\
\qquad + \frac{\psi_f \omega_e}{L_q} \sin \theta_e \\
\frac{di_\beta}{dt} = \frac{1}{L_d L_q}\{2\Delta L(R \sin 2\theta_e - \Sigma L \omega_e \cos 2\theta_e - 2\Delta L \omega_e) \cdot i_\alpha \\
\qquad -R(\Sigma L + \Delta L \cos 2\theta_e) + 2\Sigma L \Delta L \omega_e \sin 2\theta_e \cdot i_\beta\} \\
\qquad - \frac{\psi_f \omega_e}{L_q} \cos \theta_e.
\end{cases}
\tag{3.40}
$$

Figure 3.22 shows the proposed sensorless control scheme. In the diagram, the stationary frame currents are used to estimate the rotor position and speed.

Considering an estimated *d-q* frame, d_{est}-q_{est} frame, as shown in Fig. 3.23, the estimated electrical rotor position is $\hat{\theta}_e$ and the offset from the real rotor position is

$$
\tilde{\theta}_e = \theta_e - \hat{\theta}_e. \tag{3.41}
$$

Actually, the current derivatives in (3.40) can be regarded as independent variables, since the current derivatives in the stationary frame are calculated from separately sampled current groups during ZVV intervals. To eliminate terms related to the rotor position in current coefficients, the phase-locked loop method is adopted. Multiplying $\sin \hat{\theta}_e$ for terms reflected in α-axis, and multiplying $\cos \hat{\theta}_e$ for terms reflected in β-axis in (3.40), the equation can be rewritten as (3.42) and (3.43).

Fig. 3.23 Relationship of the actual frame and the estimation frame

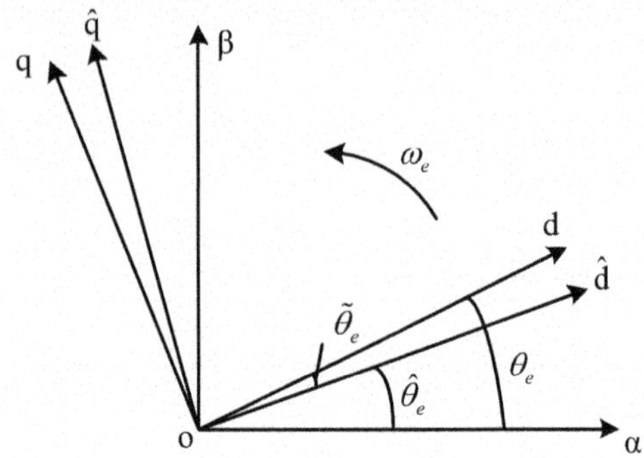

$$
\begin{aligned}
\frac{di_\alpha}{dt}\sin\hat{\theta}_e = &\frac{1}{L_dL_q}\{2\Sigma L\Delta L\omega_e \sin 2\theta_e \sin\hat{\theta}_e \cdot i_\alpha \\
&- R\left(\Sigma L \sin\hat{\theta}_e - \Delta L\cos 2\theta_e \sin\hat{\theta}_e\right)\cdot i_\alpha \\
&- 2\Delta L\omega_e(\Sigma L\cos 2\theta_e - 2\Delta L)\sin\hat{\theta}_e \cdot i_\beta \\
&+ 2R\Delta L \sin 2\theta_e \sin\hat{\theta}_e \cdot i_\beta\} \\
&+ \frac{\psi_f}{L_q}\omega_e \sin\theta_e \sin\hat{\theta}_e,
\end{aligned} \tag{3.42}
$$

$$
\begin{aligned}
\frac{di_\beta}{dt}\cos\hat{\theta}_e = &\frac{1}{L_dL_q}\{R\Delta L\sin 2\theta_e \cos\hat{\theta}_e \cdot i_\alpha \\
&- 2\Delta L\omega_e\left(\Sigma L\cos 2\theta_e \cos\hat{\theta}_e - \cos\hat{\theta}_e\right)\cdot i_\alpha \\
&- 2\Sigma L\Delta L\omega_e \sin 2\theta_e \cos\hat{\theta}_e \cdot i_\beta \\
&- R\left(\Sigma L\cos\hat{\theta}_e + \Delta L\cos 2\theta_e \cos\hat{\theta}_e\right)\cdot i_\beta\} \\
&- \frac{\psi_f\omega_e}{L_q}\cos\theta_e\cos\hat{\theta}_e.
\end{aligned} \tag{3.43}
$$

Subtracting (3.43) from (3.42), the (3.44) can be expressed by

$$
\begin{aligned}
&\frac{di_\alpha}{dt}\sin\hat{\theta}_e - \frac{di_\beta}{dt}\cos\hat{\theta}_e \\
&= \frac{1}{L_dL_q}\{2\Delta L\Sigma L\omega_e\cos\left(2\theta_e - \hat{\theta}_e\right)\cdot i_\alpha \\
&-R\left(\Sigma L\sin\hat{\theta}_e + \Delta L\sin\left(2\theta_e - \hat{\theta}_e\right)\right)\cdot i_\alpha \\
&+R\left(\Sigma L\cos\hat{\theta}_e + \Delta L\cos\left(2\theta_e - \hat{\theta}_e\right)\right)\cdot i_\beta \\
&+2\Sigma L\Delta L\omega_e\sin\left(2\theta_e - \hat{\theta}_e\right)\cdot i_\beta \\
&+2\Sigma L\Delta L\omega_e\sin\left(2\theta_e - \hat{\theta}_e\right)\cdot i_\beta \\
&+\frac{\psi_f\omega_e}{L_q}\cos\left(\theta_e - \hat{\theta}_e\right).
\end{aligned} \tag{3.44}
$$

From (3.44), it can be seen that the coefficients of stationary frame currents contain the position information. To eliminate $2\theta_e - \hat{\theta}_e$ terms, the following can be substituted into (3.44)

$$2\theta_e - \hat{\theta}_e = \hat{\theta}_e + 2\tilde{\theta}_e. \tag{3.45}$$

When the estimated rotor position approaches the real rotor position, the position error $\tilde{\theta}_e$ converges to zero. According to the Taylor expansion, the following approximate equation is valid:

$$\cos\left(\hat{\theta}_e + 2\tilde{\theta}_e\right) = \cos\hat{\theta}_e - \frac{1}{2!}\left(2\tilde{\theta}_e\right)^2 + o\left(\tilde{\theta}_e^2\right) \cong \cos\hat{\theta}_e, \tag{3.46}$$

$$\sin\left(\hat{\theta}_e + 2\tilde{\theta}_e\right) = \sin\hat{\theta}_e + 2\tilde{\theta}_e + o\left(\tilde{\theta}_e\right) \cong \sin\hat{\theta}_e. \tag{3.47}$$

By substituting (3.46) and (3.47) into (3.44), the simplified equation is rewritten as

$$\begin{aligned} &L_q\left(\frac{di_\alpha}{dt}\sin\hat{\theta}_e - \frac{di_\beta}{dt}\cos\hat{\theta}_e\right) = \\ &R\left(i_\beta\cos\hat{\theta}_e - i_\alpha\sin\hat{\theta}_e\right) \\ &+2\Delta L\left(i_\alpha\cos\hat{\theta}_e + i_\beta\sin\hat{\theta}_e\right)\cdot\omega_e \\ &+\psi_f\cos\tilde{\theta}_e\cdot\omega_e. \end{aligned} \tag{3.48}$$

For (3.48), considering the second term in the right side, when the position estimation method is applied in surface-mounted PMSM drives, it is clear that it can be neglected since the difference inductance in SPMSM is zero. However, this term is a non-zero quantity in IPMSM drive applications. Actually, the expression $i_\alpha\cos\hat{\theta}_e + i_\beta\sin\hat{\theta}_e$ can be approximately regarded as the *d*-axis current if the estimated rotor position approaches the actual rotor position. In addition, the second term is a smaller quantity compared with the permanent magnet flux linkage which is shown in the third term, and the two terms both contain rotor speed. For the third term, considering $\cos\tilde{\theta}_e \cong 1$, this condition is fulfilled in a relatively large scale of $\tilde{\theta}_e$, since the Taylor expansion of cosine function converges to unitary in second order in the immediate zero range. The actual speed in the equation can be replaced by the estimated speed when the estimation method converges,

$$\omega_e \cong \hat{\omega}_e = \frac{d\hat{\theta}_e}{dt}. \tag{3.49}$$

Therefore, (3.48) can be simplified and expressed as

$$\left(\frac{di_\alpha}{dt} + \frac{R}{L_q} i_\alpha\right) \cdot \sin\hat{\theta}_e - \left(\frac{di_\beta}{dt} + \frac{R}{L_q} i_\beta\right) \cdot \cos\hat{\theta}_e = \frac{\psi_f}{L_q}\frac{d\hat{\theta}_e}{dt}. \tag{3.50}$$

Equation (3.50) is the PMSM equation in the ZVV period. Different from the conventional motor equation which is based on states of currents and voltages, Eq. (3.50) establishes an input-output relationship among the rotor position, the stator currents, and the current derivatives. The rotor speed is a non-zero quantity as long as the motor rotates. It means that conventional position observers based on the current state equations cannot be applied into this control scheme. However, the rotor position can be calculated by integrating the left side:

$$\hat{\theta}_e^{n+1} = \hat{\theta}_e^{n} + \frac{L_q T_s}{\psi_f}\left(\frac{di_\alpha}{dt}\sin\hat{\theta}_e - \frac{di_\beta}{dt}\cos\hat{\theta}_e \quad + \frac{R}{L_q} i_\alpha \sin\hat{\theta}_e - \frac{R}{L_q} i_\beta \cos\hat{\theta}_e\right), \tag{3.51}$$

where T_s is the control period and n is the calculating index representing the time instant $t = nT_s$.

From (3.51), it can be seen that the electrical rotor position is calculated by integrating terms related to stator currents and current derivatives. As aforementioned, the four variables can be measured or calculated through measured parameters, and all the four variables are multiplied by a term pertaining to $\hat{\theta}_e$. Thus, the rotor position estimation scheme can be realized by a phase-locked loop, as shown in Fig. 3.24.

In Fig. 3.24, the dash line shows the term relating to the difference inductance. It can be neglected in the surface-mounted SPMSM drive application, since the inductance difference in SPMSMs is zero. However, it should be taken into consideration

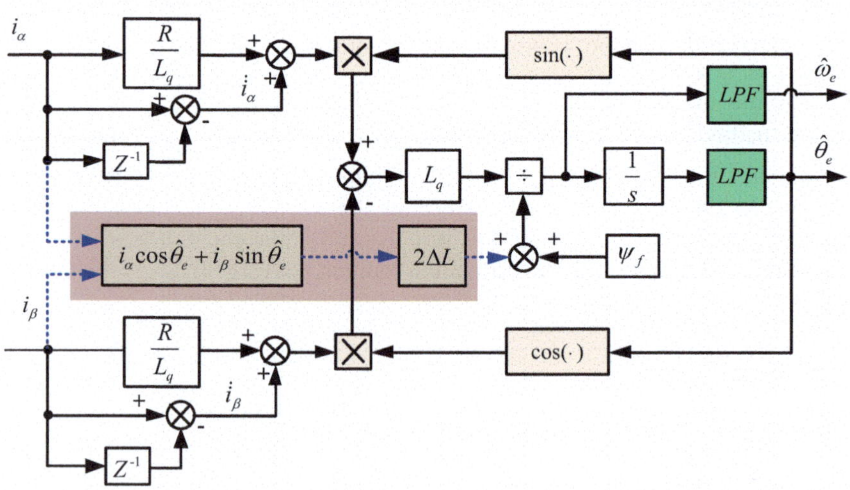

Fig. 3.24 Proposed position estimation scheme

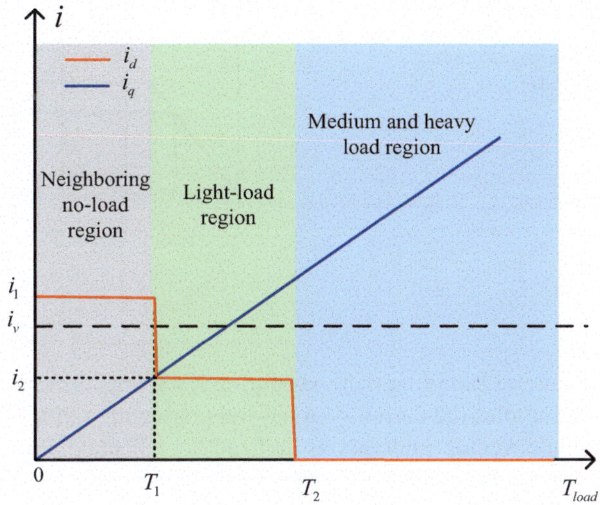

Fig. 3.25 Arrangement of dq frame current according to load torque

when applying in the IPMSM drives if the IPMSM has a large inductance difference. From (3.51) and the control scheme in Fig. 3.24, it can be seen that the implementation of the proposed position estimation method necessitates maintaining the coefficients of $\sin\hat{\theta}_e$ and $\cos\hat{\theta}_e$ nonconstant zero value. In medium- and heavy-load regions, the requirements are unconditionally satisfied as the existence of high-amplitude q-axis current. In no-load and light-load regions, to ensure the proposed method can be employed successfully, the d-axis current should be regulated to a non-zero value by modifying the reference of i_d as shown in Fig. 3.25.

3.5.2 Current Sampling Method

From (3.51), it can be seen that the accuracy of sampled currents and current derivatives will decide the level of position error of the proposed method. Figure 3.26 shows the proposed modified SVPWM method for acquiring the rotor position more effectively. To take advantage of the relatively shorter period of the effective vector, the proportions of zero voltage intervals t_0 and t_7 are tuned, as shown in Fig. 3.26. In addition, a ZVV is inserted between each carrier signal according to load torque and rotor speed as shown in Fig. 3.26. Occasionally, torque disturbance or speed variation may cause the increase of effective vectors, and as a consequence, t_0 is not enough to calculate current derivatives. Hence, the injecting of ZVV is necessary. However, the injected ZVV will not limit the bandwidth significantly since the control period has not been extended too much. The time difference Δt between two sampled current groups can be set to a desired value by modifying the length of t_{e0} for the convenience of current derivative calculation and for the sake of truncation error reduction.

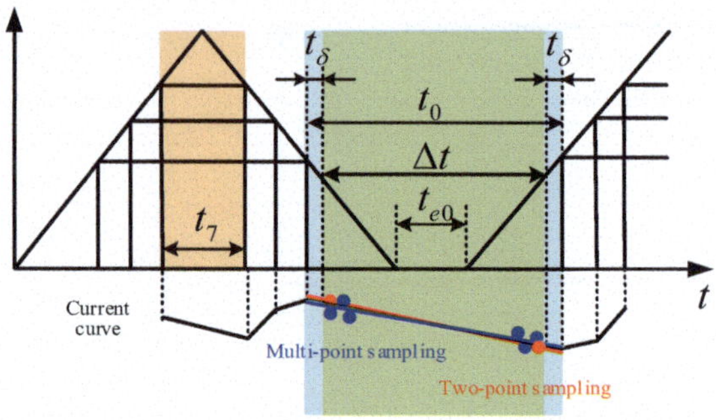

Fig. 3.26 Zero voltage period extension scheme and current sampling methods

Current sampling methods during ZVV time are analyzed in Fig. 3.26. A method that samples the currents for several consecutive times in each start and end of zero voltage vector intervals is utilized to calculate the current slope instead of a two-point method. Apparently, it can help reduce the disturbance and errors caused by hardware circuits.

The current derivative calculation formula is described as

$$\dot{i}_{sp} = \frac{\sum_{k=m+1}^{2m} i_{sp}[k] - \sum_{n=1}^{m} i_{sp}[n]}{m\Delta t}, \tag{3.52}$$

where the subscript sp denotes a, b, c. Δt represents the sampling time difference, and m is the sampling times in the start and end of each zero voltage interval.

To reduce the dead-time effect and the current ripples aroused by IGBT turn-on and turn-off, a modified current sampling method is used as shown in Fig. 3.27. As it is known, the conclusion that the sum of three phase currents is zero is valid at any moment. Therefore, stationary frame current derivatives can be calculated by two of the three-phase currents. For example, in sector I and sector VI of six FOC sectors shown in Fig. 3.27, i_b and i_c will be used to calculate current derivatives, since in sector I and sector VI, the zero voltage interval t_0 begins with the switching off of phase A. Phase B and phase C switch off earlier than phase A as the existence of effective vectors. The dead time effect and current ripple caused by IGBT turn-off will be drastically reduced if calculating current derivatives by using phase B and phase C.

3.5.3 Experimental Results for ZVV Injection

The ZVV injection-based sensorless control scheme is verified on a 1.0-kW IPMSM, supplied by a VSI as shown in Fig. 3.28. The parameters of IPMSM are shown in Table 3.3. A mechanically coupled load IPMSM is employed to produce the load torque. The sensorless control algorithm system is implemented through a low-cost ARM chip STM32F103VCT6. The unmodified PWM switching frequency of the

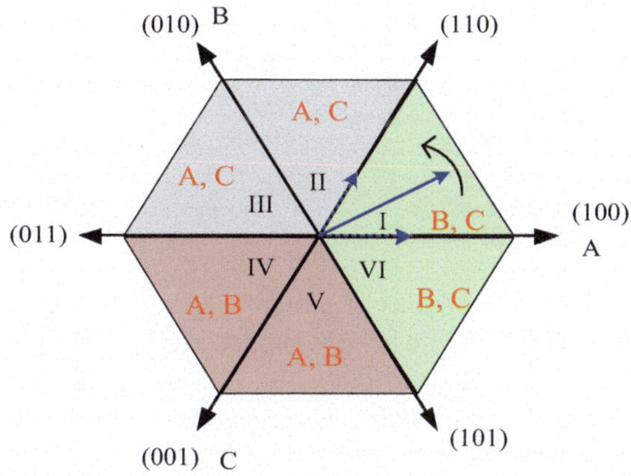

Fig. 3.27 Relationship of current vector location and corresponding current sampling selection

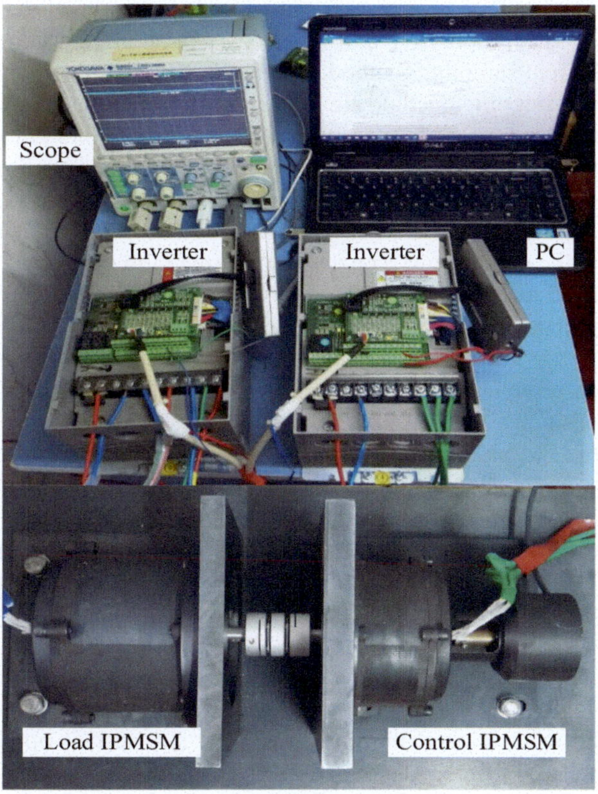

Fig. 3.28 Experimental test setup of IPMSM drive

inverter is 6 kHz. The current sampling procedure is completed through a built-in 12 bits A/D converter, and the current sampling range is restricted to −12.5A to +12.5A, in order to balance an accessible current sampling resolution.

Table 3.3 IPMSM parameters

Parameter	Value	Parameter	Value
Rated power	1.0 kW	Rated speed	3000 r/min
d-axis inductance	7.9 mH	Pole pairs	3
q-axis inductance	11.7 mH	Resistance	0.74 Ω
Torque constant	0.51 N·m/arms	Flux linkage	0.374 Wb

The comparisons of current response and the corresponding fast Fourier transformation (FFT) when the proposed ZVV injection method is applied are shown in Fig. 3.29. By modifying the injected ZVV, the zero voltage period t_0 can be maintained to desired values, i.e., 130 us in this experiment. For the aim of balancing an acceptable bandwidth, the entire control period is maintained to no more than 200 us. From the results, it can be seen that the modified control scheme has limited influence on the augmentation of current harmonics, and the ZVV period has better current slope response compared to the unmodified method.

The proposed control strategy based on the proposed two-phase-multipoint current sampling method, multipoint current sampling method, and two-point current sampling method are implemented in Fig. 3.30. The current slopes calculated by the three methods are presented in different colors in the first row of Fig. 3.30. The reference speed is set as 100 r/min under no-load condition, and the *d*-axis current is set to 1.4 A. The corresponding estimated rotor electrical position and the position error compared to the real position provided by an incremental encoder are presented in the second row. The diagrams in the third tier show the position error when 43% rated load and full-load torques are applied. From the results, it can be seen that the proposed two-phase-multipoint sampling method can reduce the calculated current slope noises effectively compared with the other two methods, and the corresponding estimated rotor position matches better the real position, and the position error is within 10°. In loading conditions, the position error estimated by the two-phase-multipoint current sampling method is within 15°.

Figure 3.31 shows the experimental results of reversal operation under rated load condition. The operating speed of the IPMSM is ±100 r/min. The load torque, the actual rotor position, the estimation errors, and the phase current are shown, respectively. From the results, it can be observed that the proposed position estimation method can operate in both motoring and regenerating mode. The position estimation error is within 15° during the whole process.

The experimental results of zero speed with rated step load disturbance sensorless control of IPMSM are shown in Fig. 3.32. As can be seen, the speed and position error fluctuate at the moments when step load is applied and removed. In steady state, both speed and position error regress to a normal level. The position error is within 20° during the entire process. Besides, the experimental results also demonstrate that the proposed position estimation method has a good robustness performance in step load condition.

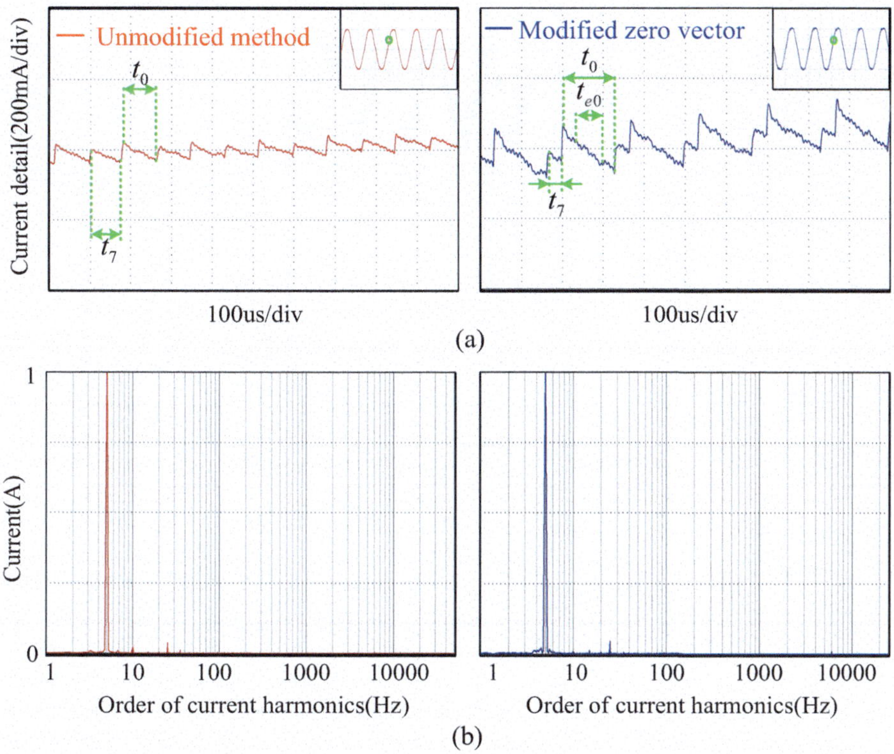

Fig. 3.29 Comparison of harmonics and current when the proposed ZVV injection method is applied. (**a**) Stator current waveforms under 100 r/min condition. (**b**) Current harmonic analysis results

3.6 Design of Position Observer

Normally, after acquiring the position estimation errors, the position observer or the state filter should be adopted to extract the precise rotor position information. The design method of the position observer will be introduced in this section [10]. The position observer is designed based on the mechanical model of PMSM, which takes both the load and motor rotation inertia into consideration. For the operation under load variation, the position observer possesses better robustness. The block diagram of the position observer is as shown in the Fig. 3.33, the input rotor position estimation errors are regulated with the PIID regulator to estimate the load torque, and feedforward electromagnetic torque is directly calculated with the dq-axis current i_{sdf} and i_{sqf} as shown in the (1.56).

The modeling of the position observer shown in Fig. 3.33 is according to the physical concept. It can be assumed as a transfer function from classic control theory aspect. It can be represented as a one-input-one-output system. By equivalent transformation, the space state model of the observer can be obtained. The state

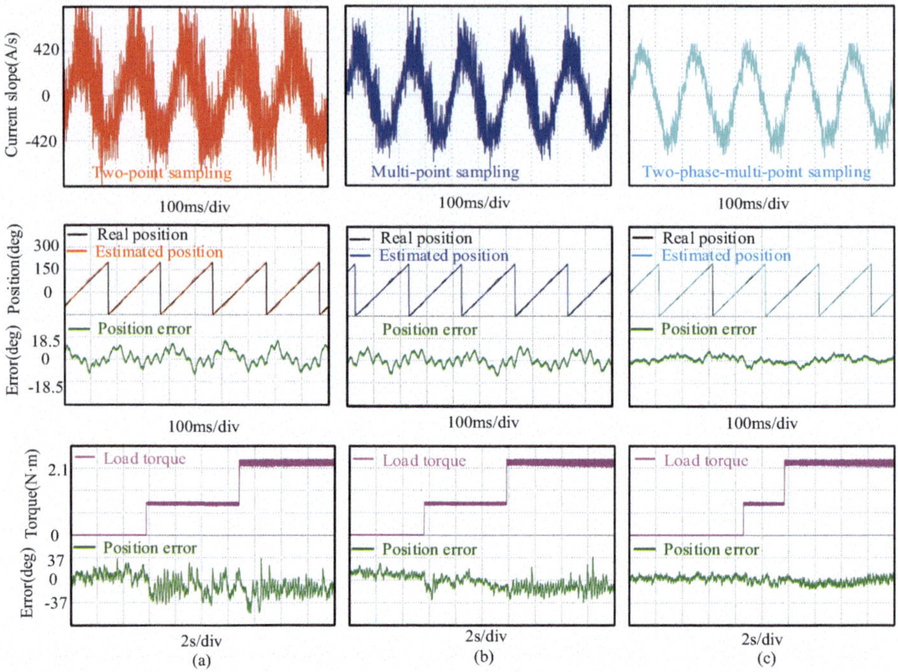

Fig. 3.30 Experimental comparison of zero voltage period current slopes and estimated position when different current sampling methods are applied under 100 r/min and no-load condition. (**a**) Two-point current sampling method. (**b**) Multipoint current sampling method. (**c**) Two-phase-multipoint current sampling method

variables of the whole process can be further expressed. The equivalent space state model is shown in the Fig. 3.34. The derivative of electromagnetic torque, the electromagnetic torque, the operation speed, and the rotor position are chosen as the state variables.

The state functions can be expressed as

$$\dot{\mathbf{X}} = \mathbf{AX} + \mathbf{B}u, \tag{3.53}$$

$$y = \mathbf{CX}, \tag{3.54}$$

where $\mathbf{X} = \begin{bmatrix} \dot{T}_e \\ T \\ \omega_e \\ \theta_e \end{bmatrix}$, $u = \ddot{T}_e$, $y = \theta_e$, $\mathbf{A} = \begin{bmatrix} 0 & 0 & 0 & 0 \\ 1 & 0 & 0 & 0 \\ 0 & \frac{1}{J_0} & 0 & 0 \\ 0 & 0 & P & 0 \end{bmatrix}$, $\mathbf{B} = \begin{bmatrix} 1 \\ 0 \\ 0 \\ 0 \end{bmatrix}$, $\mathbf{C} = \begin{bmatrix} 0 \\ 0 \\ 0 \\ 1 \end{bmatrix}^T$.

Hence the equivalent space state model can be expressed as

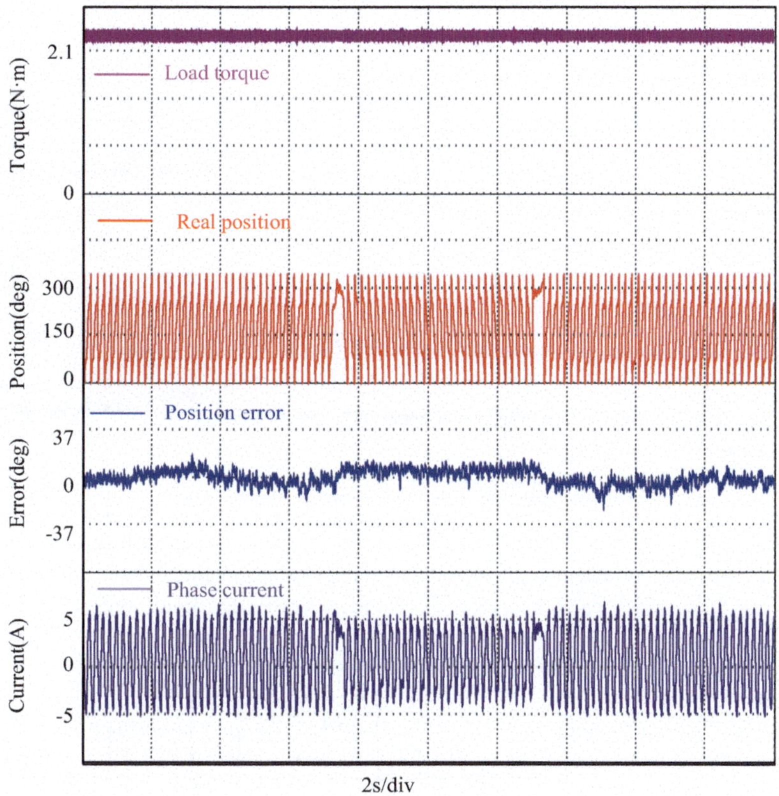

Fig. 3.31 Motoring and regenerating sensorless operation at ±100 r/min with rated load

$$\dot{\hat{\mathbf{X}}} = \begin{bmatrix} 0 & 0 & 0 & 0 \\ 1 & 0 & 0 & 0 \\ 0 & \frac{1}{J_1} & 0 & 0 \\ 0 & 0 & P & 0 \end{bmatrix} \hat{\mathbf{X}} + \begin{bmatrix} 1 \\ 0 \\ 0 \\ 0 \end{bmatrix} u + \begin{bmatrix} l_1 \\ l_2 \\ \frac{l_3}{J_1} \\ \frac{l_4}{J_1} \end{bmatrix} (y - \hat{y}), \tag{3.55}$$

where J_0 is the actual inertia and J_1 is the inertia for observer.

The relation between rotor position estimation errors and the disturbing torque can be expressed as

$$\Delta\theta(s) = \frac{-\frac{J_1}{J_0} s^2}{J_1 s^4 + l_4 s^3 + l_3 s^2 + l_2 s + l_1} T_d(s). \tag{3.56}$$

According to (3.56), it can be known that when there is a step or ramp load change, the steady estimation error of the observer can be controlled to zero, which

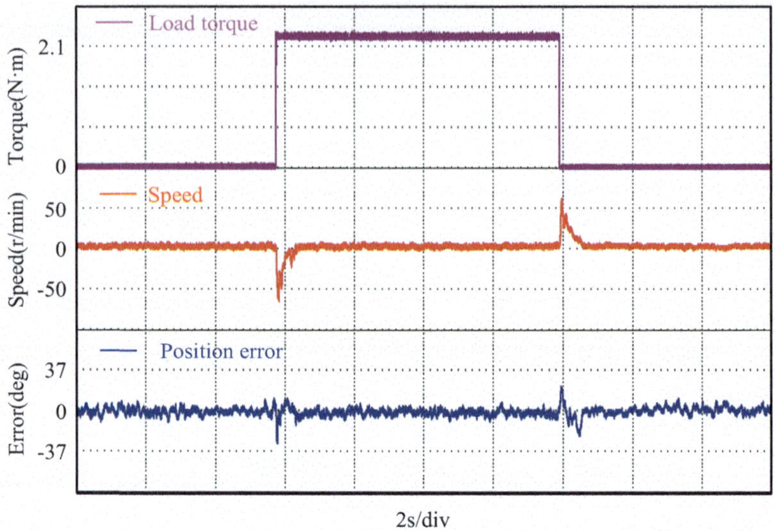

Fig. 3.32 Sensorless operation at zero speed with rated load disturbance

Fig. 3.33 The block diagram of rotor position observer

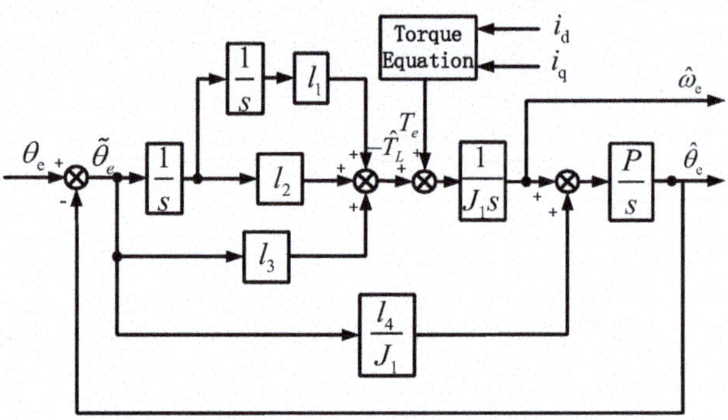

Fig. 3.34 Equivalent block diagram of the PMSM position observer

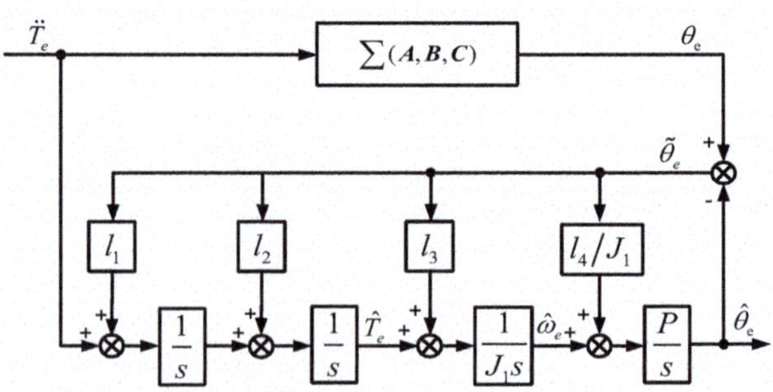

can prevent disturbing load from converging to minus 180°. It can be known that the anti-disturbance ability can be improved.

Normally, the parameters of the observer are designed based on the poles of the feedback matrix, which aims to keep the characteristic roots of the transfer function A-LC in the left half plane of the reference frame.

The characteristic polynomial can be expressed as

$$f(\lambda) = det[\lambda \mathbf{I} - (\mathbf{A} - \mathbf{LC})] = \lambda^4 + l_4\lambda^3 + l_3\lambda^2 + \frac{l_2}{\hat{J}}\lambda + \frac{l_1}{\hat{J}}, \tag{3.57}$$

where λ is the pole of the closed loop.

The arrangement of the poles is achieved by limiting maximum observation errors under load disturbance. The maximum estimation errors can be limited under the step load disturbance via limiting the maximum observation errors. This can prevent the polarity of the observed position from converging to the wrong one.

Assuming the maximum estimation error $\Delta\theta_{\max}$ is the maximum observation error under transient state in the time domain, the relation between the $\Delta\theta_{\max}$ and the poles can be deduced to meet the design goal. The step and ramp load disturbance will be discussed in the following, respectively.

Under the step load disturbance, $J_1 \approx J_0$ can be assumed, and then

$$\Delta\theta = \frac{s^2}{J_1 s^4 + l_4 s^3 + l_3 s^2 + l_2 s + l_1} \cdot \frac{T_{\mathrm{d}}}{s}. \tag{3.58}$$

The poles are set in the left half plane of the reference frame and coincident with P, and then

$$\Delta\theta = \frac{T_{\mathrm{d}} s}{J_1 \cdot (s+p)^4}. \tag{3.59}$$

The response in the time domain can be expressed as

$$\Delta\theta(t) = \frac{T_{\mathrm{d}}}{J_1} \cdot \left(-p \cdot \frac{1}{6} t^3 e^{-pt} + \frac{1}{2} t^2 e^{-pt}\right). \tag{3.60}$$

The extreme value can be obtained:

$$\Delta\theta_{\max 1} = \frac{54.146 e^{-1.268}}{p^2}. \tag{3.61}$$

Under ramp disturbance, the following can be obtained:

$$\Delta\theta = \frac{s^2}{J_1 s^4 + l_4 s^3 + l_3 s^2 + l_2 s + l_1} \cdot \frac{T_{\mathrm{d}}}{s^2}. \tag{3.62}$$

The four poles are set coincident at the negative real axis, and the response in the time domain can be expressed as

$$\Delta\theta(t) = \frac{T_{\rm d}}{J} \cdot \frac{1}{6} t^3 e^{-pt}. \tag{3.63}$$

In a similar way, the extreme value of the observation errors under ramp load disturbance can be obtained:

$$\Delta\theta_{\max 2} = \frac{525e^{-3}}{p^3}. \tag{3.64}$$

3.7 Parameters Design of PLL

3.7.1 Structure Design of the PLL

The quadrature signals $\Delta i'_\alpha$ and $\Delta i'_\beta$ containing rotor position information after using the signals processing is further processed by PLL shown in Fig. 3.35. The output rotor position angle is smoother than that of antitangent, and the influence of high frequency harmonics can be eliminated. The loop filter of the PLL adopts a PI regulator; the adjusting of the PLL can be achieved by designing the parameters properly [11].

The PLL can be equivalent to the block diagram as shown in Fig. 3.36 to analyze its characteristic in frequency domain. The transfer function can be expressed as

$$G_{PLL} = \frac{\hat{\theta}_{\rm e}}{\theta_{\rm e}} = \frac{Ek_p s + Ek_i}{s^2 + Ek_p s + Ek_i}. \tag{3.65}$$

The characteristic function can be expressed as

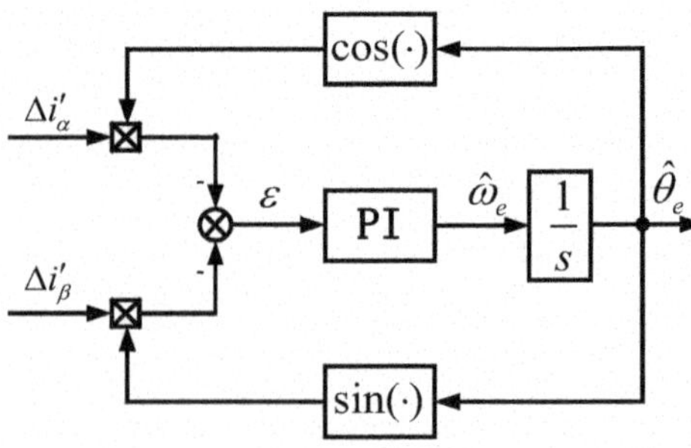

Fig. 3.35 Position information extraction of PLL

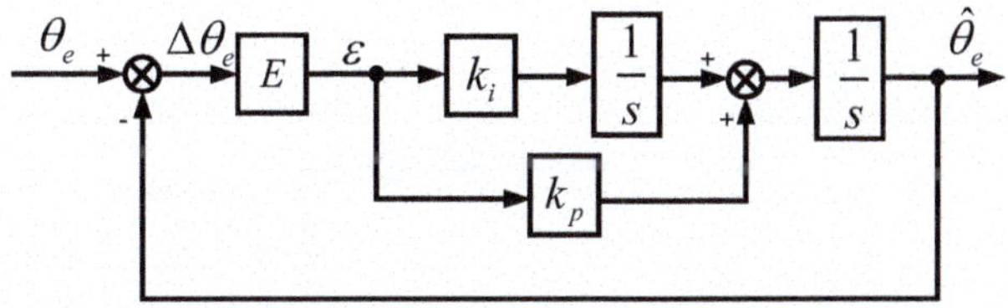

Fig. 3.36 Equivalent block diagram of the PLL

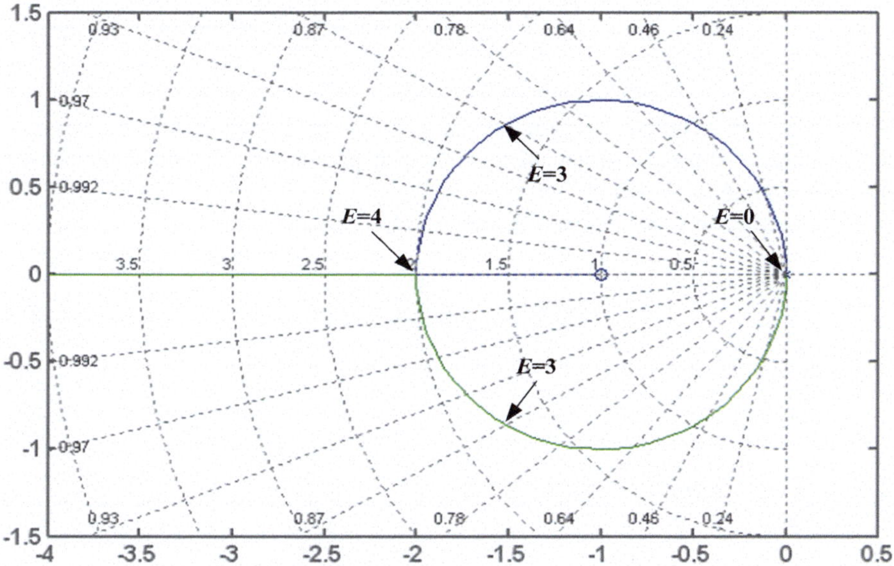

Fig. 3.37 Root path of gain coefficient E

$$s^2 + Ek_p s + Ek_i = 0, \tag{3.66}$$

where $E = \frac{\Delta V_h}{L_d}$ is the gain coefficient of the error. It varies with different injection voltage together with different d-axis inductance. The characteristic function can be equivalent to

$$1 + \frac{E(k_p s + k_i)}{s^2} = 0. \tag{3.67}$$

Normally, the root path of the gain coefficient can be obtained as shown in Fig. 3.37 after setting the parameters k_p and k_i.

When k_p and k_i are fixed, the bandwidth of the PLL is mostly influenced by the gain coefficient. The tracking performance of the PLL varies along with different injection voltage and different motor, which make it tedious to design the parameters of the PLL.

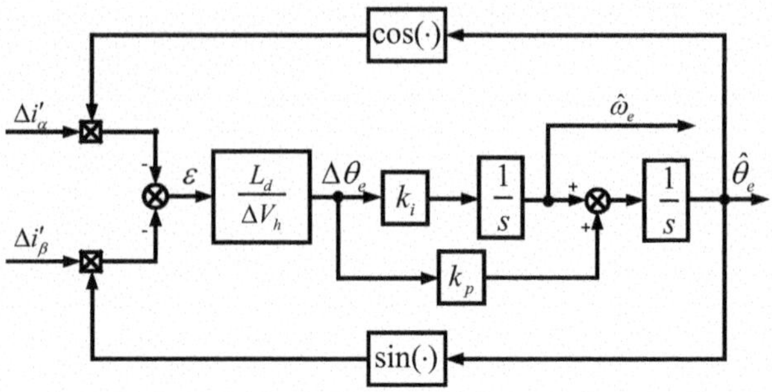

Fig. 3.38 PLL block diagram after observation error normalization

To solve this problem, the normalization of the errors can make the dynamic performance of the PLL irrelevant to the motor parameters and amplitude of the injection voltage as shown in the Fig. 3.38.

The closed-loop transfer function can be expressed as

$$G_{PLL} = \frac{\hat{\theta}_e}{\theta_e} = \frac{k_p s + k_i}{s^2 + k_p s + k_i}. \tag{3.68}$$

3.7.2 Parameter Design of PLL

For the PLL after normalization, the following can be obtained:

$$\begin{aligned} \dot{\hat{\omega}}_e &= k_i \Delta\theta_e, \\ \dot{\hat{\theta}}_e &= \hat{\omega}_e + k_p \Delta\theta_e. \end{aligned} \tag{3.69}$$

For the limitation, the poles are normally set in the left half plane of reference frame; when it is assumed as ρ, then

$$\frac{\hat{\theta}_e}{\theta_e} = \frac{2\rho s + \rho^2}{s^2 + 2\rho s + \rho^2}. \tag{3.70}$$

The results can be obtained: $k_p = 2\rho$, $k_i = \rho^2$.

Hence, the following can be met:

$$\begin{aligned} \dot{\omega}_e &= a, \\ \dot{\theta}_e &= \omega_e. \end{aligned} \tag{3.71}$$

Subtract (3.71) from (3.69):

$$\begin{aligned}\dot{\tilde{\omega}}_e &= a - k_i\Delta\theta_\mathrm{e},\\ \Delta\dot{\theta}_\mathrm{e} &= \tilde{\omega}_e - k_p\Delta\theta_\mathrm{e}.\end{aligned} \tag{3.72}$$

where $\tilde{\omega}_e = \omega_e - \hat{\omega}_e$, $\Delta\theta_e = \theta_e - \hat{\theta}_e$.

When $\dot{\tilde{\omega}}_e = \Delta\dot{\theta}_e = 0$, then

$$\begin{aligned}\tilde{\omega}_e &= \frac{ak_p}{k_i} = \frac{2a}{\rho},\\ \Delta\theta_\mathrm{e} &= \frac{a}{k_i} = \frac{a}{\rho^2}.\end{aligned} \tag{3.73}$$

The designed parameters of the PLL can be obtained by limiting ρ:

$$\rho = \sqrt{\frac{a}{\Delta\theta_{\mathrm{e\_max}}}}, \tag{3.74}$$

where $\Delta\theta_{e\_\max}$ is the maximum value of the observation error.

In conclusion, the parameters of the PLL can be obtained via limiting the maximum observation error, and the adaptive ability of the PLL can be enforced by normalization. The high frequency fluctuation can be suppressed due to the bandwidth limit of the PLL, which can make the obtained rotor position information smoother.

3.8 Summary

The saliency-based sensorless control for PMSM drives is introduced in detail in the chapter. Firstly, the HF model and the basic principle of the sensorless method based on saliency tracking are presented. Then two kinds of injection method which are sinusoidal and square-wave signal injection are further demonstrated. Two special methods based on the pulse signal injection and the fundamental PWM excitation are illustrated as well. Lastly, the position observer design method is addressed.

References

1. P.L. Jansen, R.D. Lorenz, Transducerless position and velocity estimation in induction and salient AC machines. IEEE Trans. Ind. Appl. **31**(2), 240–247 (March–April 1995)
2. N. Bianchi, S. Bolognani, J.H. Jang, S.K. Sul, Advantages of inset PM machines for zero-speed sensorless position detection. IEEE Trans. Ind. Appl. **44**(4), 1190–1198 (July/August 2008)
3. D. Raca, P. Garcia, D.D. Reigosa, F. Briz, R.D. Lorenz, Carrier-signal selection for sensorless control of PM synchronous machines at zero and very low speeds. IEEE Trans. Ind. Appl. **46** (1), 167–178 (January/February 2010)

4. Y.D. Yoon, S.K. Sul, S. Morimoto, K. Ide, High-bandwidth sensorless algorithm for AC machines based on square-wave-type voltage injection. IEEE Trans. Ind. Appl. **47**(3), 1361–1370 (May/June 2011)
5. N. Park, S. Kim, Simple Sensorless algorithm for interior permanent magnet synchronous motors based on high-frequency voltage injection method. IET Electr. Power Appl. **8**(2), 68–75 (February 2014)
6. D. Kim, Y.C. Kwon, S.K. Sul, J.H. Kim, R.S. Yu, Suppression of injection voltage disturbance for high-frequency square-wave injection Sensorless drive with regulation of induced high-frequency current ripple. IEEE Trans. Ind. Appl. **52**(1), 302–312 (January/February 2016)
7. R. Ni, D. Xu, F. Blaabjerg, K. Lu, G. Wang, G. Zhang, Square-wave voltage injection algorithm for PMSM position sensorless control with high robustness to voltage errors. IEEE Trans. Power Electron. **32**(7), 5425–5437 (July 2017).
8. G. Xie, K. Lu, S.K. Dwivedi, J.R. Rosholm, F. Blaabjerg, Minimum-voltage vector injection method for Sensorless control of PMSM for low-speed operations. IEEE Trans. Power Electron. **31**(2), 1785–1794 (February 2016)
9. G. Wang, J. Kuang, N. Zhao, G. Zhang, D. Xu, Rotor position estimation of PMSM in low-speed region and standstill using zero-voltage vector injection. IEEE Trans. Power Electron. **33**(9), 7948–7958 (September 2018)
10. G. Wang, R. Yang, D. Xu, DSP-based control of sensorless IPMSM drives for wide-speed-range operation. IEEE Trans. Ind. Electron. **60**(2), 720–727 (February 2013)
11. G. Wang, Z. Li, G. Zhang, Y. yu, D. Xu, Quadrature PLL-based high-order sliding-mode observer for IPMSM sensorless control with online MTPA control strategy. IEEE Trans. Energy Convers. **28**(1), 214–224 (March 2013)

Chapter 4
Pseudo-Random-Frequency Signal Injection for Acoustic Reduction

HF signal injection method can obtain the rotor position accurately by tracking the rotor saliency at low and zero speed. However, this method arouses the high-frequency component in the induced current, causing audible noise and electromagnetic interference (EMI) problems. The audible noise generated by injected HF signals is uncomfortable to hear, which inevitably restricts the implementation of this kind of sensorless control scheme in industrial and household applications. Adopting the idea of noise reduction in power converters, random HF signal injection method is a good way for noise suppression.

For a square-wave signal, there are three elements: frequency, amplitude, and phase. The random-signal injection method can be implemented by changing any of them. In order to maintain a suitable signal-to-noise ratio (SNR), once the injection frequency is determined, the amplitude range of the injection signal will also be limited. Thus, the frequency and amplitude cannot be changed individually. They should be considered as a whole and changed at the same time to ensure the required level of SNR in the process of random signal injection. Considering the implementation of the random-signal injection method in the practical system, the random-signal injection method can be divided into two categories, which are pseudo-random frequency-switching fixed-phase (PRFSFP) square-wave signal injection method [1, 2] and pseudo-random phase-switching fixed-frequency (PRPSFF) square-wave injection method [3].

G. Wang et al., *Position Sensorless Control Techniques for Permanent Magnet Synchronous Machine Drives*, https://doi.org/10.1007/978-981-15-0050-3_4

4.1 Pseudo-Random Frequency-Switching Fixed-Phase Square-Wave Signal Injection

4.1.1 Principle of Pseudo-Random HF Signal Injection

IPMSM can be modeled in the rotor reference frame as (3.36). Mutual inductances are not considered in this model.

If the frequency of injected signal is much higher than the fundamental frequency and the motor operates at low speed, the IPMSM can be modeled and simplified into (3.1).

For the sake of convenience, define the unit-amplitude square-wave signal as

$$\phi_{\mathrm{sqr}}(t,T_{\mathrm{i}})=\begin{cases}1 & 0<t_{\mathrm{r}}(t,T_{\mathrm{i}})\leq T_{\mathrm{i}}/2\\ -1 & T_{\mathrm{i}}/2<t_{\mathrm{r}}(t,T_{\mathrm{i}})\leq T_{\mathrm{i}}\end{cases} \tag{4.1}$$

and the unit-amplitude triangle-wave signal is set as

$$\phi_{\mathrm{saw}}(t,T_{\mathrm{i}})=\begin{cases}-1+\dfrac{4t_{\mathrm{r}}(t,T_{\mathrm{i}})}{T} & 0<t_{\mathrm{r}}(t,T_{\mathrm{i}})\leq T_{\mathrm{i}}/2\\ 3-\dfrac{4t_{\mathrm{r}}(t,T_{\mathrm{i}})}{T} & T_{\mathrm{i}}/2<t_{\mathrm{r}}(t,T_{\mathrm{i}})\leq T_{\mathrm{i}},\end{cases} \tag{4.2}$$

where T_{i} denotes the period of the unit-amplitude square-wave and unit-amplitude triangle-wave signals, and parameter $t_{\mathrm{r}}(t, T_{\mathrm{i}})$ is the remainder of the division of t by T_{i}.

Figure 4.1 illustrates the implementation of the overall scheme. The random signal generator produces two kinds of square-wave signals, the injection and demodulation signals. The injection signal is added to the output of current regulator, and the demodulation signal whose phase angle is orthogonal to the injected one is used as the input of the demodulator. Both the fundamental and the random frequency components are contained in the stator current. The former obtained by a LPF is used as a feedback of the current regulator. The latter containing the rotor position information is extracted by a HPF and then input to the demodulator. The output of the demodulator is used as the input of the normalizer, which can extract the position error more accurately. By controlling the position error ε to zero with a Luenberger or PLL observer, the rotor position can be obtained.

The detailed diagram of the random signal generator is shown in Fig. 4.2. The subscript "R" means the random value. Four signals are generated including two injection signals and two demodulation signals. The periods of the injection signals are different from each other: one is T_{i1} and the other is T_{i2}. The two demodulation

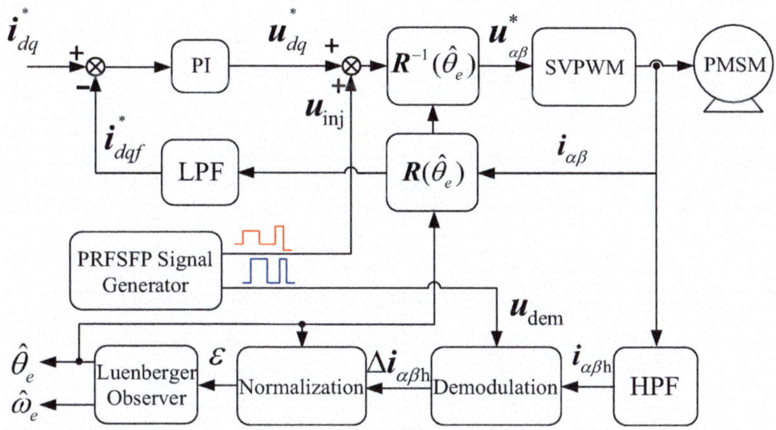

Fig. 4.1 Block diagram of the PRFSFP square-wave voltage injection method [2]

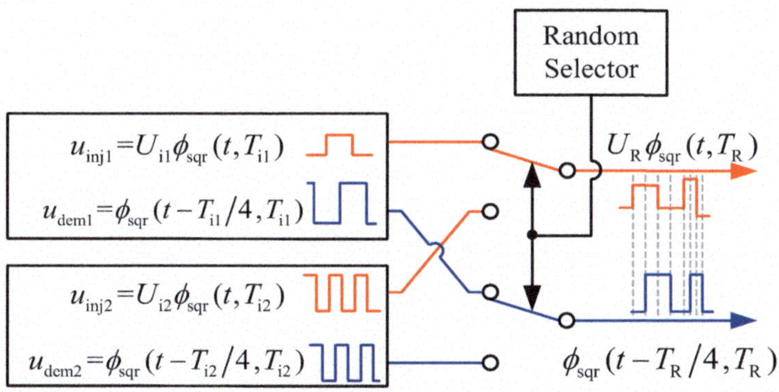

Fig. 4.2 Random signal generator for PRFSFP square-wave voltage injection

signals are orthogonal to their corresponding injection signals, respectively. The random selector chooses one of the injection signals and its corresponding demodulation signal at the same time. Two kinds of fixed-frequency and fixed-amplitude signals are selected periodically at random, since only discretely distributed frequency signal can be implemented. The random-frequency square-wave voltage signal injected into the estimated rotor reference frame can be expressed as follows:

$$\begin{bmatrix} u_{dh} \\ u_{qh} \end{bmatrix} = \begin{bmatrix} U_R\phi_{sqr}(t, T_R) \\ 0 \end{bmatrix}. \quad (4.3)$$

4.1.2 Demodulation

According to (4.3), the induced HF current in the SRF can be obtained as follows:

$$\begin{aligned}\begin{bmatrix} i_{\alpha \mathrm{h}} \\ i_{\beta \mathrm{h}} \end{bmatrix} &= \mathbf{R}^{-1}(\theta_e) \begin{bmatrix} L_d & 0 \\ 0 & L_q \end{bmatrix}^{-1} \mathbf{R}(\tilde{\theta}_e) \cdot \int \begin{bmatrix} u_{d\mathrm{h}} \\ u_{q\mathrm{h}} \end{bmatrix} \mathrm{d}t \\ &= \frac{U_{\mathrm{R}} T_{\mathrm{R}} \phi_{\mathrm{saw}}(t, T_{\mathrm{R}})}{4 L_d L_q} \begin{bmatrix} L_q \cos\tilde{\theta}_e \cos\theta_e + L_d \sin\tilde{\theta}_e \sin\theta_e \\ L_q \cos\tilde{\theta}_e \sin\theta_e - L_d \sin\tilde{\theta}_e \cos\theta_e \end{bmatrix} \\ &= K_1 \phi_{\mathrm{saw}}(t, T_{\mathrm{R}}) \begin{bmatrix} \Delta i_{\alpha\_\mathrm{pu}} \\ \Delta i_{\beta\_\mathrm{pu}} \end{bmatrix}, \end{aligned} \tag{4.4}$$

where

$$K_1 = \frac{U_{\mathrm{R}} T_{\mathrm{R}} \sqrt{\left(L_q \cos\tilde{\theta}_e\right)^2 + \left(L_d \sin\tilde{\theta}_e\right)^2}}{4 L_d L_q}, \tag{4.5}$$

$$\begin{bmatrix} \Delta i_{\alpha\_\mathrm{pu}} \\ \Delta i_{\beta\_\mathrm{pu}} \end{bmatrix} = \begin{bmatrix} \cos\left(\theta_e - \tilde{\theta}_{e1}\right) \\ \sin\left(\theta_e - \tilde{\theta}_{e1}\right) \end{bmatrix} \tag{4.6}$$

$$\tan\tilde{\theta}_{e1} = \frac{L_d}{L_q} \tan\tilde{\theta}_e. \tag{4.7}$$

Then (4.7) can be simplified into (4.8) on the assumption of $\tilde{\theta}_e \approx 0$:

$$\tilde{\theta}_{e1} = \frac{L_d}{L_q} \tilde{\theta}_e. \tag{4.8}$$

In (4.4), the frequency of $\Delta i_{\alpha\beta\_\mathrm{pu}}$ associates with the rotor speed which is much lower than that of $\phi_{\mathrm{saw}}(t, T_{\mathrm{R}})$. Consequently, $\phi_{\mathrm{saw}}(t, T_{\mathrm{R}})$ is modulated by $\Delta i_{\alpha\beta\_\mathrm{pu}}$ and the amplitude is subject to K_1. To ensure the amplitudes of HF currents induced by the two different frequency voltage signals are identical, $U_{\mathrm{R}} T_{\mathrm{R}}$ should maintain constant. The relationship of the random frequency voltage and current is shown in Fig. 4.3.

In order to extract the rotor position error, the effect of the fluctuation from $\phi_{\mathrm{saw}}(t, T_{\mathrm{R}})$ in (4.4) should be suppressed. Multiply (4.4) by the demodulation signal $\phi_{\mathrm{sqr}}(t- T_{\mathrm{R}}/4, T_{\mathrm{R}})$, (4.9) can be obtained:

$$\begin{bmatrix} \Delta i_{\alpha \mathrm{h}} \\ \Delta i_{\beta \mathrm{h}} \end{bmatrix} = \phi_{\mathrm{sqr}}\left(t - \frac{T_{\mathrm{R}}}{4}, T_{\mathrm{R}}\right) \begin{bmatrix} i_{\alpha \mathrm{h}} \\ i_{\beta \mathrm{h}} \end{bmatrix} = K_1 K_2 \begin{bmatrix} \cos\left(\theta_e - \tilde{\theta}_{e1}\right) \\ \sin\left(\theta_e - \tilde{\theta}_{e1}\right) \end{bmatrix}, \tag{4.9}$$

where

$$K_2 = \phi_{\mathrm{saw}}(t, T_\mathrm{R}) \cdot \phi_{\mathrm{sqr}}\left(t - \frac{T_\mathrm{R}}{4}, T_\mathrm{R}\right) = \frac{1}{2}\left[1 + \phi_{\mathrm{saw}}\left(t + \frac{T_\mathrm{R}}{4}, \frac{T_\mathrm{R}}{2}\right)\right]. \tag{4.10}$$

The parameter K_2 fluctuates between 0 and 1 with the frequency twice of the injected voltage. Now the sign of $\Delta i_{\alpha\beta\mathrm{h}}$ is only determined by $\Delta i_{\alpha\beta\_\mathrm{pu}}$. The aim of the LPFs in Fig. 4.4 is not to eliminate the noise $\phi_{\mathrm{saw}}(t + T_\mathrm{R}/4, T_\mathrm{R}/2)$ completely, but just to slightly reduce its amplitude in order to guarantee the sign of K_2 to be positive. Thus, the cutoff frequency of LPFs can be set high enough, which will not cause obvious phase lag to the position estimation.

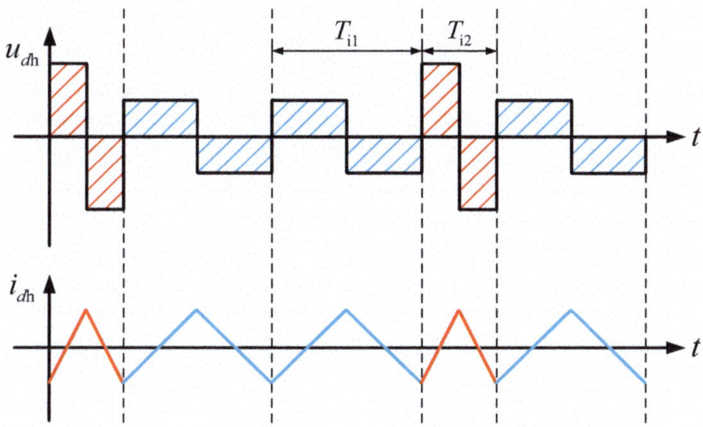

Fig. 4.3 Relationship of the random HF voltage and the induced current

Fig. 4.4 Digital signal processing for obtaining position information with the pseudo-random HF signal injection

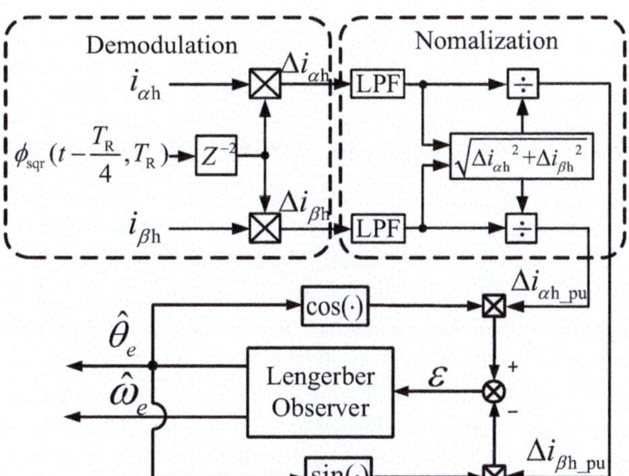

4.1.3 Normalization Process

The effect of fluctuation in K_2 can be eliminated with the help of normalization shown in Fig. 4.4. As given in (4.11), the absolute value of K_1K_2 can be calculated. If the sign of K_1K_2 is guaranteed to be positive, it can be removed from (4.9). Dividing $\Delta i_{\alpha\beta\text{h}}$ by K_1K_2, a pair of orthogonal sinusoidal signals $\Delta i_{\alpha\_\text{pu}}$ and $\Delta i_{\beta\_\text{pu}}$ obtained in (4.12) can be used to extract the position error:

$$|K_1K_2| = \sqrt{\Delta i_{\alpha\text{h}}{}^2 + \Delta i_{\beta\text{h}}{}^2}, \tag{4.11}$$

$$\begin{bmatrix} \Delta i_{\alpha\_\text{pu}} \\ \Delta i_{\beta\_\text{pu}} \end{bmatrix} = \text{PU}\left(\phi_{\text{sqr}}\left(t - \frac{T_\text{R}}{4}, T_\text{R}\right)\begin{bmatrix} i_{\alpha\text{h}} \\ i_{\beta\text{h}} \end{bmatrix}\right) = \begin{bmatrix} \cos\left(\theta_e - \tilde{\theta}_{e1}\right) \\ \sin\left(\theta_e - \tilde{\theta}_{e1}\right) \end{bmatrix}. \tag{4.12}$$

Due to the division calculation in Fig. 4.4, some noise signal will be produced when K_2 is around zero. That is why LPF is used to slightly reduce the amplitude of noise signal $\phi_{\text{saw}}(t + T_\text{R}/4, T_\text{R}/2)$ in K_2 to avoid the division of zeros. In this way, the effect caused by the fluctuation of K_2 can be suppressed. Actually, with the aid of normalization, the evaluated position only depends on the relative scale between $\Delta i_{\alpha\_\text{pu}}$ and $\Delta i_{\beta\_\text{pu}}$.

The position error signal can be extracted from (4.13). By controlling ε to zero, the rotor position can be obtained:

$$\begin{aligned} \varepsilon &= \cos\left(\hat{\theta}_e\right)\sin\left(\theta_e - \tilde{\theta}_{e1}\right) - \sin\left(\hat{\theta}_e\right)\cos\left(\theta_e - \tilde{\theta}_{e1}\right) \\ &= \sin\left(\tilde{\theta}_e - \tilde{\theta}_{e1}\right) \approx \tilde{\theta}_e \frac{L_q - L_d}{L_q} \qquad \left(\tilde{\theta}_e \approx 0\right). \end{aligned} \tag{4.13}$$

4.1.4 Compensation Strategy for Digital Time-Delay Effect

In the digital control system of IPMSM drive, there is an inevitable execution time delay, which is one sampling period. Furthermore, the digital PWM logic also produces around half sampling period time delay [4]. This one-and-a-half sampling period time delay is always neglected in fundamental frequency which is much lower than the PWM switching frequency. However, since the frequency of the injected HF signal is close to the PWM switching frequency, the time-delay effect cannot be neglected. The actual HF voltage injected into the stator lags behind the HF signal reference in the realization process. The time-delay effect will produce an additional position estimation error. In this part, the time-delay effect is discussed, and a compensation scheme is introduced.

The actual injected HF voltage considering one-and-a-half sampling period time delay is shown as

$$\begin{bmatrix} u_{dh}' \\ u_{qh}' \end{bmatrix} = \begin{bmatrix} U_i \phi_{sqr}\left(t - \frac{3}{2}T_s, T_R\right) \\ 0 \end{bmatrix}, \tag{4.14}$$

where the superscript "′" denotes the value in which time delay is considered, and T_s denotes the sampling period. Then (4.9) can be transformed into

$$\begin{bmatrix} \Delta i_{\alpha h}' \\ \Delta i_{\beta h}' \end{bmatrix} = \phi_{sqr}\left(t - \frac{T_i}{4}, T_i\right) \begin{bmatrix} i_{\alpha h}' \\ i_{\beta h}' \end{bmatrix} = K_1 K_2' \begin{bmatrix} \cos\left(\theta_e - \tilde{\theta}_{e1}\right) \\ \sin\left(\theta_e - \tilde{\theta}_{e1}\right) \end{bmatrix}, \tag{4.15}$$

$$K_2' = \phi_{saw}\left(t - \frac{3}{2}T_s, T_R\right) \cdot \phi_{sqr}\left(t - \frac{T_R}{4}, T_R\right) . \tag{4.16}$$

As can be seen in (4.16), sometimes K_2' is negative, and the HF current signal cannot be demodulated well.

To attenuate the time-delay effect, a compensation strategy by adding phase lag to the demodulation signal is adopted as shown in Fig. 4.4. In realization, two sampling periods for compensating the phase lag are added to the demodulation signal since that only integral multiples of T_s can be implemented.

Figure 4.5 shows the digital signal processing in which the time-delay effect is taken into consideration. From top to bottom, the random HF voltage reference in program, the actual HF voltage reference considering digital time delay, the induced

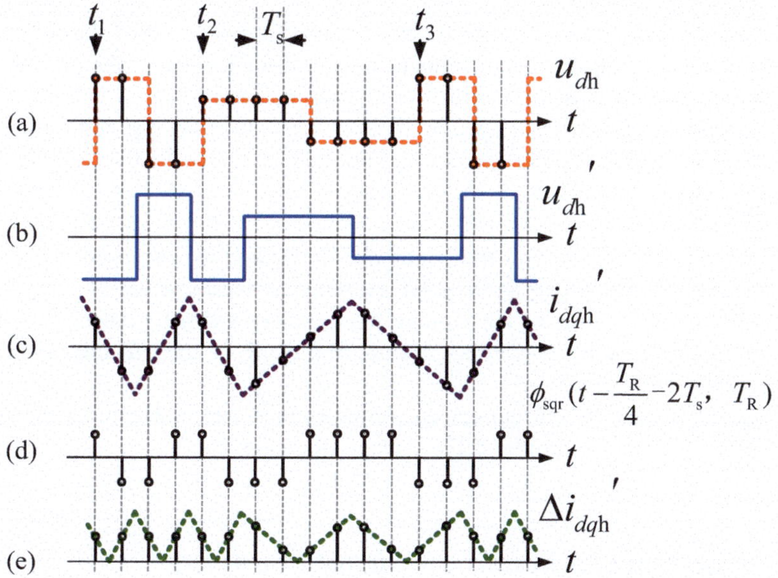

Fig. 4.5 Digital signal processing when digital time delay is taken into account. (**a**) Random HF voltage reference. (**b**) Actual HF voltage reference with digital time delay. (**c**) Induced HF stator current. (**d**) Demodulation signal with digital time-delay compensation. (**e**) Demodulated signal

current in SRF, the demodulation signal with time-delay compensation, and the demodulated signal are given, respectively. Since the period of HF signal can only be integral multiples of sampling period, four times and eight times situations are discussed here. In this way, the sign of demodulated signal only depends on the rotor position, and the position estimation error can be decreased.

4.1.5 Selection of the Injected Frequency and Voltage

The method to determine the frequency and the voltage of the injection signal is a key issue for the pseudo-random HF injection method. Here, the selection of injected frequency and voltage is discussed as follows:

1. Frequency Selection
   (a) The period of the injection signal must be integral multiples of the PWM period, for example, $2T_s$, $3T_s$, $4T_s$, $5T_s$ ... where $T_s = 1/f_s$. So, frequency division method is used to produce the HF signals.
   (b) If a higher frequency is selected, the corresponding voltage should be increased to guarantee the SNR of HF current to be high enough to extract the position information.
   (c) With higher injection frequency, the digital time-delay effect would cause larger phase lag in the injected signal. Then the influence on position estimation will be greater.
   (d) If a lower frequency is adopted, lower cutoff frequency LPFs should be used to extract the fundamental-frequency current for feedback; the performance of the position sensorless drive system will be deteriorated.
   (e) Considering the audible noise reduction, when the two injected frequencies are too close, the power spectra will not be extended effectively.
2. Voltage Selection
   (a) The ratio of voltage and frequency should be a constant, to ensure that the magnitudes of the two induced HF currents are the same.
   (b) A higher magnitude of HF voltage would cause a shriller noise, although it can achieve better performance since the SNR of the HF current can be guaranteed.
   (c) The additional HF current causes torque ripples, especially under the condition of higher magnitude of HF voltage injection.

4.1.6 Experimental Results for Signal Processing

The PRFSFP square-wave voltage injection based on the sensorless control strategy is verified on the platform with a 2.2-kW IPMSM as shown in Fig. 3.18. A low-cost

ARM chip STM32F103VCT6 is adopted to realize the sensorless control algorithm. A linear congruential generator (LCG) is adopted to produce a series of pseudo-random numbers [5]. High-accuracy current sampling resistors are used to obtain the stator currents. The PWM switching frequency of the inverter and current sampling frequency are both 6 kHz, the DC bus voltage is 510 V, and the dead time of the inverter is set as 3.2 μs. The magnitudes of the injected HF voltages are 46 V for 750 Hz and 92 V for 1.5 kHz. All experimental results are obtained by sensorless control.

Figure 4.6 shows the experimental results of the digital signal processing for the pseudo-random frequency signal injection at 100 r/min with 50% rated load. The waveforms are the HF current in SRF $i_{\alpha\beta h}$, the demodulated HF current $\Delta i_{\alpha\beta h}$, the normalized current $\Delta i_{\alpha\beta h\_pu}$, the actual position, the estimated position, and the *a*-phase current. From the digital signal processing and the estimated position, it can be seen that the proposed demodulation and normalization methods are effective for this PRFSFP square-wave voltage injection.

4.2 Comparison of Full-Period Switching Mode and Half-Period Switching Mode

4.2.1 Full-Period-Switch PRFSFP Square-Wave Signal Injection

The Full-Period-Switch (FPS) PRFSFP square-wave signal injection scheme is to inject a random-frequency voltage signal to the estimated *d*-axis [2]. Two square-wave voltage signals with fixed-frequency and fixed-amplitude are randomly selected once per full period. Figure 4.7a, b illustrate the two single-period voltage signals and their induced currents. The parameters T_{i1} and T_{i2} denote the periods of the injected signals, and the corresponding frequencies are f_{i1} and f_{i2}, respectively. The amplitude of current should be guaranteed identical between the two kinds of frequencies. According to (4.3), the current is the integration of the injected voltage. Thus, the volt-second area of the voltage should be the same.

Define the unit-area square-wave function as

$$u_{D_1,D_2}(t) = \begin{cases} 1/D_1 & 0 < t \leq D_1 \\ -1/D_2 & D_1 < t \leq D_1 + D_2 \\ 0 & \text{elsewhere} \end{cases}, \tag{4.17}$$

and the unit-amplitude saw-tooth-wave function as

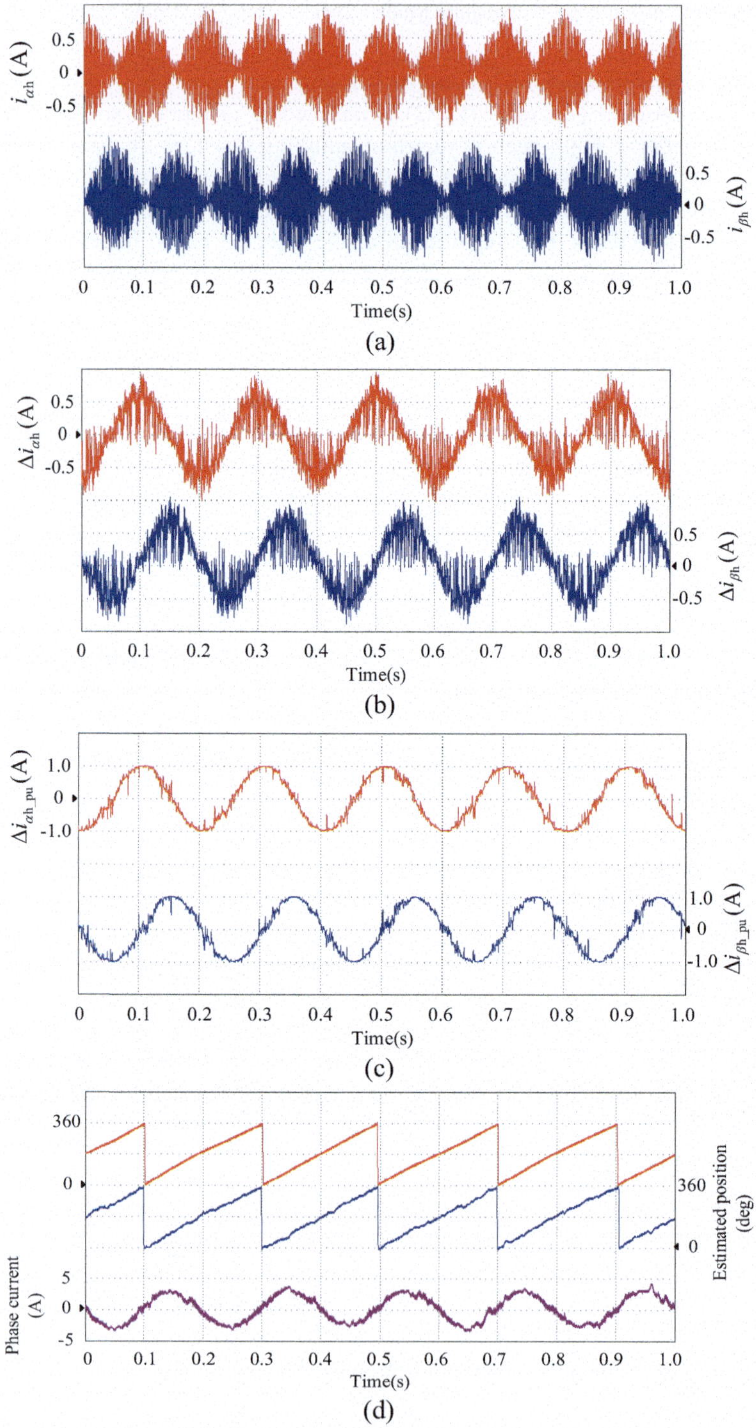

Fig. 4.6 Experimental results of digital signal processing at 100 r/min with half rated load. (**a**) HF current in SRF. (**b**) Demodulated signal. (**c**) Normalized signal. (**d**) Actual position, estimated position, and *a*-phase current. (**e**) Zoomed results of *a*-phase current

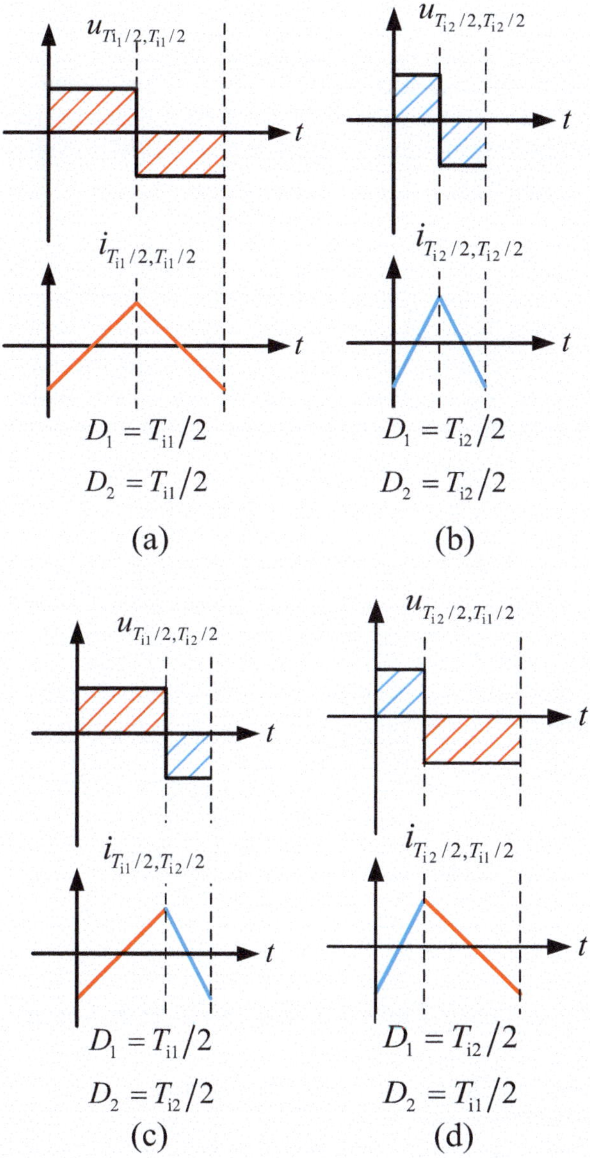

Fig. 4.7 Four voltage injection modes and the corresponding current. (**a**) $u_{T_{i1}/2,\ T_{i1}/2}$ and $i_{T_{i1}/2,\ T_{i1}/2}$. (**b**) $u_{T_{i2}/2,\ T_{i2}/2}$ and $i_{T_{i2}/2,\ T_{i2}/2}$. (**c**) $u_{T_{i1}/2,\ T_{i2}/2}$ and $i_{T_{i1}/2,\ T_{i2}/2}$. (**d**) $u_{T_{i2}/2,\ T_{i1}/2}$ and $i_{T_{i2}/2,\ T_{i1}/2}$

$$i_{D_1,D_2}(t) = \begin{cases} -1 + 2t/D_1 & 0 \le t \le D_1 \\ 1 + 2(D_1 - t)/D_2 & D_1 < t \le D_1 + D_2 \\ 0 & \text{elsewhere} \end{cases}. \tag{4.18}$$

The relationship between (4.17) and (4.18) can be obtained:

$$2u_{D_1,D_2}(t) = \frac{\mathrm{d}}{\mathrm{d}t} i_{D_1,D_2}(t). \tag{4.19}$$

The parameters D_1 and D_2 illustrate the duration of the positive and negative voltages, respectively. For FPS-PRFSFP square-wave signal injection scheme, the injected pseudo-random high-frequency (PRHF) voltage in the estimated d-axis can be written as

$$\begin{aligned} u_{d\mathrm{h}}(t) &= U_{\mathrm{R}} \cdot \sum\nolimits_{k=1}^{\infty} u_{\mathrm{FPS}}(t - t_k) \\ u_{\mathrm{FPS}}(t) &= \Re\left[u_{T_{\mathrm{i1}}/2,\ T_{\mathrm{i1}}/2}(t),\ u_{T_{\mathrm{i2}}/2,\ T_{\mathrm{i2}}/2}(t)\right], \end{aligned} \tag{4.20}$$

where t_k denotes the kth cycle of injected voltage and U_{i} is the magnitude of injected voltage. The operator R[] indicates random selection in the bracket. If (4.20) is utilized, the induced PRHF current in q-axis can be calculated as

$$\begin{aligned} i_{q\mathrm{h}}(t) &= U_{\mathrm{R}} \cdot \frac{L_q - L_d}{4L_d L_q} \sin 2\tilde{\theta}_e \cdot \sum\nolimits_{k=1}^{\infty} i_{\mathrm{FPS}}(t - t_k), \\ i_{\mathrm{FPS}}(t) &= \Re\left[i_{T_{\mathrm{i1}}/2,\ T_{\mathrm{i1}}/2}(t),\ i_{T_{\mathrm{i2}}/2,\ T_{\mathrm{i2}}/2}(t)\right]. \end{aligned} \tag{4.21}$$

Notice that the signal $i_{q\mathrm{h}}(t)$ fluctuates around zero, and the magnitude varies with the position error $\tilde{\theta}_e$. When $\tilde{\theta}_e$ is close to zero, $i_{q\mathrm{h}}(t)$ is proportional to $\tilde{\theta}_e$. Thus, it contains the equivalent position error information. By multiplying with $i_{q\mathrm{h}}(t)$ and the demodulation signal ϕ_{FPS}, the equivalent position error signal ε can be obtained:

$$\begin{aligned} \phi_{\mathrm{FPS}} &= sign\left(\sum\nolimits_{k=1}^{\infty} i_{\mathrm{FPS}}(t - t_k)\right), \\ \varepsilon &= i_{q\mathrm{h}}(t) \cdot \phi_{\mathrm{FPS}}. \end{aligned} \tag{4.22}$$

4.2.2 *Half-Period-Switch PRFSFP Square-Wave Injection*

Similar with the FPS- PRFSFP voltage injection, the half-period-switch (HPS) PRFSFP scheme also selects from two kinds of fixed-frequency and fixed-amplitude square-wave voltage signals [2]. Then it injects the PRHF voltage signal to the estimated d-axis. The difference is that, not every full period but every half period, the signal switches from one to the other. It can be equivalent to the random combination of four voltage modes shown in Fig. 4.7a–d. In HPS-PRFSFP scheme, the injected PRHF voltage can be expressed as

$$u_{dh}(t) = U_{\mathrm{R}} \cdot \sum_{k=1}^{\infty} u_{\mathrm{HPS}}(t - t_k),$$
$$u_{\mathrm{HPS}}(t) = \Re\left[u_{T_{i1}/2,\ T_{i1}/2}(t),\ u_{T_{i2}/2,\ T_{i2}/2}(t),\ u_{T_{i1}/2,\ T_{i2}/2}(t),\ u_{T_{i2}/2,\ T_{i1}/2}(t)\right]. \tag{4.23}$$

With the voltage signal (4.23) injected into the estimated d-axis, the induced current in the q-axis can be expressed as

$$i_{qh}(t) = U_{\mathrm{R}} \cdot \frac{L_q - L_d}{4L_d L_q} \sin 2\tilde{\theta}_e \cdot \sum_{k=1}^{\infty} i_{\mathrm{HPS}}(t - t_k),$$
$$i_{\mathrm{HPS}}(t) = \Re\left[i_{T_{i1}/2,\ T_{i1}/2}(t),\ i_{T_{i2}/2,\ T_{i2}/2}(t),\ i_{T_{i1}/2,\ T_{i2}/2}(t),\ i_{T_{i2}/2,\ T_{i1}/2}(t)\right]. \tag{4.24}$$

The demodulation signal ϕ_{HPS} and the equivalent position error signal ε in HPS-PRFSFP scheme can be obtained as

$$\begin{aligned} \phi_{\mathrm{HPS}} &= sign\left(\sum_{k=1}^{\infty} i_{\mathrm{HPS}}(t - t_k)\right), \\ \varepsilon &= i_{qh}(t) \cdot \phi_{\mathrm{HPS}}. \end{aligned} \tag{4.25}$$

By multiplying with $i_{qh}(t)$ and the demodulation signal ϕ_{HPS}, the equivalent position error can be obtained. Then the rotor position can be estimated through controlling the error to zero.

4.2.3 Experimental Results for PRFSFP Injection

The platform with a 2.2-kW IPMSM test machine is shown in Fig. 3.18. The PWM switching frequency and the current sampling frequency are both 6 kHz, and the DC bus voltage is 510 V. The magnitudes of the injected HF voltages are 90 V/1.5 kHz, 60 V/1 kHz, 45 V/750 Hz, and 36 V/600 Hz. The LCG is adopted to produce pseudo-random numbers. The probability parameters in FPS-PRFSFP injection schemes are set as $\boldsymbol{P}_{\mathrm{FPS}} = [0.5\ 0.5]$ and in HPS-PRFSFP injection schemes $\boldsymbol{P}_{\mathrm{HPS}} = [0.25\ 0.25\ 0.25\ 0.25]$.

Figure 4.8 compares the four sensorless control operation tests at 100 r/min with rated load, including two fixed-frequency signal injection tests and two PRFSFP signal injection tests. Not only the fixed-frequency current can be observed but also the PRHF current can be used to extract the rotor position information. The position estimation error almost remains the same in the four tests, within 15° during the rated load operation. The detailed current waveforms in the dashed blocks are shown under the figure, in which the comparison between the FPS-PRFSFP and the

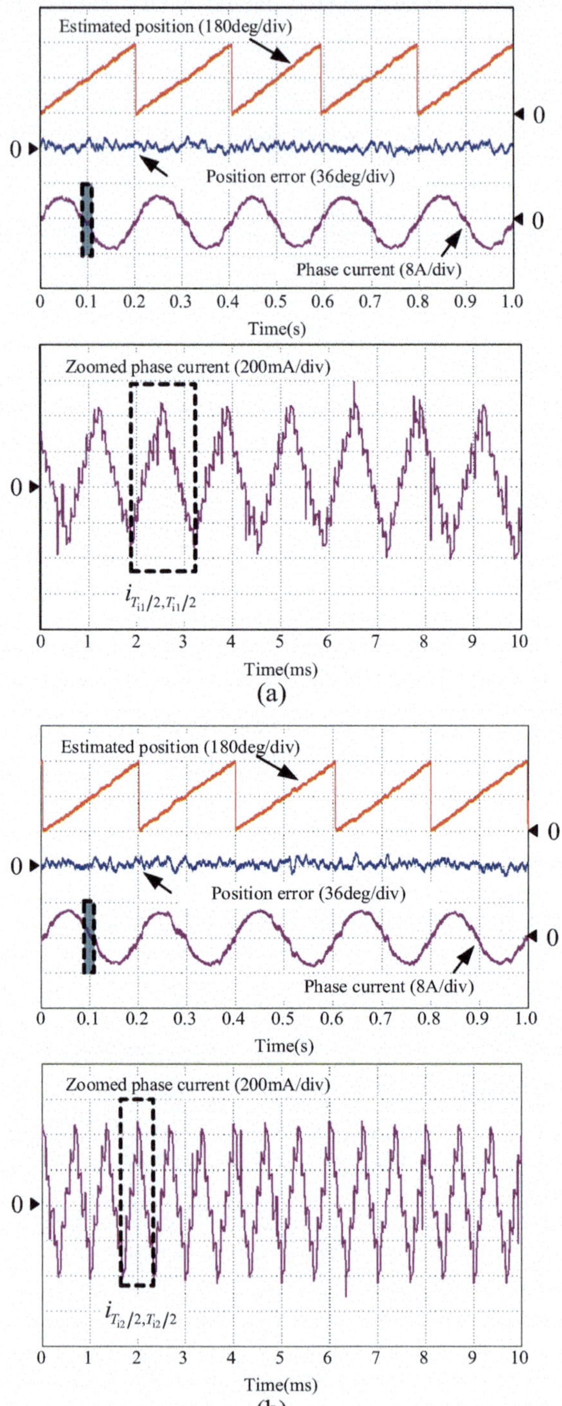

Fig. 4.8 Sensorless operation at 100 r/min with rated load using different injection schemes. (**a**) Fixed 750 Hz signal injection. (**b**) Fixed 1.5 kHz signal injection. (**c**) FPS-PRFSFP 750 Hz and 1.5 kHz signal injection. (**d**) HPS-PRFSFP 750 Hz and 1.5 kHz signal injection

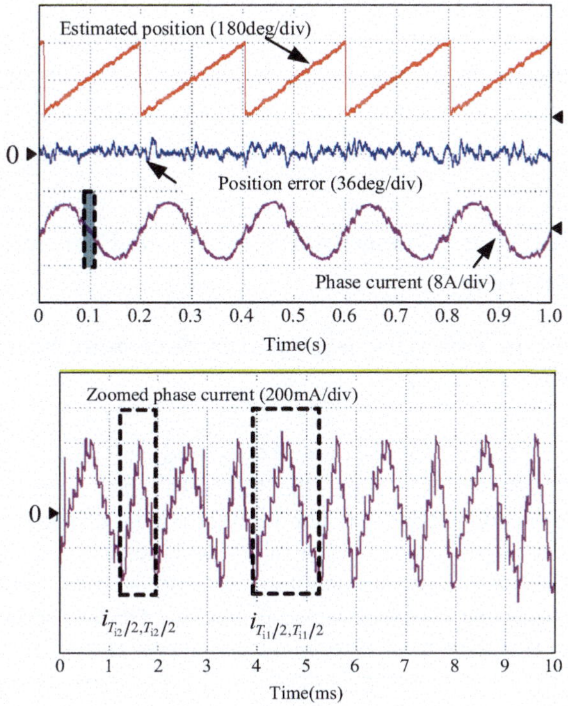

(c)

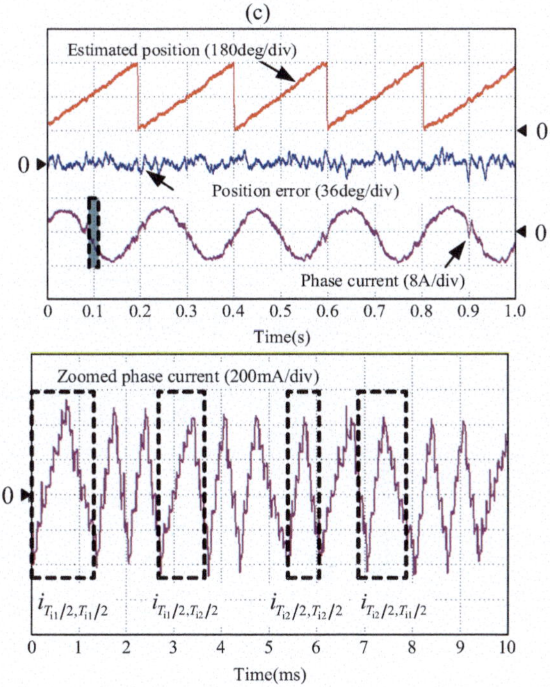

(d)

Fig. 4.8 (continued)

HPS-PRFSFP signal injection can be made. Only two modes ($i_{T_{i1}/2},\ T_{i1}/2$ and $i_{T_{i2}/2},\ T_{i2}/2$) can be observed in the FPS- PRFSFP scheme. However, in the HPS-PRFSFP, there are four modes ($i_{T_{i1}/2},\ T_{i1}/2$, $i_{T_{i2}/2},\ T_{i2}/2$, $i_{T_{i2}/2},\ T_{i1}/2$, and $i_{T_{i2}/2},\ T_{i2}/2$).

4.3 Pseudo-Random Phase-Switching Fixed-Frequency Signal Injection

4.3.1 Structure of PRPSFF Signal Injection Scheme-Based Sensorless Control

Figure 4.9 shows the block diagram of the PRPSFF square-wave voltage injection [3]. It mainly consists of three parts: the PRPSFF signal generator, the signal processing, and the position observer. The PRPSFF signal generator produces two square-wave signals, namely, the injection signal and the demodulation signal. The demodulation signal lags one current sampling cycle behind the injection one, which is used as an input of the signal processing section. The HF induced current is extracted and converted to position error in the signal processing. By converging the position error to zero with a Luenberger observer, the rotor position can be obtained.

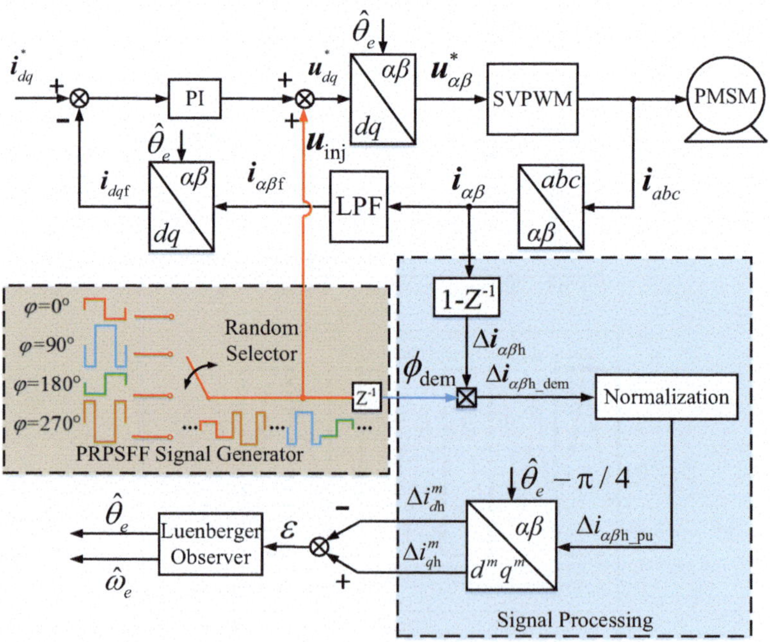

Fig. 4.9 Block diagram of the PRPSFF square-wave voltage injection method

4.3.2 DC Bias in Selection of Different Phase Signals

Figure 4.10 illustrates the selected four basic injected voltages and the corresponding induced currents. For the sake of convenience, define the unit-amplitude square-wave signal as

$$\begin{aligned}\phi_{\mathrm{sqr}}(t,T_{\mathrm{i}},\varphi) = \{&\phi_{\mathrm{sqr}}(t,T_{\mathrm{i}},0^{\circ}),\ \phi_{\mathrm{sqr}}(t,T_{\mathrm{i}},90^{\circ}),\\ &\phi_{\mathrm{sqr}}(t,T_{\mathrm{i}},180^{\circ}),\phi_{\mathrm{sqr}}(t,T_{\mathrm{i}},270^{\circ})\}.\end{aligned} \tag{4.26}$$

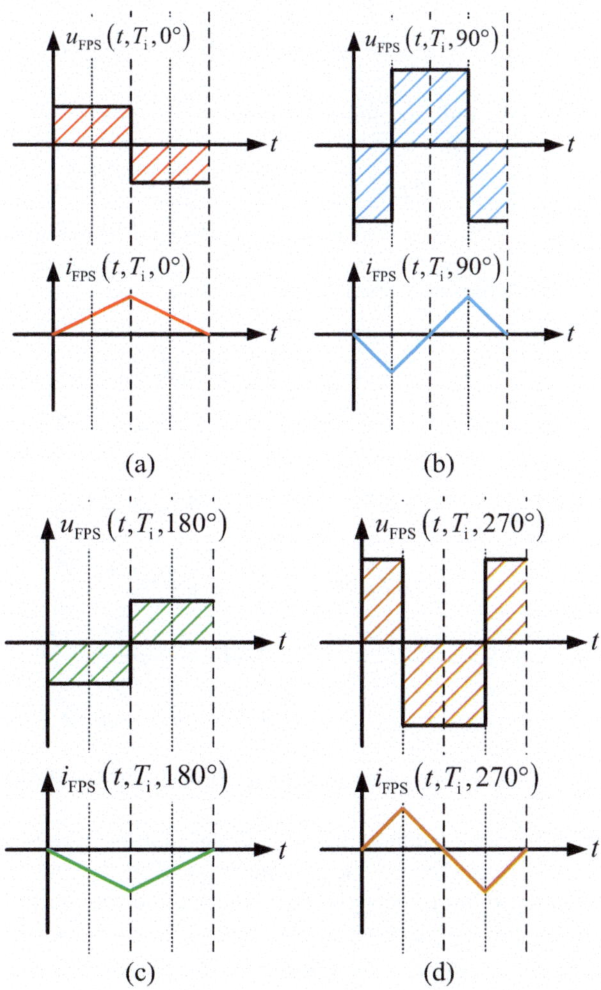

Fig. 4.10 Four basic injection voltages and corresponding currents. (**a**) $\varphi = 0°$. (**b**) $\varphi = 90°$. (**c**) $\varphi = 180°$. (d) $\varphi = 270°$.

The expression of each element in (4.26) is

$$\phi_{\text{sqr}}\left(t, T_{\text{i}}, 0^{\circ}\right)=\begin{cases}1, & 0<t_{\text{r}}(t, T_{\text{i}}) \leq \dfrac{T_{\text{i}}}{2} \\ -1, & \dfrac{T_{\text{i}}}{2}<t_{\text{r}}(t, T_{\text{i}}) \leq T_{\text{i}}\end{cases}, \tag{4.27}$$

$$\phi_{\text{sqr}}\left(t, T_{\text{i}}, 90^{\circ}\right)=\begin{cases}1, & \dfrac{T_{\text{i}}}{4}<t_{\text{r}}(t, T_{\text{i}}) \leq \dfrac{3T_{\text{i}}}{4} \\ -1, & 0<t_{\text{r}}(t, T_{\text{i}}) \leq \dfrac{T_{\text{i}}}{4}, \dfrac{3T_{\text{i}}}{4}<t_{\text{r}}(t, T_{\text{i}}) \leq T_{\text{i}}\end{cases}, \tag{4.28}$$

$$\phi_{\text{sqr}}\left(t, T_{\text{i}}, 180^{\circ}\right)=-\phi_{\text{sqr}}\left(t, T_{\text{i}}, 0^{\circ}\right), \tag{4.29}$$

$$\phi_{\text{sqr}}\left(t, T_{\text{i}}, 270^{\circ}\right)=-\phi_{\text{sqr}}\left(t, T_{\text{i}}, 90^{\circ}\right). \tag{4.30}$$

And the unit-amplitude triangle-wave signal is set as

$$\begin{aligned}\phi_{\text{tri}}(t, T_{\text{i}}, \varphi)=&\{\phi_{\text{tri}}\left(t, T_{\text{i}}, 0^{\circ}\right), \phi_{\text{tri}}\left(t, T_{\text{i}}, 90^{\circ}\right), \\ &\phi_{\text{tri}}\left(t, T_{\text{i}}, 180^{\circ}\right), \phi_{\text{tri}}\left(t, T_{\text{i}}, 270^{\circ}\right)\}.\end{aligned} \tag{4.31}$$

The expression of each element in (4.31) is

$$\phi_{\text{tri}}\left(t, T_{\text{i}}, 0^{\circ}\right)=\begin{cases}\dfrac{2t_{\text{r}}(t, T_{\text{i}})}{T_{\text{i}}}, & 0<t_{\text{r}}(t, T_{\text{i}}) \leq \dfrac{T_{\text{i}}}{2} \\ 2-\dfrac{2t_{\text{r}}(t, T_{\text{i}})}{T_{\text{i}}}, & \dfrac{T_{\text{i}}}{2}<t_{\text{r}}(t, T_{\text{i}}) \leq T_{\text{i}}\end{cases}, \tag{4.32}$$

$$\phi_{\text{tri}}\left(t, T_{\text{i}}, 90^{\circ}\right)=\begin{cases}-\dfrac{4t_{\text{r}}(t, T_{\text{i}})}{T_{\text{i}}} & 0<t_{\text{r}}(t, T_{\text{i}}) \leq \dfrac{T_{\text{i}}}{4} \\ -2+\dfrac{4t_{\text{r}}(t, T_{\text{i}})}{T_{\text{i}}} & \dfrac{T_{\text{i}}}{4}<t_{\text{r}}(t, T_{\text{i}}) \leq \dfrac{3T_{\text{i}}}{4} \\ 4-\dfrac{4t_{\text{r}}(t, T_{\text{i}})}{T_{\text{i}}} & \dfrac{3T_{\text{i}}}{4}<t_{\text{r}}(t, T_{\text{i}}) \leq T_{\text{i}}\end{cases}, \tag{4.33}$$

$$\phi_{\text{tri}}\left(t, T_{\text{i}}, 180^{\circ}\right)=-\phi_{\text{tri}}\left(t, T_{\text{i}}, 0^{\circ}\right), \tag{4.34}$$

$$\phi_{\text{tri}}\left(t, T_{\text{i}}, 270^{\circ}\right)=-\phi_{\text{tri}}\left(t, T_{\text{i}}, 90^{\circ}\right), \tag{4.35}$$

where T_{i} and φ denote the period and phase of the square-wave and triangle-wave signals, respectively. The parameter $t_{\text{r}}(t, T_{\text{i}})$ is the remainder of the division of t by T_{i}. The start point of triangle-wave signals is set as zero.

Considering an extreme case, when the same signal (0° or 180° phase signal) is selected for several consecutive cycles, the average value of the d-axis HF induced current will not equal to zero, resulting in a DC bias in the d-axis, as shown in Fig. 4.11a. Due to the existence of the DC bias, the output of the d-axis current

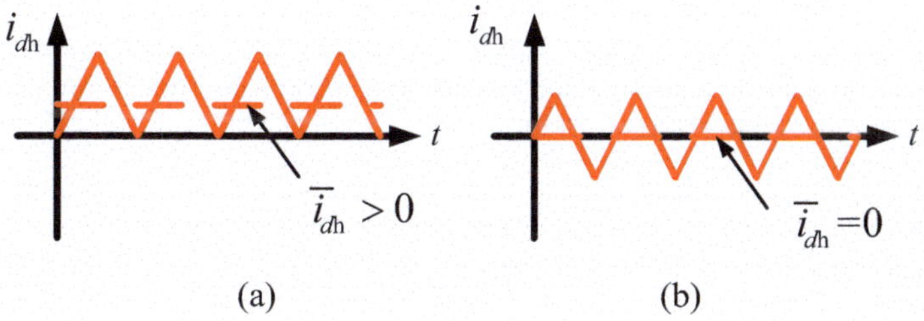

Fig. 4.11 The HF induced current when the same signal is consecutively injected. (a) $\varphi = 0°$. (b) $\varphi = 270°$

regulator will contain an additional voltage component, which can be regarded as a disturbance of the system. It makes the actual amplitude of the injection signal no longer equal to the given value. As a result, the actual injection signal will be uncontrollable. When the 90° or 270° phase signal is injected for several consecutive cycles, the average of the HF induced current will always maintain at zero, and the DC bias does not exist, as shown in Fig. 4.11b. Thus, 0° and 180° phase signals cannot be directly used for injection.

4.3.3 PRPSFF Square-Wave Voltage Injection Scheme

To deal with the DC bias, a limitation for PRPSFF injection is set. Since the *d*-axis DC bias only exists when 0° or 180° phase signal is injected successively, it can make 0° and 180° phase signals injected by pairs, and 90° and 270° phase signals between them are arbitrary. Then the DC bias can be eliminated.

The injected voltage in the estimated rotor reference frame can be expressed as

$$\begin{bmatrix} u_{dh}^{\hat{r}}(t) \\ u_{qh}^{\hat{r}}(t) \end{bmatrix} = U_{\mathrm{i}} \begin{bmatrix} u_{\mathrm{inj}}(t, T_{\mathrm{i}}, \varphi_{\mathrm{R}}) \\ 0 \end{bmatrix}, \tag{4.36}$$

$$u_{\mathrm{inj}}(t, T_{\mathrm{i}}, \varphi_{\mathrm{R}}) = \sum_{k=1}^{\infty} u_{\mathrm{sam}}(t - kT_{\mathrm{i}}, T_{\mathrm{i}}, \varphi_{\mathrm{R}}), \tag{4.37}$$

$$\begin{aligned} u_{\mathrm{sam}}(t, T_{\mathrm{i}}, \varphi_{\mathrm{R}}) =& \Re\left[0.5\phi_{\mathrm{sqr}}(t, T_{\mathrm{i}}, 0^{\circ}), \phi_{\mathrm{sqr}}(t, T_{\mathrm{i}}, 90^{\circ}),\right. \\ & \left. 0.5\phi_{\mathrm{sqr}}(t, T_{\mathrm{i}}, 180^{\circ}), \phi_{\mathrm{sqr}}(t, T_{\mathrm{i}}, 270^{\circ})\right], \end{aligned} \tag{4.38}$$

where U_{i} is the amplitude of the injected voltage, the subscript "R" represents the result of the random operation, and the subscript "h" denotes the HF component. $u_{\mathrm{sam}}(t, T_{\mathrm{i}}, \varphi_{\mathrm{R}})$ is the injected square-wave voltage sample. The operator R[] indicates the random selection.

When the motor operates at low speed (usually under 5% rated speed), and the injection frequency is set sufficiently high, the voltage drops on the stator resistance and the terms associated with ω_e can be neglected. Therefore, the HF induced current in the SRF can be obtained as

$$\begin{bmatrix} i_{\alpha h} \\ i_{\beta h} \end{bmatrix} = K i_{\text{inj}}(t, T_{\text{i}}, \varphi_{\text{R}}) \begin{bmatrix} L_q \cos\theta_e \cos\tilde{\theta}_e + L_d \sin\theta_e \sin\tilde{\theta}_e \\ L_q \sin\theta_e \cos\tilde{\theta}_e - L_d \cos\theta_e \sin\tilde{\theta}_e \end{bmatrix}, \tag{4.39}$$

where

$$K = \frac{U_{\text{i}} T_{\text{i}}}{4 L_d L_q}, \tag{4.40}$$

$$i_{\text{inj}}(t, T_{\text{i}}, \varphi_{\text{R}}) = \sum_{k=1}^{\infty} i_{\text{sam}}(t - kT_{\text{i}}, T_{\text{i}}, \varphi_{\text{R}}), \tag{4.41}$$

$$\begin{aligned} i_{\text{sam}}(t, T_{\text{i}}, \varphi_{\text{R}}) =& \Re\left[\phi_{\text{tri}}\left(t, T_{\text{i}}, 0^{\circ}\right), \phi_{\text{tri}}\left(t, T_{\text{i}}, 90^{\circ}\right),\right. \\ &\left.\phi_{\text{tri}}\left(t, T_{\text{i}}, 180^{\circ}\right), \phi_{\text{tri}}\left(t, T_{\text{i}}, 270^{\circ}\right)\right], \end{aligned} \tag{4.42}$$

and θ_e is the actual rotor position and $\tilde{\theta}_e$ $(\tilde{\theta}_e = \theta_e - \hat{\theta}_e)$ is the estimated position error.

According to (4.40), in order to ensure the same amplitude of HF induced currents, the volt-second area of each injection signal should be identical. Thus, the amplitude of 0° and 180° phase signals should be half of the value of 90° and 270° phase signals. The injected voltages and induced currents for PRPSFF injection scheme are shown in Fig. 4.12.

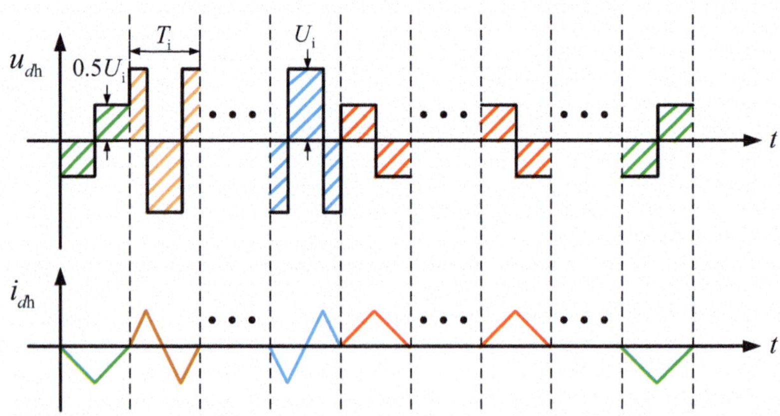

Fig. 4.12 Injected voltages and corresponding induced currents for PRPSFF injection scheme

4.3.4 Signal Processing and Extraction of Rotor Position

According to (4.39), the rotor position information is included in the envelope of the HF induced current. Adopting a HPF is a feasible method to extract the HF induced current. However, when the load is changed abruptly, the fundamental component of the induced current will change as well to balance the load torque, producing a loaded HF component. This loaded HF component cannot be effectively separated from the injection of HF component by HPFs. As a result, it may decrease the accuracy of the estimated position in the transient. In [6], the HF induced current was extracted without HPFs, but it needs to keep the current sampling frequency twice as high as the injection one, which restricts the scope of the injection frequency.

The processing of the rotor position extraction can be explained in Fig. 4.13, where T_s denotes the current sampling period. Figure 4.13a shows the injected voltage in the estimated rotor reference frame. The corresponding induced current contains both HF and fundamental components, as shown in Fig. 4.13c. Since the speed is low enough (usually under 5% rated speed, i.e., several hertz), the variation of the fundamental component between two consecutive sampling points (6 kHz) is very small. For the sake of convenience, the fundamental components at two consecutive sampling points are regarded as a constant. Then the increment of the HF induced current can be obtained as

$$\begin{cases} i_{\alpha,\beta}(k) = i_{\alpha,\beta\mathrm{h}}(k) + i_{\alpha,\beta\mathrm{f}}(k) \\ i_{\alpha,\beta}(k-1) = i_{\alpha,\beta\mathrm{h}}(k-1) + i_{\alpha,\beta\mathrm{f}}(k-1)\,, \\ i_{\alpha,\beta\mathrm{f}}(k) \approx i_{\alpha,\beta\mathrm{f}}(k-1) \end{cases} \tag{4.43}$$

$$\Delta i_{\alpha,\beta\mathrm{h}} \approx i_{\alpha,\beta}(k) - i_{\alpha,\beta}(k-1), \tag{4.44}$$

where the subscript “f” denotes the fundamental component.

By substituting (4.39) in (4.44), the expression of the increment of the HF induced current shown in Fig. 4.13d can be calculated as

$$\begin{bmatrix} \Delta i_{\alpha\mathrm{h}} \\ \Delta i_{\beta\mathrm{h}} \end{bmatrix} = K u_{\mathrm{inj}}(t - T_{\mathrm{s}}, T_{\mathrm{i}}, \varphi_{\mathrm{R}}) \cdot \begin{bmatrix} L_q \cos\theta_e \cos\tilde{\theta}_e + L_d \sin\theta_e \sin\tilde{\theta}_e \\ L_q \sin\theta_e \cos\tilde{\theta}_e - L_d \cos\theta_e \sin\tilde{\theta}_e \end{bmatrix}. \tag{4.45}$$

According to (4.45), $\Delta i_{\alpha,\beta\mathrm{h}}$ has a signal processing delay behind the injection signal. In [7, 8], the envelope of $\Delta i_{\alpha,\beta\mathrm{h}}$ was directly extracted by unit injected voltage, as (4.46). It will cause some oscillations in $\Delta i_{\alpha,\beta\mathrm{h}}$ at each alternating edge of the injection signal.

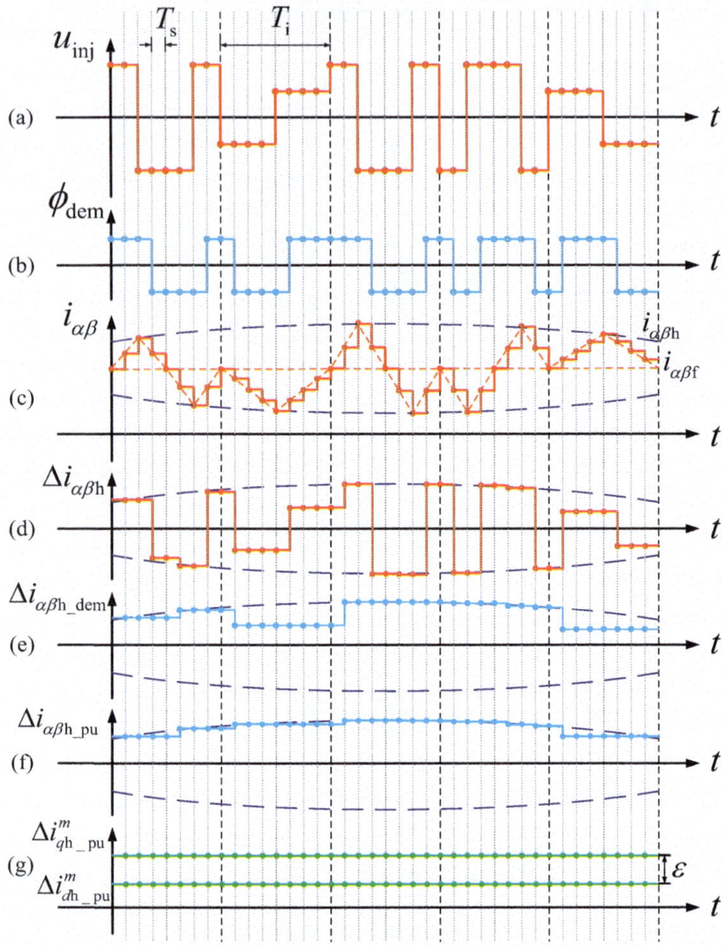

Fig. 4.13 Processing of the position extraction for PRPSFF injection scheme. (**a**) HF voltage injection signal. (**b**) Demodulation signal. (**c**) Induced current in the SRF. (**d**) Increment of the HF induced current. (**e**) Demodulated HF induced current increment. (**f**) Normalized HF induced current increment. (**g**) Normalized HF induced current increment in $d^m q^m$-axes

$$\phi_{dem0}(t, T_i, \varphi_R) = \sum_{k=1}^{\infty} \Re\left[\phi_{sqr}(t - kT_i, T_i, \varphi)\right]. \tag{4.46}$$

Defining the demodulation signal as

$$\phi_{dem}(t, T_i, \varphi_R) = \sum_{k=1}^{\infty} \Re\left[\phi_{sqr}(t - kT_i - T_s, T_i, \varphi)\right]. \tag{4.47}$$

The demodulation signal can be regarded as a unit-amplitude injection signal with one current sampling cycle delay, as shown in Fig. 4.13b. Multiplying $\Delta i_{\alpha,\beta \text{h}}$ by $\phi_{\text{dem}}(t, T_{\text{i}}, \varphi_{\text{R}})$, the demodulated HF induced current increment $\Delta i_{\alpha,\beta \text{h\_dem}}$ shown in Fig. 4.13e can be expressed as

$$\begin{bmatrix} \Delta i_{\alpha \text{h\_dem}} \\ \Delta i_{\beta \text{h\_dem}} \end{bmatrix} = K\phi_{\text{sqr\_dem}}(t, T_{\text{i}}, \varphi_{\text{R}}) \cdot \begin{bmatrix} L_q \cos\theta_e \cos\tilde{\theta}_e + L_d \sin\theta_e \sin\tilde{\theta}_e \\ L_q \sin\theta_e \cos\tilde{\theta}_e - L_d \cos\theta_e \sin\tilde{\theta}_e \end{bmatrix}, \tag{4.48}$$

$$\phi_{\text{sqr\_dem}}(t, T_{\text{i}}, \varphi_{\text{R}}) = \Re[0.5, 1, 0.5, 1]. \tag{4.49}$$

To reduce the effect of variations of the motor parameters and the amplitude of the injected voltage, the normalization is implemented as

$$\begin{aligned} \begin{bmatrix} \Delta i_{\alpha \text{h\_pu}} \\ \Delta i_{\beta \text{h\_pu}} \end{bmatrix} &= \frac{1}{\sqrt{(\Delta i_{\alpha \text{h\_dem}})^2 + (\Delta i_{\beta \text{h\_dem}})^2}} \begin{bmatrix} \Delta i_{\alpha \text{h\_dem}} \\ \Delta i_{\beta \text{h\_dem}} \end{bmatrix}, \\ &= \begin{bmatrix} \cos(\theta_e - \delta) \\ \sin(\theta_e - \delta) \end{bmatrix} \end{aligned} \tag{4.50}$$

$$\tan\delta = \frac{L_d}{L_q} \tan\tilde{\theta}_e, \delta \approx \frac{L_d}{L_q}\tilde{\theta}_e, (\tilde{\theta}_e \approx 0), \tag{4.51}$$

where δ is the intermediate variable in the operation of trigonometric functions, which simplifies the expression of formulas. $\Delta i_{\alpha,\beta \text{h\_pu}}$ is the normalized HF induced current increment. As shown in Fig. 4.13f, the shape of $\Delta i_{\alpha,\beta \text{h\_pu}}$ is similar to the envelope of the HF induced current.

In order to obtain the position error, define the measuring rotor reference frame as $d^m q^m$-axes, which has a phase lag of π/4 compared with the estimated rotor reference frame. As shown in Fig. 4.14, dq-axes are the actual rotor reference frame and $\hat{d}\hat{q}$-axes are the estimated rotor reference frame.

Then the normalized HF induced current increment in the $d^m q^m$-axes can be calculated from (4.50) as

$$\begin{bmatrix} \Delta i^m_{d \text{h\_pu}} \\ \Delta i^m_{q \text{h\_pu}} \end{bmatrix} = \boldsymbol{R}\left(\hat{\theta}_e - \frac{\pi}{4}\right) \begin{bmatrix} \Delta i_{\alpha \text{h\_pu}} \\ \Delta i_{\beta \text{h\_pu}} \end{bmatrix} = \begin{bmatrix} \cos\left(\tilde{\theta}_e - \delta + \frac{\pi}{4}\right) \\ \sin\left(\tilde{\theta}_e - \delta + \frac{\pi}{4}\right) \end{bmatrix}, \tag{4.52}$$

where $\boldsymbol{R}(\theta) = \begin{bmatrix} \cos(\theta) & \sin(\theta) \\ -\sin(\theta) & \cos(\theta) \end{bmatrix}$ is the reference frame transformation matrix, and the superscript "m" represents the measuring rotor reference frame.

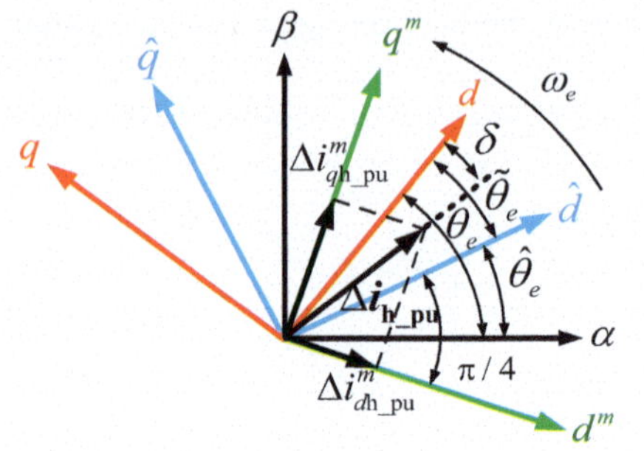

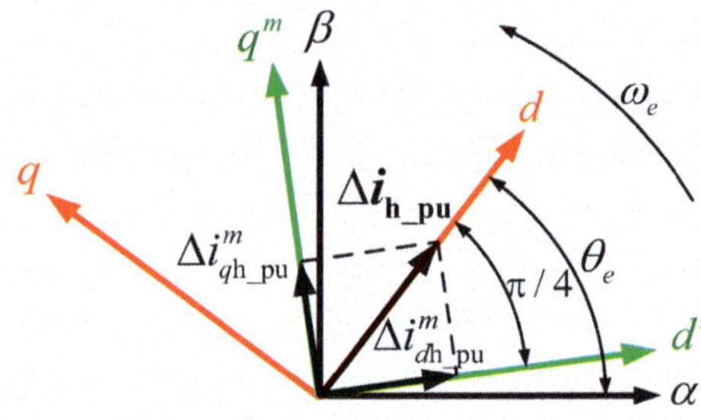

Fig. 4.14 Coordinate distribution of $d^m q^m$-axes. (**a**) $\tilde{\theta}_e \neq 0$. (b) $\tilde{\theta}_e \approx 0$

The equivalent position deviation ε can be calculated as

$$\varepsilon = \Delta i^m_{qh\_pu} - \Delta i^m_{dh\_pu} \approx \sqrt{2}\frac{L_q - L_d}{L_q}\tilde{\theta}_e, \left(\tilde{\theta}_e \approx 0\right). \tag{4.53}$$

By converging ε to zero with a Luenberger observer [9], the rotor position can be obtained. The block diagram of the rotor position extraction method is shown in the dashed box of Fig. 4.9.

4.3.5 Experimental Results for PRPSFF Injection

As mentioned above, the DC-bias problem in the d-axis exists when the 0° or 180° phase signal is injected for several consecutive cycles. In order to verify the limitation of signal injection is advantageous, a simplified PRPSFF injection experiment is carried out. The injection signals are selected only from two different phases (0° and 180°, 90°, and 270°), and then the probability of consecutive injection of the same signal can be increased. The experimental results are shown in Fig. 4.15.

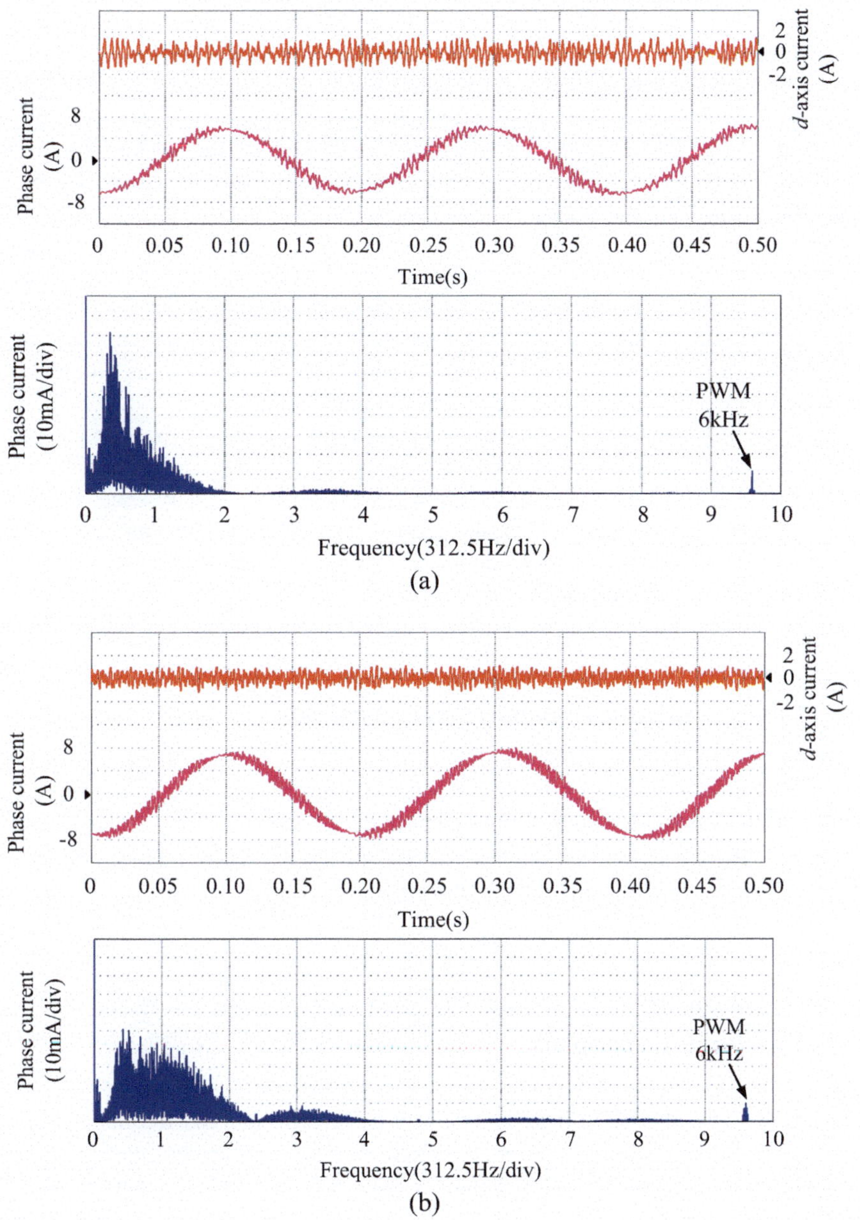

Fig. 4.15 Comparison of *d*-axis current, *a*-phase current, and harmonics at 100 r/min for PRPSFF injection (two-phase switching) with rated load. (**a**) 0° and 180° phase voltage injection. (**b**) 90° and 270° phase voltage injection

In Fig. 4.15, it can be easily observed that some oscillations exist in the d-axis current for 0° and 180° phase voltage injection, leading to some ripples in the torque. According to the harmonic analysis, the harmonic distribution of 90° and 270° phase injection is wider than that of 0° and 180° phase injection. Thus, setting the limitation of the injected voltage is propitious to reduce the probability of the consecutive appearance of 0° or 180° phase signal and improve the spectra of the induced current.

Figure 4.16 shows the experimental results of $\Delta i_{\alpha,\beta \mathrm{h-pu}}$ for PRPSFF injection. With the unit-injection signal demodulation (4.45), there are many oscillations in normalized signals. It is inappropriate to use the unit-injection signal for demodulation since the $\Delta i_{\alpha,\beta \mathrm{h}}$ has a signal processing delay behind the injection signal, as shown in (4.46). When the introduced demodulation signal (4.47) is adopted, the envelope of the HF induced current can be extracted smoothly.

Figure 4.17 shows the estimated rotor position, position error, and a-phase current at 100 r/min with rated load. The proposed scheme can extract the rotor position effectively, and the position error is within 15° in steady state. From the details of a-phase current shown in the dashed box, 0°, 90°, 180°, and 270° phase signals can be observed. It should be noted that the 0° and 180° phase signals are injected by pairs to reduce the d-axis DC bias.

4.4 Power Spectra Density (PSD) Analysis

PSD, which is the Fourier transform of the autocorrelation of a signal, is an important tool for random signal analysis. In this section, the current PSD for different kinds of injection schemes (fixed-frequency-phase injection, FPS-PRFSFP injection, HPS-PRFSFP injection, and PRPSFF injection) are analyzed.

4.4.1 Current PSD in Fixed-Frequency HF Square-Wave Voltage Injection

Figure 4.18 shows the HF voltage and current waveforms in fixed-frequency voltage injection. For analysis simplification, the fixed-frequency HF square-wave voltage is assumed to be injected into the actual d-axis, and the induced current can be expressed as

$$i_{d\mathrm{h}}(t) = \sum\nolimits_{k=1}^{\infty} i_{D_1,\ D_2}(t - t_k). \tag{4.54}$$

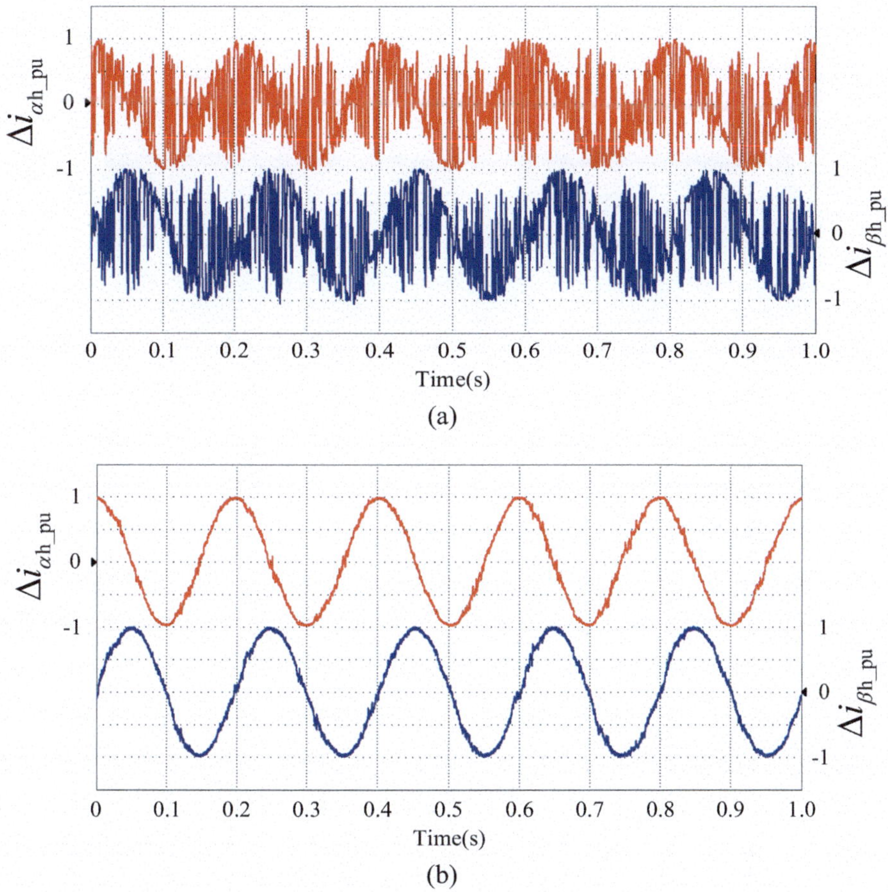

Fig. 4.16 Experimental results of normalized HF induced current increment at 100 r/min for PRPSFF injection with rated load. (**a**) With injection signal demodulated. (**b**) With proposed signal demodulated

Here, unit magnitude current is discussed. The PSD of the fixed-frequency current can be calculated as

$$S(f) = \begin{cases} \dfrac{1}{(D_1 + D_2)^2} \left| I_{D_1,\ D_2}(f) \right|^2 & f = n\dfrac{1}{D_1 + D_2} \\ 0 & \text{elsewhere,} \end{cases} \tag{4.55}$$

where $n(0,1,2,3\ldots)$ is the integer, and $I_{D_1,\ D_2}(f)$ is the Fourier transform of one cycle of the current

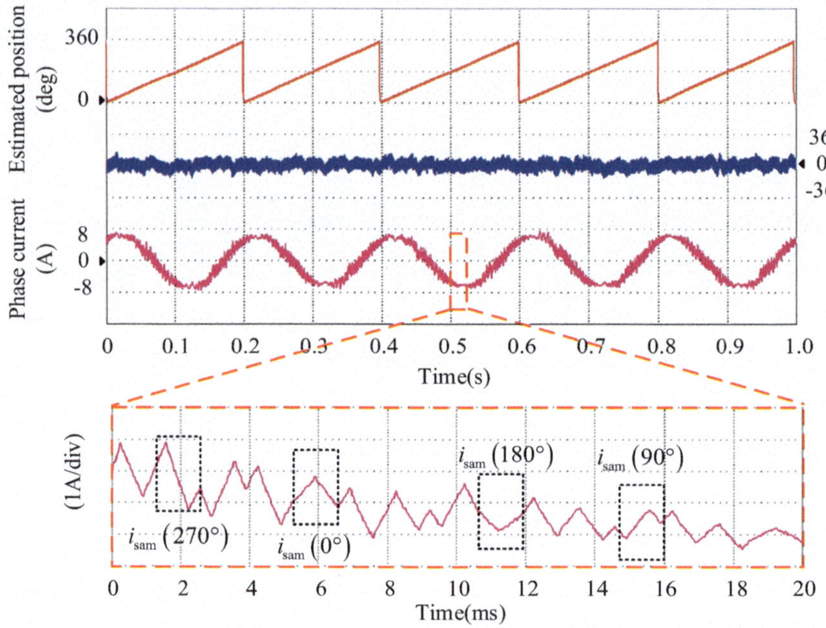

Fig. 4.17 Sensorless operation at 100 r/min with rated load using PRPSFF injection scheme

Fig. 4.18 Injected voltage and induced current in fixed-frequency voltage injection

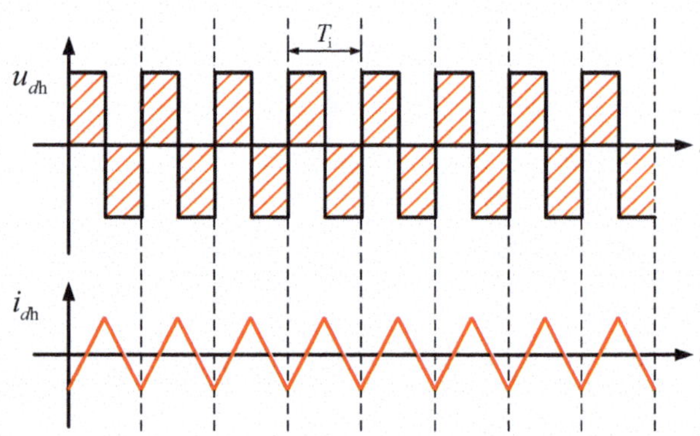

$$
\begin{aligned}
I_{D_1,\ D_2}(f) &= \int_0^{D_1} \left(-1 + \frac{2}{D_1} t\right) e^{-j2\pi ft} dt + \int_{D_1}^{D_1+D_2} \left(1 + \frac{2}{D_2}(D_1 - t)\right) e^{-j2\pi ft} dt \\
&= \frac{e^{-jD_1 2\pi f}}{2\pi^2 f^2} \left[\frac{1 - e^{jD_1 2\pi f}(1 - j\pi f D_1)}{D_1} + \frac{1 - e^{-jD_2 2\pi f}(1 + j\pi f D_2)}{D_2}\right].
\end{aligned} \tag{4.56}
$$

As can be seen in (4.55), only discrete harmonic components are contained. Usually, in fixed-frequency voltage injection, the positive and the negative voltages have the same duration, where $D_1 = D_2 = T/2$, and (4.56) can be transformed into

$$I_{T/2,\ T/2}(f) = \frac{2 - 2\cos(f\pi T) - f\pi T\sin(f\pi T)}{f^2\pi^2 T} e^{-j\pi fT}. \tag{4.57}$$

By substituting $f = n/T$ into (4.57), the following result can be obtained:

$$I_{T/2,\ T/2}\left(\frac{n}{T}\right) = \begin{cases} 0 & \text{when } n \text{ is even} \\ -\dfrac{4T}{n^2\pi^2} & \text{when } n \text{ is odd} \end{cases}. \tag{4.58}$$

From the results of (4.58), the conclusion can be obtained: in the fixed-frequency square-wave voltage injection scheme, only discrete harmonics are contained in the current PSD. Furthermore, it concentrates on the frequencies of the injected signal and its odd multiples.

4.4.2 Current PSD in Full-Period-Switch PRFSFP Square-Wave Voltage Injection

Figure 4.19 shows the PRHF voltage and current using the FPS-PRFSFP voltage injection. The induced current is a series of successive random-frequency saw-tooth-wave currents:

$$i_{dh}(t) = \sum\nolimits_{k=1}^{\infty} i_{FPS}(t - t_k). \tag{4.59}$$

According to [1], the PSD result $S(f)$ of the PRHF current (4.59) can be calculated as follows.

For $E[e^{j2\pi fT}] \neq 1$:

$$S(f) = \frac{1}{E[T]}\left\{E\left[|I(f)|^2\right] + 2\,\mathrm{Re}\left(\frac{E\left[I(f)e^{j2\pi fT}\right] \cdot E[I^*(f)]}{1 - E\left[e^{j2\pi fT}\right]}\right)\right\}. \tag{4.60}$$

And for $E[e^{j2\pi fT}] = 1$:

$$S(f) = \frac{1}{\{E[T]\}^2}\{E[|I(f)|]\}^2, \tag{4.61}$$

where E[] denotes the mathematical expectation operator. The expression of the PSD consists of two parts, the continuous spectrum (4.60) and the discrete spectrum (4.61). If a signal has only discrete components, the distribution concentrates

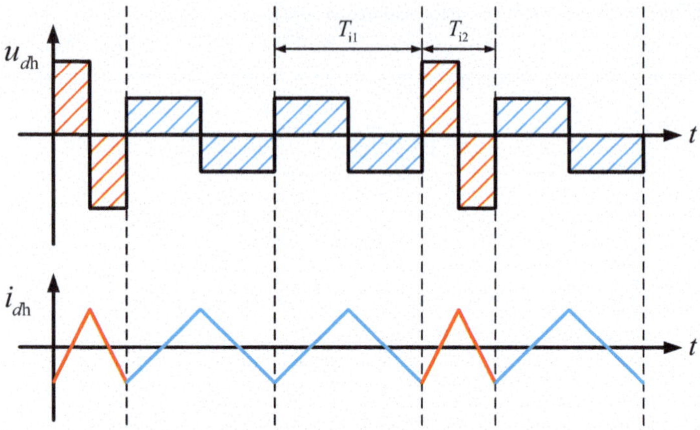

Fig. 4.19 Injected voltage and induced current in FPS-PRFSFP voltage injection

excessively. Discrete harmonics cause peaks in the spectrum, which means terrible noises. Continuous components are expected compared with the discrete harmonics. Therefore, the key problem of the noise reduction is to eliminate the discrete components.

From (4.61), the discrete harmonics may present at the points where f satisfies the following condition:

$$\mathrm{E}\left[e^{j2\pi\bar{f}T}\right] = \mathbf{P}_{\mathrm{FPS}} \cdot \left[e^{j2\pi\bar{f}T_1} e^{j2\pi\bar{f}T_2}\right]^{\mathrm{T}} = 1. \tag{4.62}$$

Let $\bar{f}$ denote the root of (4.62). And $\mathbf{P}_{\mathrm{FPS}} = [p_1 p_2]$ is the probability matrix and p_1 and p_2 are the probability of the injection mode in Fig. 4.7a, b, respectively. Furthermore, if $\bar{f}$ satisfies the equation

$$\mathrm{E}\left[\left|I(\bar{f})\right|\right] \neq 0 \qquad \left(\text{ when } \mathrm{E}\left[e^{j2\pi\bar{f}T}\right] = 1\right), \tag{4.63}$$

the spectrum peaks would come out.

Considering that $e^{j2\pi fT_{i1}}$ and $e^{j2\pi fT_{i2}}$ are both less than or equal to 1, if (4.62) holds, the following statements are equivalent:

1. The equation $e^{j2\pi fT_{i1}} = e^{j2\pi fT_{i2}} = 1$ meets.
2. $f_{i1}T_{i1}$ and $f_{i2}T_{i2}$ are both integers.
3. f is the common multiple of f_{i1} and f_{i2}.

The ratio between f_{i1} and f_{i2} is

$$\frac{f_{i1}}{f_{i2}} = \frac{m_1}{m_2}, \tag{4.64}$$

where m_1 and m_2 are the numerator and denominator of the irreducible fraction, respectively. The least common multiple (LCM) of f_{i1} and f_{i2} can be expressed as $f_{12} = f_{i1}m_1 = f_{i2}m_2$. Therefore, when f is the integral multiple of f_{12}, or $\bar{f} = n \cdot f_{12}$, Eq. (4.62) holds. Here, n is an integer. When n is even or odd, the result (4.61) will be different.

1. *When n is even*

When f is the integral multiple of f_{12}, the following relation can be obtained:

$$\begin{aligned} \bar{f}T_{i1} &= nm_1 f_{i1} T_{i1} = nm_1, \\ \bar{f}T_{i2} &= nm_2 f_{i2} T_{i2} = nm_2. \end{aligned} \tag{4.65}$$

As can be seen in (4.65), when n is even, nm_1 and nm_2 are also even. By substituting (4.65) into (4.57),

$$I_{T_1/2,\ T_1/2}(nf_{12}) = I_{T_2/2,\ T_2/2}(nf_{12}) = 0. \tag{4.66}$$

Equation (4.68) can be obtained. It is obvious that the Eq. (4.69) is satisfied if n is even.

$$\begin{aligned} \mathrm{E}[|I(nf_{12})|] &= \mathbf{P}_{\mathrm{FPS}} \cdot \left[\left| I_{T_{i1}/2,\ T_{i1}/2}(nf_{12}) \right| \left| I_{T_{i2}/2,\ T_{i2}/2}(nf_{12}) \right| \right]^{\mathrm{T}}. \\ &= 0 \quad (\text{when } n \text{ is even}) \end{aligned} \tag{4.67}$$

Therefore, the peaks will not present on the points where f is the even multiple of the LCM of f_{i1} and f_{i2}.

2. *When n is odd*

It is impossible that both m_1 and m_2 are even simultaneously. As can be seen in (4.65), when n is odd, there is at least one odd number in $\bar{f}T_{i1}$ and $\bar{f}T_{i2}$. Obviously, the following relation can be obtained as

$$\begin{aligned} \mathrm{E}[|I(nf_{12})|] &= \mathbf{P}_{\mathrm{FPS}} \cdot \left[\left| I_{T_{i1}/2,\ T_{i1}/2}(nf_{12}) \right| \left| I_{T_{i2}/2,\ T_{i2}/2}(nf_{12}) \right| \right]^{\mathrm{T}} \\ &\neq 0 \quad (\text{when } n \text{ is odd}), \end{aligned} \tag{4.68}$$

since the terms in the matrix cannot be zero simultaneously. As a consequence, Conclusion II can be obtained: In the FPS-PRHF square-wave voltage injection scheme, both continuous and discrete spectrum are contained in the current PSD. The spectrum peaks present at the odd multiples of the LCM of the two frequencies.

4.4.3 Current PSD in Half-Period-Switch PRFSFP Square-Wave Voltage Injection

Although the induced current in FPS-PRFSFP voltage injection has a relatively spread power spectra, some spectrum peaks still exist. To solve this problem, a new PRHF voltage injection method is introduced as follows. Figure 4.20 shows the voltage and induced current in the HPS-PRFSFP scheme. The PRHF current in *d*-axis can be expressed as

$$i_{dh}(t) = \sum\nolimits_{k=1}^{\infty} i_{\mathrm{HPS}}(t - t_k). \tag{4.69}$$

The PRHF current $i_{\mathrm{HPS}}(t)$ is chosen randomly from four current modes as shown in Fig. 4.7.

The PSD results (4.60) and (4.61) are general expression. For HPS-PRFSFP scheme, the condition equation is

$$\mathrm{E}\left[e^{j2\pi\bar{f}T}\right] = \mathbf{P}_{HPS} \cdot \left[e^{j2\pi\bar{f}T_{i1}} e^{j2\pi\bar{f}T_{i2}} e^{j\pi\bar{f}(T_{i1}+T_{i2})} e^{j\pi\bar{f}(T_{i1}+T_{i2})}\right]^{\mathrm{T}} = 1, \tag{4.70}$$

where

$$\mathbf{P}_{\mathrm{HPS}} = [p_1 p_2 p_3 p_4]. \tag{4.71}$$

Equation (4.71) is the probability matrix and p_1 to p_4 are the probability parameters of the injection mode Fig. 4.7a–d, respectively.

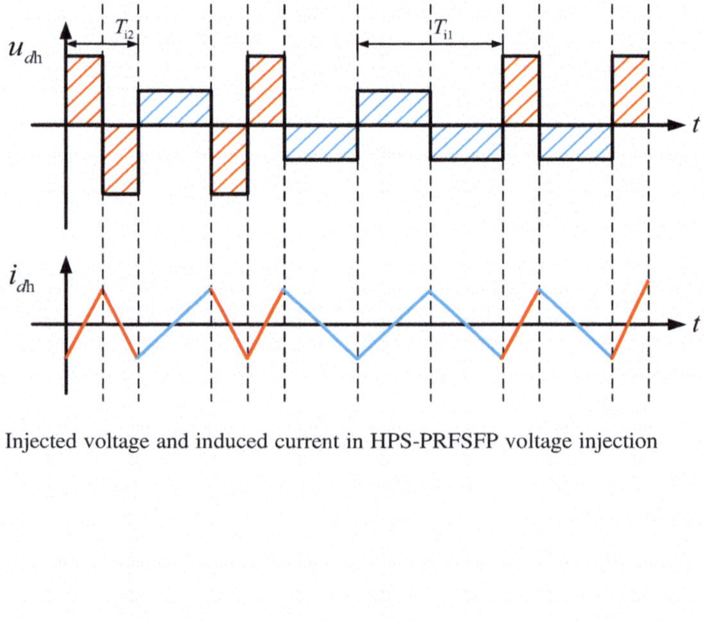

Fig. 4.20 Injected voltage and induced current in HPS-PRFSFP voltage injection

Similar with the analysis procedure above, to find the roots of (4.70) is to make the equation

$$\bar{f}\frac{(T_{\mathrm{i}1}+T_{\mathrm{i}2})}{2}=\frac{\bar{f}}{2}\left(\frac{nm_1}{\bar{f}}+\frac{nm_2}{\bar{f}}\right)=\frac{n}{2}(m_1+m_2) \tag{4.72}$$

an integer. Specifically, it can be classified into the following two cases:

1. *Both m_1 and m_2 are odd.*

If both m_1 and m_2 are odd, then $m_1 + m_2$ is even. According to (4.72), as long as n is an integer, $n(m_1 + m_2)/2$ is an integer and (4.70) holds. Now, the roots of (4.70) are $\bar{f}=nf_{12}$ $(n=1,2,3\ldots)$. By substituting the roots into (4.56), the following term can be obtained:

$$\begin{aligned}I_{T_{\mathrm{i}1}/2,\ T_{\mathrm{i}2}/2}(\bar{f})&=I_{T_{\mathrm{i}2}/2,\ T_{\mathrm{i}1}/2}(\bar{f})\\&=\begin{cases}0 & (\text{when } n \text{ is even})\\ -\dfrac{2(T_{\mathrm{i}1}+T_{\mathrm{i}2})}{T_{\mathrm{i}1}T_{\mathrm{i}2}\pi^2\bar{f}^2} & (\text{when } n \text{ is odd})\end{cases}.\end{aligned} \tag{4.73}$$

Combine (4.67), (4.68) with (4.73), and then

$$\begin{aligned}&\mathrm{E}[|I(nf_{12})|]=\mathbf{P}_{\mathrm{HPS}}\cdot\\&\left[\left|I_{T_{\mathrm{i}1}/2,\ T_{\mathrm{i}1}/2}(nf_{12})\right|\ \left|I_{T_{\mathrm{i}2}/2,\ T_{\mathrm{i}2}/2}(nf_{12})\right|\ \left|I_{T_{\mathrm{i}1}/2,\ T_{\mathrm{i}2}/2}(nf_{12})\right|\ \left|I_{T_{\mathrm{i}2}/2,\ T_{\mathrm{i}1}/2}(nf_{12})\right|\right]^{\mathrm{T}}\\&\begin{cases}=0 & (\text{when } n \text{ is even})\\ \neq 0 & (\text{when } n \text{ is odd})\end{cases}\end{aligned} \tag{4.74}$$

can be obtained.

2. *One is odd and the other is even.*

If one odd and one even are in m_1 and m_2, then $m_1 + m_2$ is odd. According to (4.72), only when n is even, $n(m_1 + m_2)/2$ is an integer and (4.70) holds. $\bar{f}=nf_{12}$ $(n=2,4,6\ldots)$are the roots of (4.70). According to (4.73), if n is even, we can get both $I_{T_{\mathrm{i}1}/2,\ T_{\mathrm{i}2}/2}(\bar{f})$ and $I_{T_{\mathrm{i}2}/2,\ T_{\mathrm{i}1}/2}(\bar{f})$ are zero. Furthermore, $\mathrm{E}[|I(nf_{12})|]=0$ can be obtained on the basis of (4.74). Therefore, the discrete harmonics will not appear in this case, and the peaks can be suppressed.

Therefore, Conclusion III can be obtained: in the HPS-PRFSFP square-wave voltage injection scheme, if both the numerator and denominator in the irreducible fraction of the two frequencies are odd, the discrete harmonics would come out at the odd multiples of the LCM of the two frequencies. However, if one is odd and the other is even, the discrete harmonics will not occur. In this case, the spectrum peaks would not appear.

4.4.4 Current PSD in PRPSFF Square-Wave Voltage Injection

According to the analytical model proposed in [3] the current PSD for PRPSFF injection can be expressed as

$$\begin{aligned} S(f) = & f_{\mathrm{i}} \{ \overbrace{E_{\varphi}\left[|I(f)|^2\right] - |E_{\varphi}[I(f)]|^2}^{\text{continuous term}} \\ & + \overbrace{f_{\mathrm{i}} |E_{\varphi}[I(f)]|^2 \sum_{k=-\infty}^{+\infty} \delta(f - kf_{\mathrm{i}})}^{\text{discrete term}} \}, \end{aligned} \tag{4.75}$$

where the operator $E[]$ denotes the mathematical expectation operation, $I(f)$ is the Fourier transform of the $i^{r}_{dh}(t)$, f_{i} is the injection frequency, and $\delta(f)$ is the unit impulse function.

According to (4.75), the current PSD for PRPSFF injection contains both continuous and discrete harmonics. The discrete harmonic spectrum can cause a loud audible noise, which is the undesired component.In PRPSFF injection, a limitation of the injected voltage is set to reduce the influence of the d-axis DC bias. The concern of current PSD is the statistical expectation of the induced current, but not the random distribution of the actual induced current. Under the limitation of the injected voltage, it can be considered that the induced current is composed of 90° and 270° phase currents with different frequencies. The relationship between two frequencies is shown as

$$f_{\mathrm{i}1} = 0.5 f_{\mathrm{i}}, \tag{4.76}$$

where $f_{\mathrm{i}1}$ is the equivalent frequency. The equivalent injected voltage and induced current are shown in Fig. 4.21.

The expression of the equivalent induced current is

$$i_{d\mathrm{h}}(t) = \sum_{k=1}^{\infty} [i_{\mathrm{sam}1}(t - kT_{\mathrm{R}}, T_{\mathrm{R}}, \varphi_{\mathrm{R}})], \tag{4.77}$$

$$\begin{aligned} i_{\mathrm{sam}1}(t, T_{\mathrm{R}}, \varphi_{\mathrm{R}}) = & \Re\left[\phi_{\mathrm{tri}}(t, T_{\mathrm{i}}, 90^{\circ}), \phi_{\mathrm{tri}}(t, T_{\mathrm{i}}, 270^{\circ}), \right. \\ & \left. \phi_{\mathrm{tri}}(t, 2T_{\mathrm{i}}, 90^{\circ}), \phi_{\mathrm{tri}}(t, 2T_{\mathrm{i}}, 270^{\circ})\right]. \end{aligned} \tag{4.78}$$

According to (4.77) and (4.78), both frequency and phase of equivalent current are selected randomly. Therefore, PRPSFF injection can be regarded as a dual-random injection scheme.

The Fourier transform for different induced currents are calculated as

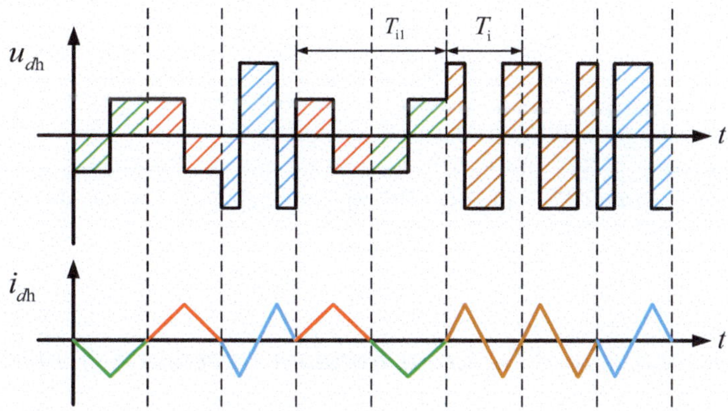

Fig. 4.21 The equivalent injected voltage and induced current for PRPSFF injection

$$I_{90^\circ}(f) = \frac{-2}{\pi^2 f^2 T_R}\left[2je^{-j\pi f T_R}\sin\frac{\pi}{2}fT_R + \frac{1}{2}\left(e^{-j2\pi f T_R} - 1\right)\right], \tag{4.79}$$

$$I_{270^\circ}(f) = \frac{2}{\pi^2 f^2 T_R}\left[2je^{-j\pi f T_R}\sin\frac{\pi}{2}fT_R + \frac{1}{2}\left(e^{-j2\pi f T_R} - 1\right)\right]. \tag{4.80}$$

According to (4.79) and (4.80), with the same injection cycle, the amplitudes of 90° and 270° phase current are identical, but the signs of them are opposite because a 180° phase difference exists in the induced current waveform. By substituting $f = nf_R$ into (4.79) and (4.80), the following results can be obtained.

$$I_{90^\circ}(f) = \frac{4}{\pi^2 n^2 f_R} e^{-j\frac{n\pi}{2}}, \tag{4.81}$$

$$I_{270^\circ}(f) = \frac{-4}{\pi^2 n^2 f_R} e^{-j\frac{n\pi}{2}}. \tag{4.82}$$

The Fourier transform of the current samples for fixed-frequency-phase injection and FPS-PRFSFP injection is calculated as

$$I(f) = \begin{cases} 0 & \text{when } n \text{ is even} \\ -\dfrac{4}{n^2\pi^2 f_R} & \text{when } n \text{ is odd.} \end{cases} \tag{4.83}$$

Compared with (4.79), (4.80), and(4.83), although the distribution of discrete spectra for PRPSFF injection is twice than that of FPS-PRFSFP injection and fixed-frequency-phase injection schemes, it can be canceled with the phase difference. Supposing the system has a good random ability, the probabilities of the four signals are equal, which can be obtained as

$$E_{\varphi}[I(f)] = \sum_{k=-\infty}^{+\infty} P(\varepsilon_k) I_k(f) = 0, \tag{4.84}$$

where $P(\varepsilon_{\mathrm{k}})$ is the probability of the k phase signal.

By substituting (4.84) into (4.75), the second and the third terms in (4.75) are zero; hence the current PSD of PRPSFF injection scheme only has the continuous component.

4.4.5 *Comparison of Different Random Injection Schemes*

From the perspective of the injected principle and current PSD distribution, PRFSFP and PRPSFF injection schemes are compared, and the following conclusions can be obtained.

1. The current samples for PRFSFP injection and PRPSFF injection are different. PRFSFP injection uses 0° or 180° phase signal as the injection signal, while PRPSFF injection injects four different phase signals randomly.
2. The principle of eliminating discrete harmonic spectra is different. PRFSFP uses the characteristic of 0° or 180° phase induced current of which discrete spectra at even multiples of the injection frequency points are not existent. The distribution range of the discrete spectrum expands to odd multiples of the LCM of the two injection frequencies, and then the discrete spectra in low frequency can be eliminated. PRPSFF injection method utilizes the symmetry of the different phase induced current to offset the discrete harmonic spectrum in the whole range. Besides, since there are two different frequencies in the equivalent induced current, it can also extend the discrete spectrum between f_{i} and $0.5f_{\mathrm{i}}$, and integer multiples of its LCM.

4.4.6 *Verification of the Analysis Results of Current PSD*

Figure 4.22 shows the theoretical and experimental current PSD results for different kinds of injection schemes (fixed-frequency-phase injection, FPS-PRFSFP injection, HPS-PRFSFP injection, and PRPSFF injection). The experimental results are consistent with the theoretical analysis. Compared with fixed-frequency-phase injection, both PRFSFP injection and PRPSFF injection have a broad continuous spectrum. However, some discrete harmonic spectra still remain in PRPSFF injection, as shown in the dashed box in Fig. 4.22e. It can be explained as follows. During the switch of phases, the average current in d-axis has a DC bias. Under the control of the current regulator, the amplitude of the injected voltage also contains a bias, and it makes the amplitude of the HF induced current not symmetric anymore. According to (4.75), the second and third terms are no longer equal to zero, causing the

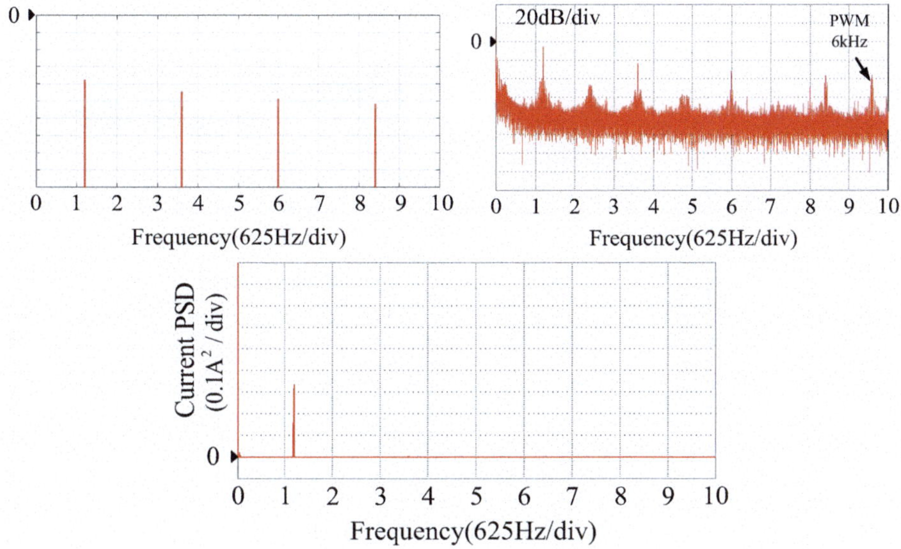

(a)

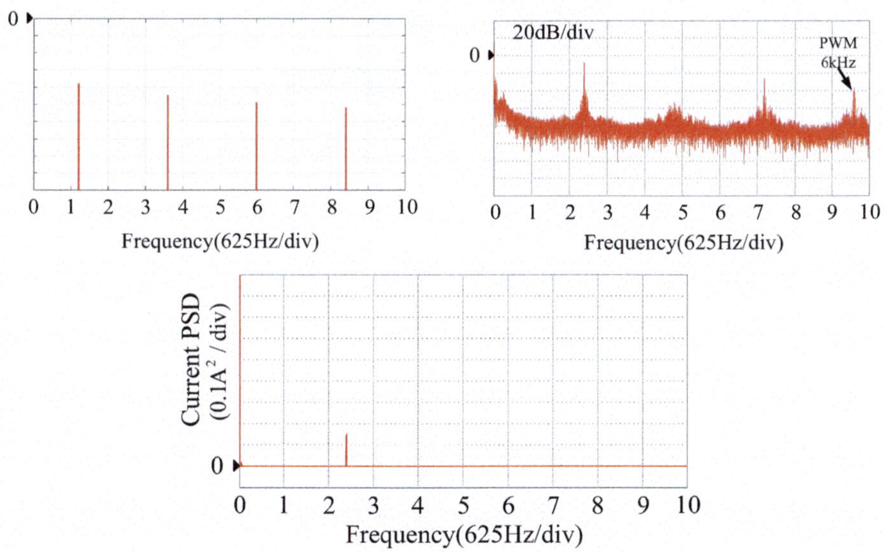

(b)

Fig. 4.22 Theoretical and experimental current PSD results for different injection schemes. (**a**) 750 Hz fixed-frequency-phase injection. (**b**) 1500 Hz fixed-frequency-phase injection. (**c**) 750 Hz/1500 Hz full-period-switch PRFSFP injection. (**d**) 750 Hz/1500 Hz half-period-switch PRFSFP injection. (**e**) 750 Hz PRPSFF injection. (**f**) 1500 Hz PRPSFF injection

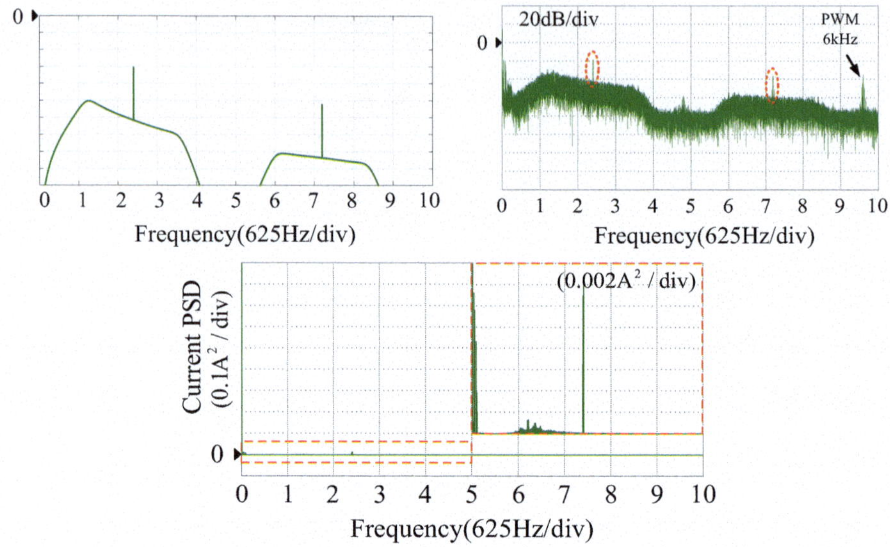

(c)

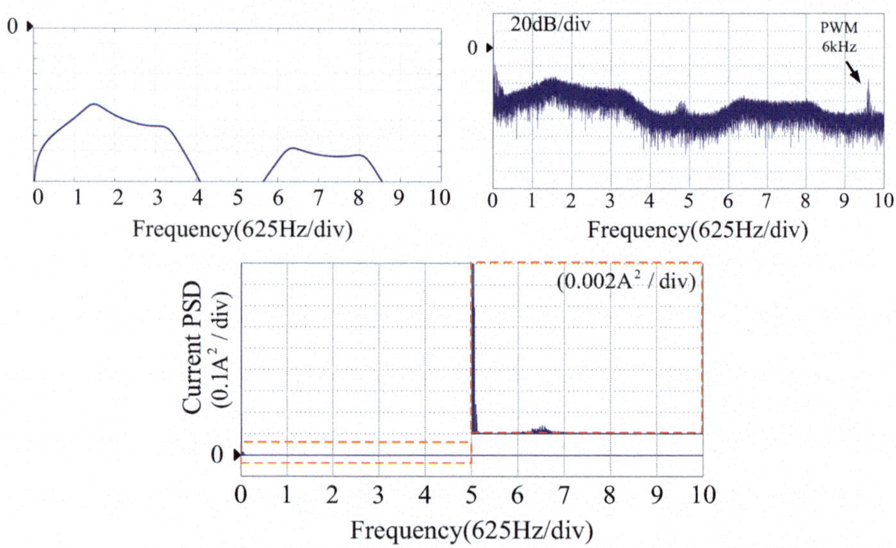

(d)

Fig. 4.22 (continued)

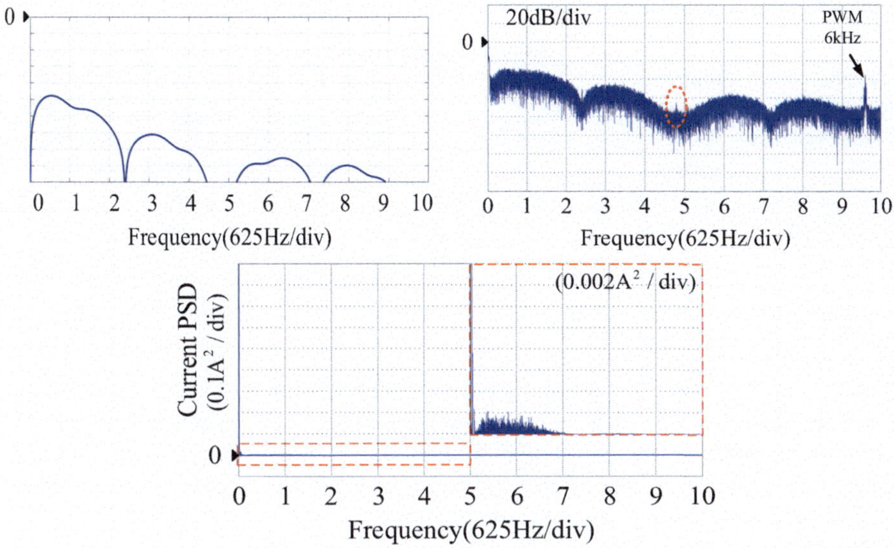

(e)

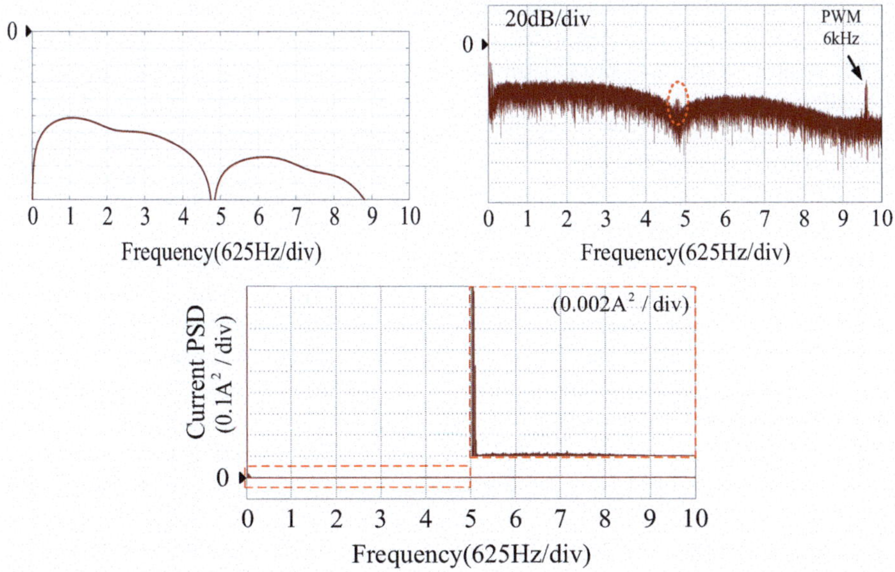

(f)

Fig. 4.22 (continued)

Table 4.1 Comparison of Current psd

Injection scheme	Injection frequency (Hz)	Spectral peaks (dB)							
		0.75	1.5	2.25	3.0	3.75	4.5	5.25	6.0 PWM
Fixed-frequency -phase	750	**−4.7**	−46	**−22**	−58	**−31**	−63	**−36**	**−35**
	1500	−77	**−8**	−86	−52	−92	**−25**	−83	**−35**
FPS-PRFSFP	750/1500	−52	**−18**	−77	−57	−76	**−36**	−76	**−35**
HPS-PRFSFP	750/1500	−47	−49	−65	−59	−74	−68	−75	**−35**
PRPSFF	750	−40	−50	−63	−69	−63	−74	−73	**−35**
	1500	−39	−59	−43	−55	−51	−78	−71	**−35**

existence of the discrete harmonic spectra in the current PSD. As the injection frequency increases, the distribution range of the continuous spectrum of PRPSFF injection becomes wider, as shown in Fig. 4.22e, f. Compared with PRFSFP injection method, the continuous spectrum of PRPSFF injection is smoother, which is conducive to reducing the audible noise.

The presence of additional signals causes a harmonic pollution in the current, and they will increase the losses of PMSM drives. That is an inherent drawback for the sensorless control method based on the additional signal injection. In general, the losses of PMSM can be categorized into copper loss and iron loss. The former is mainly related to the magnitude of induced current. If the volt-second areas of the injected signals are equal, the copper loss for different schemes will be approximately identical. The latter is the mainly HF induced loss, which can be regarded as a function of frequency [10–13].

The PRPSFF method can extend the discrete peaks between injection frequency and half of injection frequency again on the basis of offsetting the peaks by induced current with different phases. Therefore, the peaks of the injection frequency and its multiples decrease significantly. According to the linear current PSD shown in Fig. 4.22, the overall power spectra increase for PRPSFF and PRFSFP is so light compared with the decrease in peaks. In consequence, compared with fixed-frequency-phase injection scheme, the proposed scheme can reduce the iron losses caused by additional HF signals.

The quantification of current PSD between 0.75 and 6.0 kHz is shown in Table 4.1. The red bold font represents the discrete spectrum. Compared with fixed-frequency-phase injection, the discrete spectrum of PRPSFF injection decreases approximately 35 dB for 750 Hz and 50 dB for 1500 Hz injection. Although the peak values of continuous spectrum for PRPSFF injection are increased, it can extend the electromagnetic noise to the white noise within a limited bandwidth under the same total decibel.

4.5 Summary

This chapter introduces some pseudo-random HF voltage injection-based sensorless control schemes for PMSM drive. The introduced methods can attenuate the audible noise effectively by extending the electromagnetic noise to the white noise within a limited bandwidth. Both theoretical and experimental results of the current PSDs among the four injection schemes are compared to evaluate the audible noise and HF losses. All introduced methods are evaluated by 2.2 kW IPMSM platform.

References

1. G. Wang, L. Yang, G. Zhang, X. Zhang, D. Xu, Comparative investigation of pseudo-random high frequency signal injection schemes for sensorless IPMSM drives. IEEE Trans. Power Electron **32**(3), 2123–2132 (March 2017)
2. G. Wang, L. Yang, B. Yuan, B. Wang, G. Zhang, D. Xu, Pseudo-random hgh-Frequency Square-wave voltage injection based sensorless control of IPMSM drives for audible noise reduction. IEEE Trans. Ind. Electron. **63**(12), 7423–7433 (December 2016)
3. G. Wang, H. Zhou, N. Zhao, C. Li, D. Xu, Sensorless control of IPMSM drives using a pseudo-random phase-switching fixed-frequency signal injection scheme. IEEE Trans. Ind. Electron. **65** (10), 7660–7671 (January 2018)
4. B.H. Bae, S.K. Sul, A compensation method for time delay of full-digital synchronous frame current regulator of PWM AC drives. IEEE Trans. Ind. Appl. **39**(3), 802–810 (May/June 2003)
5. R.L. Kirlin, M.M. Bech, A.M. Trzynadlowski, Analysis of power and power spectral density in PWM inverters with randomized switching frequency. IEEE Trans. Ind. Electron. **49**(2), 486–499 (April 2002)
6. S. Kim, J.I. Ha, S.K. Sul, PWM switching frequency signal injection Sensorless method in IPMSM. IEEE Trans. Ind. Appl. **48**(5), 1576–1587 (September/October 2012)
7. N.C. Park, S.H. Kim, Simple Sensorless algorithm for interior permanent magnet synchronous motors based on high-frequency voltage injection method. IET Electr. Power Appl. **8**(2), 68–75 (February 2014)
8. S. Murakami, T. Shiota, M. Ohto, K. Ide, M. Hisatsune, Encoderless servo drive with adequately designed IPMSM for pulse-voltage-injection-based position detection. IEEE Trans. Ind. Appl. **48**(6), 1922–1930 (November/December 2012)
9. P. Vadstrup and R. D. Lorenz, Robust Estimator Design for Signal Injection-Based IPM Synchronous Machine Drives, in *IEEE Industry Applications Conference*, October 2004, pp. 957–963
10. N. Limsuwan, T. Kato, C.Y. Yu, J. Tamura, D.D. Reigosa, K. Akatsu, R.D. Lorenz, Secondary resistive losses with high-frequency injection-based self-sensing in IPM machines. IEEE Trans. Ind. Appl. **49**(4), 1499–1507 (July/August 2013)
11. X. Wang, R. Kennel, W. Xie and D. Gerling, Analysis of losses in a novel IPMSM resulting from high-frequency injection for sensorless control, in *IEEE International Symposium on Sensorless Control for Electrical Drives and Predictive Control of Electrical Drives and Power Electronics*, October 2013, pp. 1–5
12. S. C. Yang, R. D. Lorenz, Analysis of iron and magnet losses in surface-permanent-magnet machines resulting from injection-based self-Ssensing position Estimation, IEEE Trans. Ind. Appl., 48, 6, 1901–1910, November/December 2012
13. S.C. Yang, R.D. Lorenz, Comparison of resistance-based and inductance-based self-sensing controls for surface permanent-magnet machines using high-frequency signal injection. IEEE Trans. Ind. Appl. **48**(3), 977–986 (May/June 2012)

Chapter 5
Model-Based Sensorless Control for PMSM Drives

Model-based method is used to achieve position sensorless control of the IPMSM drive system when the motor runs at middle or high speed. The model method estimates the rotor position by the back-EMF or the flux linkage model of the fundamental frequency excitation. Since the rotor saliency and the additional test signal are not needed, the rotor position information can be estimated only by the motor model. This model-based method is relatively simple to implement, so the method is widely used in the vector control system for sensorless IPMSM drive.

5.1 Classification of Model-Based Methods

The model-based methods mainly include model reference adaptive method, extended Kalman filter method, sliding mode observer method, etc.

1. Model reference adaptive method

The model reference adaptive system (MRAS) includes two models: one is the reference model and the other is the adjustable model (the controlled object). First, design a reference model with the same order according to the performance requirements of the adjustable model, and then connect them in parallel. The two deviations $e(t)$ are compared under the same reference input r(t). The adjustable model parameters are adjusted by the adaptive mechanism to make e(t) converge to zero.

Based on the stability theory, the rotor speed estimation reference adaptive mechanism can be obtained. The asymptotic stability of the model reference adaptive system can be verified by Lyapunov stability theory and Popov hyperstability theory [1]. The application of the model reference adaptive method in fault-tolerant control is generally based on the motor model to establish an observer to obtain motor position and current information. According to the difference between the reference model and the adjustable model, different model reference adaptive

G. Wang et al., *Position Sensorless Control Techniques for Permanent Magnet Synchronous Machine Drives*, https://doi.org/10.1007/978-981-15-0050-3_5

algorithms can be constructed. Among them, the model reference adaptive method based on back-EMF is used more commonly.

The model reference adaptive method is simpler and easier to implement. However, the estimation accuracy of the model reference adaptive method is greatly affected by the system parameters, especially the reference model and the adjustable model.

2. Extended Kalman filter method

The Kalman filter method is an efficient recursive estimation algorithm based on the minimum variance. The main idea of this method is to predict and correct, that is, to correct the system state variable prediction value based on the observation data and then obtain the optimal estimation of the state variable [2]. In the specific implementation process, the Kalman filter method considers the influence of the inherent model error and the measurement error on the estimated value of the system state variable and thus has high precision.

The extended Kalman filter is one of the Kalman filters suitable for nonlinear systems. The basic idea is to near the filter value, and the Taylor expansion algorithm is applied to expand the nonlinear system where the higher order terms above the second order are omitted. The original nonlinear system is approximated into a linear system, and the standard Kalman filter algorithm is used to filter the linearized model of the system. The advantage of this method is that it can adjust the gain matrix of the observer based on the noise to reduce its adverse effects and is robust to the change of the system parameters. However, the disadvantage is that many random error parameters are needed, and the calculation is complex and massive, which brings difficulties to implement.

3. Flux observer method

The flux observer method can be easily divided into the open-loop flux observer method, the closed-loop flux observer method, and the comprehensive method. The open-loop flux observer method uses the stator voltage model of the motor to calculate the permanent magnet excitation flux linkage directly and obtain the rotor position information. This method requires high accuracy of the motor parameters and less robustness to parameter changes, especially at low speed. However, the open-loop flux observer method is easy to implement and has a fast dynamic response, which is an effective method for systems with low performance requirements. To broaden the application range of the flux observer, the composite flux observer method with variable structure is proposed in [3]. The composite voltage-current model is used at low speed, and the closed-loop voltage model is used at high speed and switching the two models at specified speed points (variable structure). The method effectively broadens the speed regulation range of the flux observer method, but its performance depends on the estimation accuracy of the operation speed and the motor parameters. Based on the concept of active flux linkage, the reference [4] proposed a flux linkage observer method that the speed information is not needed. This method combines the voltage model in the stationary coordinate system and the current model in the synchronous rotating coordinate system. It can take the advantages of voltage and current models at high speed and low speed. The combination of the models in two coordinate systems can effectively eliminate the influence of the estimated speed on the accuracy of the system and improve the performance of the system. The introduction of the active flux linkage concept makes the observer suitable for all AC motor drive systems. However, this method is still affected by changes in motor parameters, and the performance of the observer depends on the accuracy of the motor parameters.

4. Sliding mode observer method

The sliding mode observer is an observer-design method based on sliding mode control. Its essence is a kind of nonlinear control, which is manifested as the discontinuity of its control. The characteristic is that the structure of the system is not fixed but purposefully changed according to the current state of the system (such as deviation and its derivatives), forcing the system to move according to the predetermined sliding mode trajectory. Its advantages include fast response, strong robustness, and insensitivity to both parameter changes and disturbances. The disadvantage of this method is that when the state trajectory reaches the sliding surface, the chattering will be generated.

5.2 Active Flux Modeling and Extended EMF Modeling for IPMSM

5.2.1 *Active Flux Modeling for IPMSM*

Based on the idea of generating the flux linkage [5, 6], the active flux linkage model can transform the salient AC motor into the nonsalient one, which simplifies the analysis and design of the position sensorless control system. For IPMSM, the active flux linkage is defined as shown in (5.1):

$$\begin{cases} T_{\mathrm{e}} = \frac{3}{2} P_{\mathrm{n}} \left[\psi_{\mathrm{f}} + \left(L_{\mathrm{d}} - L_{\mathrm{q}} \right) i_{\mathrm{d}} \right] i_{\mathrm{q}} \\ \psi_{\mathrm{f}}^{\mathrm{a}} = \psi_{\mathrm{f}} + \left(L_{\mathrm{d}} - L_{\mathrm{q}} \right) i_{\mathrm{d}}, \end{cases} \tag{5.1}$$

where T_e is the electromagnetic torque, P_n is the number of pole pairs, and ψ_f^a is the active flux linkage. The active flux linkage consists of two parts: the permanent magnet flux linkage and the flux linkage generated by the rotor saliency. It is noted that the active flux linkage direction is in the same direction as the d-axis, so the active flux linkage model can fit for any mode of operation.

Based on the idea of active flux linkage, the state equation of IPMSM using the stator current as the state variable in the α-β stationary axis frame can be obtained:

$$p\mathbf{i}_{\alpha\beta} = \mathbf{A}\mathbf{i}_{\alpha\beta} + \mathbf{B}\left(\mathbf{u}_{\alpha\beta} - \mathbf{e}_{\alpha\beta}\right), \tag{5.2}$$

where $\mathbf{u}_{\alpha\beta} = [u_\alpha, u_\beta]^{\mathrm{T}}$, $\mathbf{i}_{\alpha\beta} = [i_\alpha, i_\beta]^{\mathrm{T}}$, $\mathbf{e}_{\alpha\beta} = [e_\alpha, e_\beta]^{\mathrm{T}}$, $A = -(R/L_q)I$, and $\mathbf{B} = (1/L_q)\mathbf{I}$.

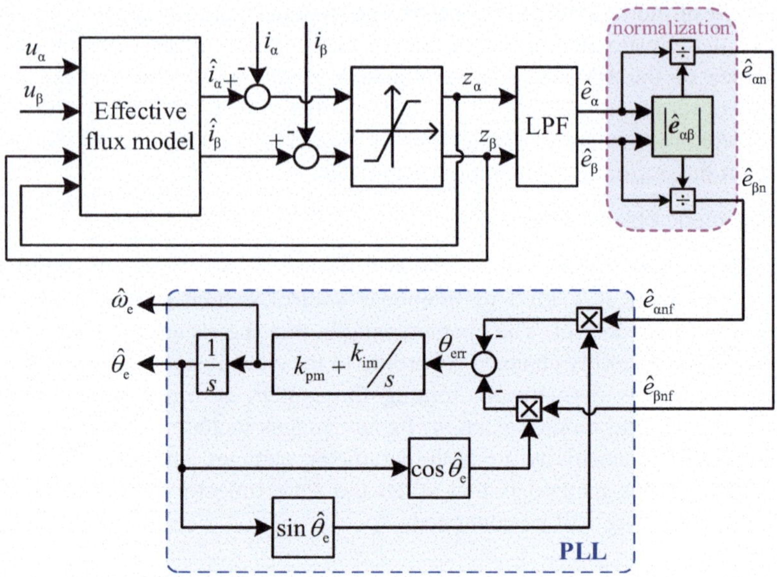

Fig. 5.1 Stator current sliding mode observer

The equivalent back-EMF based on the active flux linkage can be defined as

$$\mathbf{e}_{\alpha\beta} = \begin{bmatrix} e_\alpha \\ e_\beta \end{bmatrix} = p \begin{bmatrix} \psi_{f\alpha}^{a} \\ \psi_{f\beta}^{a} \end{bmatrix} = \psi_f^a \omega_e \begin{bmatrix} -\sin\theta_e \\ \cos\theta_e \end{bmatrix}. \tag{5.3}$$

It is known from reference [7] that the IPMSM is observable when it is operated in a nonstationary state. Therefore, according to the basic principle of sliding mode variable structure and the mathematical model of IPMSM in the α-β axis (5.3), the stator current sliding mode observer can be constructed as shown in Fig. 5.1.

$$p\mathbf{i}_{\alpha\beta} = \mathbf{A}\mathbf{i}_{\alpha\beta} + \mathbf{B}\left(\mathbf{u}_{\alpha\beta} - \mathbf{e}_{\alpha\beta}\right), \tag{5.4}$$

where $\hat{\mathbf{i}}_{\alpha\beta}$ is the observed stator current in the α-β axes; $\mathbf{z}_{\alpha\beta} = [z_\alpha, z_\beta]^T$ is the sliding mode control function and can be expressed as

$$\mathbf{z}_{\alpha\beta} = \begin{cases} k_{SM}/\delta \cdot \mathbf{S}, & |\mathbf{S}| < \delta_{SM} \\ k_{SM} \cdot \mathit{sgn}\,(\mathbf{S}), & |\mathbf{S}| > \delta_{SM}, \end{cases} \tag{5.5}$$

where $sgn(\mathbf{S})$ is the switching function, k_{SM} is the switching function gain, and δ_{SM} is the boundary layer thickness. Define the stator current error as the sliding surface, i.e., $\mathbf{S} = \hat{\mathbf{i}}_{\alpha\beta} - \mathbf{i}_{\alpha\beta} = \mathbf{0}$.

By subtracting (5.2) from (5.4), the dynamic error equation of the stator current sliding mode observer can be obtained.

$$p\mathbf{S} = \mathbf{AS} + \mathbf{B}\left(\mathbf{e}_{\alpha\beta} - \mathbf{z}_{\alpha\beta}\right). \tag{5.6}$$

In order to determine the stability of the sliding mode observer, the *Lyapunov* stability equation is established as follows:

$$pV = \mathbf{S}^{\mathrm{T}} p\mathbf{S} < 0. \tag{5.7}$$

And then,

$$k_{\mathrm{SM}} > \max\left\{\left|e_{\alpha}\right|, \left|e_{\beta}\right|\right\}. \tag{5.8}$$

Furthermore, (5.9) can be expressed as follows, considering the influence of modeling uncertainty:

$$p\hat{\mathbf{i}}_{\alpha\beta} = \mathbf{A}\hat{\mathbf{i}}_{\alpha\beta} + \mathbf{B}\left(\mathbf{u}_{\alpha\beta} - \mathbf{z}_{\alpha\beta}\right) + \Delta\mathbf{f}_{\alpha\beta}, \tag{5.9}$$

where $\Delta\mathbf{f}_{\alpha\beta}$ is the bounded modeling error, and $\Delta\mathbf{f}_{\alpha\beta} < \xi_{err}, \xi_{err}$ is a positive constant.

Based on the *Lyapunov* stability analysis, there are positive numbers.

$$k_{\mathrm{SM}} > \max\left\{\left|e_{\alpha}\right|, \left|e_{\beta}\right|\right\} + \xi_{\mathrm{err}}. \tag{5.10}$$

Therefore, the sliding mode observer convergence condition is established. Based on the equivalent control theory, the equivalent back EMF observation can be obtained by low-pass filter the switching function containing the HF switching signal, i.e.,

$$\hat{\mathbf{e}}_{\alpha\beta} = LPF\left(\mathbf{z}_{\alpha\beta}\right), \tag{5.11}$$

where $\hat{\mathbf{e}}_{\alpha\beta}$ is the back-EMF observation in the α-β axes and LPF() refers to the low-pass filter function.

After obtaining the equivalent back-EMF information through the stator current sliding mode observer based on the active flux linkage model, it is normalized, and then the rotor position/speed is tracked by quadrature PLL. The algorithm can quickly track position changes and filter the HF harmonics and noise, which is conducive to improving the accuracy of position observation [9].

5.2.2 *Extended EMF Modeling for IPMSM*

In the synchronous reference frame, the mathematical model of IPMSM can be expressed as (3.36). By transforming this equation into the α-β stationary reference frame, the mathematical model of IPMSM in the stationary coordinate system can be obtained as

$$\begin{bmatrix} u_\alpha \\ u_\beta \end{bmatrix} = \begin{bmatrix} R_s + pL_\alpha & pM_{\alpha\beta} \\ pM_{\beta\alpha} & R_s + pL_\beta \end{bmatrix} \begin{bmatrix} i_\alpha \\ i_\beta \end{bmatrix} + \omega_e \psi_f \begin{bmatrix} -\sin\theta_e \\ \cos\theta_e \end{bmatrix}. \tag{5.12}$$

According to (5.12), the rotor position information of the IPMSM is contained not only in the back-EMF but also in the inductance matrix due to the existence of the rotor saliency.In addition, the inductance matrix contains θ term and 2θ term, which increase the difficulty of rotor position detection. Therefore, it is not possible for IPMSM to directly use the voltage equation of the α-β axis system for position estimation like SPMSMs.

In order to decouple the rotor position informationof IPMSM from the voltage equation in the α-β axis system, (5.12) is rewritten to make the coefficient matrix symmetrical, and the following equivalent transformation is performed:

$$\begin{bmatrix} u_d \\ u_q \end{bmatrix} = \begin{bmatrix} R_s + pL_d & -\omega_e L_q \\ \omega_e L_q & R_s + pL_d \end{bmatrix} \begin{bmatrix} i_d \\ i_q \end{bmatrix} + \begin{bmatrix} 0 \\ (L_d - L_q)(\omega_e i_d - \bullet i_q) + \omega_e \psi_f \end{bmatrix}. \tag{5.13}$$

Define E_{ex} as extended electromotive force (EEMF) [10]:

$$E_{ex} = (L_d - L_q)(\omega_e i_d - \dot{i}_q)\omega_e \psi_f. \tag{5.14}$$

Thus, the EEMF E_{ex} includes not only the traditional counter EMF term but also the term generated by saliency of the polarity of IPMSM. The concept of EEMF can be extended to the entire synchronous motor category. When $L_d = L_q$, it is the back-EMF model of SPMSM. When $\psi_f = 0$, it is the reluctance motor EMF model.

5.3 Sliding Mode Observer Based on EEMF

5.3.1 *Reduced-Order Position Observer*

In general, the mathematical model in α-β axis system is usually used to construct the PMSM position observer instead of the d-q axis system mathematical model.

Because the mathematical model in d-q axis system is obtained by Park transformation, where the rotor position information θ_e is necessary. Thus, (5.13) is transformed into the α-β stationary coordinate system to obtain the voltage equations based on the EEMF model:

$$\begin{bmatrix} u_\alpha \\ u_\beta \end{bmatrix} = \begin{bmatrix} R_s + pL_d & \omega_e(L_d - L_q) \\ -\omega_e(L_d - L_q) & R_s + pL_d \end{bmatrix} \begin{bmatrix} i_\alpha \\ i_\beta \end{bmatrix} + \begin{bmatrix} e_\alpha \\ e_\beta \end{bmatrix}, \tag{5.15}$$

where $e_\alpha = - E_{ex} \sin \theta_e$ and $e_\beta = E_{ex} \cos \theta_e$ are components of EEMF in the α-β axis.

According to (5.15), the rotor position information can be decoupled from the inductance matrix by means of the equivalent transformation and the introduction of the EEMF concept, so that the EEMF is the only term that contains the rotor pole position information. And then the EEMF phase information can be directly used to realize the rotor position observation.

Rewrite the IPMSM voltage Eq. (5.16) as a state equation using the stator current as a state variable:

$$\begin{bmatrix} \dot{i}_\alpha \\ \dot{i}_\beta \end{bmatrix} = \frac{1}{L_d} \begin{bmatrix} -R_s & -\omega_e(L_d - L_q) \\ \omega_e(L_d - L_q) & -R_s \end{bmatrix} \begin{bmatrix} i_\alpha \\ i_\beta \end{bmatrix} + \frac{1}{L_d} \begin{bmatrix} u_\alpha - e_\alpha \\ u_\beta - e_\beta \end{bmatrix}. \tag{5.16}$$

Since the stator current is the only physical quantity that can be directly measured, the sliding surface is selected on the stator current path:

$$\mathbf{s}(x) = \begin{bmatrix} \tilde{i}_\alpha \\ \tilde{i}_\beta \end{bmatrix} = \begin{bmatrix} \hat{i}_\alpha - i_\alpha \\ \hat{i}_\beta - i_\beta \end{bmatrix} = 0, \tag{5.17}$$

where the superscript "~" indicates the variable error, which refers to the difference between the observed value and the actualone.

The traditional reduced-order sliding mode observer is constructed. The mathematical model is shown in (5.18) and the block diagram is shown in Fig. 5.2.

$$\begin{bmatrix} \dot{\hat{i}}_\alpha \\ \dot{\hat{i}}_\beta \end{bmatrix} = \frac{1}{L_d} \begin{bmatrix} -R_s & -\hat{\omega}_e(L_d - L_q) \\ \hat{\omega}_e(L_d - L_q) & -R_s \end{bmatrix} \begin{bmatrix} \hat{i}_\alpha \\ \hat{i}_\beta \end{bmatrix} + \frac{1}{L_d} \begin{bmatrix} u_\alpha - \hat{e}_\alpha - z_\alpha \\ u_\beta - \hat{e}_\beta - z_\beta \end{bmatrix}, \tag{5.18}$$

where superscripts "^" indicates the estimated value and z_α and z_β are sliding mode feedback components:

$$\begin{bmatrix} z_\alpha \\ z_\beta \end{bmatrix} = k \begin{bmatrix} \operatorname{sgn}(\hat{i}_\alpha - i_\alpha) \\ \operatorname{sgn}(\hat{i}_\beta - i_\beta) \end{bmatrix}, \tag{5.19}$$

where k is the constant sliding mode gain designed by *Lyapunov* stability analysis.

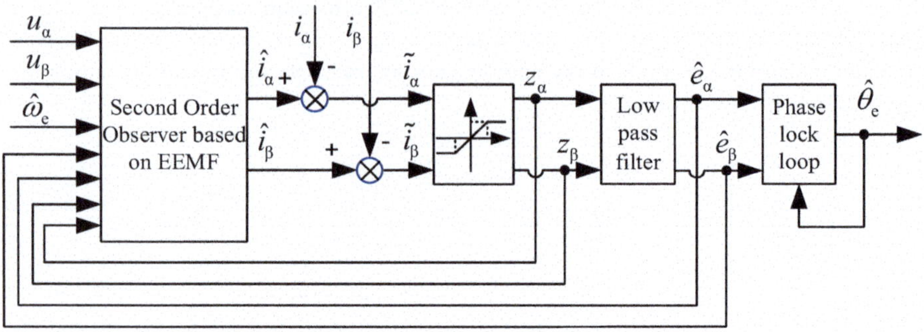

Fig. 5.2 Block diagram of traditional second-order sliding mode observer

The estimated value of EEMF in α-β axes ($\hat{e}_\alpha$, $\hat{e}_\beta$) can be obtained by low-pass filteri from the discontinuous switching signals z_α and z_β:

$$\begin{bmatrix} \hat{e}_\alpha \\ \hat{e}_\beta \end{bmatrix} = \frac{\omega_c}{s + \omega_c} \begin{bmatrix} z_\alpha \\ z_\beta \end{bmatrix}, \tag{5.20}$$

where ω_c is the cutoff frequency of the LPF, which is usually selected according to the fundamental frequency of the stator current. It is ensured that the fundamental component is not affected while filtering out the HF components. However, the addition of the LPF causes the phase lag and amplitude attenuation of the EEMF estimation, which affects the performance of the position sensorless PMSM vector control system.

A dynamic model of the current observation error can be obtained by subtracting the (5.18) from (5.16):

$$\begin{aligned} \dot{\tilde{i}}_\alpha &= -\frac{R_s}{L_d}\tilde{i}_\alpha - \frac{L_d - L_q}{L_d}\left(\hat{\omega}_e\hat{i}_\beta - \omega_e i_\beta\right) - \frac{1}{L_d}\tilde{e}_\alpha - \frac{1}{L_d}k\,\mathrm{sgn}\left(\hat{i}_\alpha - i_\alpha\right), \\ \dot{\tilde{i}}_\beta &= -\frac{R_s}{L_d}\tilde{i}_\beta + \frac{L_d - L_q}{L_d}\left(\hat{\omega}_e\hat{i}_\alpha - \omega_e i_\alpha\right) - \frac{1}{L_d}\tilde{e}_\beta - \frac{1}{L_d}k\,\mathrm{sgn}\left(\hat{i}_\beta - i_\beta\right), \end{aligned} \tag{5.21}$$

where $\tilde{e}_\alpha = \hat{e}_\alpha - e_\alpha$ and $\tilde{e}_\beta = \hat{e}_\beta - e_\beta$ are the α-β components of EEMF.

Normally, when the stator current reaches the sliding mode surface, the observed speed can converge to the actual speed, i.e., $\hat{\omega}_e = \omega_e$. Therefore, (5.21) can be further simplified to

$$\begin{aligned} \dot{\tilde{i}}_\alpha &= -\frac{R_s}{L_d}\tilde{i}_\alpha - \frac{L_d - L_q}{L_d}\omega_e\tilde{i}_\beta - \frac{1}{L_d}\tilde{e}_\alpha - \frac{1}{L_d}k\,\mathrm{sgn}\left(\hat{i}_\alpha - i_\alpha\right), \\ \dot{\tilde{i}}_\beta &= -\frac{R_s}{L_d}\tilde{i}_\beta + \frac{L_d - L_q}{L_d}\omega_e\tilde{i}_\alpha - \frac{1}{L_d}\tilde{e}_\beta - \frac{1}{L_d}k\,\mathrm{sgn}\left(\hat{i}_\beta - i_\beta\right). \end{aligned} \tag{5.22}$$

In order to analyze the stability of the reduced-order sliding mode observer, define the positive definite Lyapunov function:

$$V = \frac{1}{2}\mathbf{s}^{\mathrm{T}}\mathbf{s} = \frac{1}{2}\left(\tilde{i}_{\alpha}^{2} + \tilde{i}_{\beta}^{2}\right). \tag{5.23}$$

In order to enhance the robustness of observers to disturbance and parameter variation, and to ensure the convergence, it can be known from the conditionality of the sliding mode variable structure that $\dot{V}$ must be negative, i.e., $\dot{V} < 0$,

$$\begin{aligned}
\dot{V} =& \mathbf{s}^{\mathrm{T}}\dot{\mathbf{s}} \\
=& \tilde{i}_{\alpha} \cdot \dot{\tilde{i}}_{\alpha} + \tilde{i}_{\beta} \cdot \dot{\tilde{i}}_{\beta} \\
=& \tilde{i}_{\alpha}\left\{-\frac{R_{\mathrm{s}}}{L_{\mathrm{d}}}\tilde{i}_{\alpha} - \frac{L_{\mathrm{d}} - L_{\mathrm{q}}}{L_{\mathrm{d}}}\hat{\omega}_{\mathrm{e}}\tilde{i}_{\beta} - \frac{1}{L_{\mathrm{d}}}\tilde{e}_{\alpha} - \frac{1}{L_{\mathrm{d}}}k\,\mathrm{sgn}\left(\tilde{i}_{\alpha}\right)\right\} \\
&+ \tilde{i}_{\beta}\left\{-\frac{R_{\mathrm{s}}}{L_{\mathrm{d}}}\tilde{i}_{\beta} + \frac{L_{\mathrm{d}} - L_{\mathrm{q}}}{L_{\mathrm{d}}}\hat{\omega}_{\mathrm{e}}\tilde{i}_{\alpha} - \frac{1}{L_{\mathrm{d}}}\tilde{e}_{\beta} - \frac{1}{L_{\mathrm{d}}}k\,\mathrm{sgn}\left(\tilde{i}_{\beta}\right)\right\} \\
=& -\frac{R_{\mathrm{s}}}{L_{\mathrm{d}}}\left(\tilde{i}_{\alpha}^{2} + \tilde{i}_{\beta}^{2}\right) - \frac{1}{L_{\mathrm{d}}}\left\{\tilde{i}_{\alpha}\left(\tilde{e}_{\alpha} + k\,\mathrm{sgn}\left(\tilde{i}_{\alpha}\right)\right)\right\} - \frac{1}{L_{\mathrm{d}}}\left\{\tilde{i}_{\beta}\left(\tilde{e}_{\beta} + k\,\mathrm{sgn}\left(\tilde{i}_{\beta}\right)\right)\right\} \\
=& V_{1} - V_{2},
\end{aligned} \tag{5.24}$$

where

$$V_{1} = -\frac{R_{\mathrm{s}}}{L_{\mathrm{d}}}\left(\tilde{i}_{\alpha}^{2} + \tilde{i}_{\beta}^{2}\right), \tag{5.25}$$

$$\begin{aligned}
V_{2} =& \frac{1}{L_{\mathrm{d}}}\left\{\tilde{i}_{\alpha}\left(\tilde{e}_{\alpha} + k\,\mathrm{sgn}\left(\tilde{i}_{\alpha}\right)\right)\right\} + \frac{1}{L_{\mathrm{d}}}\left\{\tilde{i}_{\beta}\left(\tilde{e}_{\beta} + k\,\mathrm{sgn}\left(\tilde{i}_{\beta}\right)\right)\right\} \\
=& \frac{1}{L_{\mathrm{d}}}\left\{\tilde{i}_{\alpha}\left(\frac{s + 2\omega_{\mathrm{c}}}{s + \omega_{\mathrm{c}}}k\,\mathrm{sgn}\left(\tilde{i}_{\alpha}\right) - e_{\alpha}\right) + \tilde{i}_{\beta}\left(\frac{s + 2\omega_{\mathrm{c}}}{s + \omega_{\mathrm{c}}}k\,\mathrm{sgn}\left(\tilde{i}_{\beta}\right) - e_{\beta}\right)\right\}.
\end{aligned} \tag{5.26}$$

Equation (5.25) shows that V_1 is always negative. The cutoff frequency of the low-pass filter is relatively large, $\frac{s+2\omega_{\mathrm{c}}}{s+\omega_{\mathrm{c}}} \approx 2$. If k is large enough and satisfy (5.27),

$$k > \frac{1}{2}\max\left(\left|e_{\alpha}\right|,\ \left|e_{\beta}\right|\right). \tag{5.27}$$

Then the sliding mode convergence condition can be met. When V converges to zero, i_a and i_b will be zero and the estimated current will converge to the actual value.

5.3.2 Full-Order Position Observer

Normally, the mechanical time constant of the system is much larger than its electromagnetic time constant. In the PWM control period, the angular velocity can be considered as a constant, i.e., $\dot{\omega}_e = 0$. Under these circumstances, the EEMF and its derivative satisfy the following relationship:

$$\begin{bmatrix} \dot{e}_\alpha \\ \dot{e}_\beta \end{bmatrix} = \omega_e \begin{bmatrix} -e_\beta \\ e_\alpha \end{bmatrix}. \tag{5.28}$$

The full-order state equation of IPMSM using the stator currents and the extended back-EMFs as state variables can be derived:

$$\begin{bmatrix} \dot{\mathbf{i}} \\ \dot{\mathbf{e}} \end{bmatrix} = \begin{bmatrix} \mathbf{A}_{11} & \mathbf{A}_{12} \\ \mathbf{0} & \mathbf{A}_{22} \end{bmatrix} \begin{bmatrix} \mathbf{i} \\ \mathbf{e} \end{bmatrix} + \begin{bmatrix} \mathbf{B}_1 \\ \mathbf{0} \end{bmatrix} \mathbf{u}, \tag{5.29}$$

where

$$\begin{aligned} &\mathbf{i} = \left[i_\alpha \ i_\beta\right]^{\mathrm{T}}, \ \mathbf{e} = \left[e_\alpha e_\beta\right]^{\mathrm{T}}, \ \mathbf{u} = \left[u_\alpha u_\beta\right]^{\mathrm{T}}, \\ &\mathbf{A}_{11} = -R/L_d \cdot \mathbf{I} + \omega_e \left(L_d - L_q\right)/L_d \cdot \mathbf{J}, \\ &\mathbf{A}_{12} = -1/L_d \cdot \mathbf{I}, \ \mathbf{A}_{22} = \omega_e \cdot \mathbf{J}, \\ &\mathbf{B}_1 = 1/L_d \cdot \mathbf{I}, \end{aligned}$$

$$\mathbf{I} = \begin{bmatrix} 1 & 0 \\ 0 & 1 \end{bmatrix}, \ \mathbf{J} = \begin{bmatrix} 0 & -1 \\ 1 & 0 \end{bmatrix}.$$

According to the full-order state Eq. (5.29), the full-order sliding mode observer can be established as shown in Eq. (5.30). Figure 5.3 shows the block diagram of the full-order sliding mode observer based on EEMF model.

$$\begin{bmatrix} \dot{\hat{\mathbf{i}}} \\ \dot{\hat{\mathbf{e}}} \end{bmatrix} = \begin{bmatrix} \hat{\mathbf{A}}_{11} & \mathbf{A}_{12} \\ \mathbf{0} & \hat{\mathbf{A}}_{22} \end{bmatrix} \begin{bmatrix} \hat{\mathbf{i}} \\ \hat{\mathbf{e}} \end{bmatrix} + \begin{bmatrix} \mathbf{B}_1 \\ \mathbf{0} \end{bmatrix} \mathbf{u} - \frac{1}{L_d} \mathbf{K} \operatorname{sgn}(\mathbf{s}). \tag{5.30}$$

where

$$\begin{aligned} \hat{\mathbf{i}} &= \left[\hat{i}_\alpha \hat{i}_\beta\right]^{\mathrm{T}}, \\ \hat{\mathbf{e}} &= \left[\hat{e}_\alpha \hat{e}_\beta\right]^{\mathrm{T}}, \\ \mathbf{s} &= \hat{\mathbf{i}} - \mathbf{i}, \end{aligned}$$

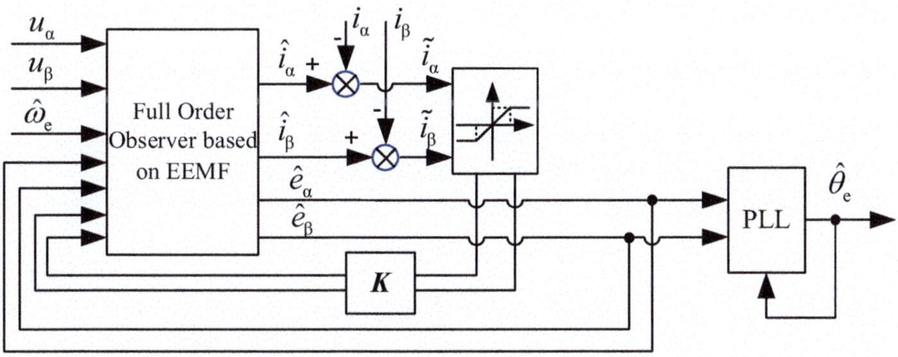

Fig. 5.3 Block diagram of full-order sliding mode observer

$$\hat{\mathbf{A}}_{11} = -R/L_d \cdot \mathbf{I} + \hat{\omega}_e(L_d - L_q)/L_d \cdot \mathbf{J},$$

$$\hat{\mathbf{A}}_{22} = \hat{\omega}_e \cdot \mathbf{J},$$

$$\mathbf{K} = \begin{bmatrix} k & 0 \\ 0 & k \\ -m & 0 \\ 0 & -m \end{bmatrix},$$

where $\boldsymbol{K}$ is the feedback gain matrix and k and m are the sliding mode observer gains that can obtained from *Lyapunov* stability analysis.

By using the idea of equivalent control, the stability of the full-order sliding mode observer can be analyzed.

Step 1: Refering to the stability analysis of the traditional second-order sliding mode observer, as shown in (5.19)–(5.27). The difference is that the full-order sliding mode observer does not need the low-pass filter which causes the phase lag in the rotor position observation. However, its derivation process is basically the same.

Step 2: When the stator current reaches the sliding mode region, the stator current observation value converges to its actual value, i.e., $\dot{\tilde{i}}_\alpha = \dot{\tilde{i}}_\beta = 0$. At this time, (5.22) can be rewritten as follows:

$$\begin{aligned} \tilde{e}_\alpha &= -k\,\mathrm{sgn}\left(\hat{i}_\alpha - i_\alpha\right), \\ \tilde{e}_\beta &= -k\,\mathrm{sgn}\left(\hat{i}_\beta - i_\beta\right). \end{aligned} \tag{5.31}$$

Let (5.30) be subtracted from (5.29); the error dynamic equation of EEMF can be obtained as

$$\begin{aligned}\dot{\tilde{e}}_{\alpha} &= -\tilde{e}_{\beta}\hat{\omega}_{e} + \frac{1}{L_{d}} m \operatorname{sgn}\left(\hat{i}_{\alpha} - i_{\alpha}\right), \\ \dot{\tilde{e}}_{\beta} &= \tilde{e}_{\alpha}\hat{\omega}_{e} + \frac{1}{L_{d}} m \operatorname{sgn}\left(\hat{i}_{\beta} - i_{\beta}\right).\end{aligned} \tag{5.32}$$

According to (5.31) and (5.32), the results can be obtained:

$$\begin{aligned}\dot{\tilde{e}}_{\alpha} &= -\tilde{e}_{\beta}\hat{\omega}_{e} - \frac{1}{L_{d}} m/k \cdot \tilde{e}_{\alpha}, \\ \dot{\tilde{e}}_{\beta} &= \tilde{e}_{\alpha}\hat{\omega}_{e} - \frac{1}{L_{d}} m/k \cdot \tilde{e}_{\beta}.\end{aligned} \tag{5.33}$$

Equation (5.33) shows that the EEMF error dynamic equations have two basic terms: the prediction and the correction of Kalman filter. EEMF observation equations are the prediction terms, and the feedback correction composed of the gain coefficient $\frac{1}{L_{d}} m/k$ is the correction term. Therefore, the abovementioned full-order sliding mode observer can remove the LPF used in the traditional reduced-order sliding mode observer, thereby avoiding the phase lag.

Solve the differential (5.33) and obtain its characteristic equation:

$$\alpha(s) = s^{2} + 2/L_{d} \cdot m/k \cdot s + \left(\hat{\omega}_{e}^{2} + 1/L_{d}^{2} \cdot (m/k)^{2}\right) = 0, \tag{5.34}$$

where '*s*' is the Laplace operator. Solve the equation and the eigenvalue can be obtained:

$$s_{1,2} = \frac{-m/k \pm jL_{d}\omega_{e}}{L_{d}}. \tag{5.35}$$

Equation (5.35) shows that the eigenvalues s_1 and s_2 are conjugate complex roots, which are located on the left half plane. And the system is asymptotically stable.

Equation (5.34) can be further rewritten as the following standard form:

$$s^{2} + 2\xi\omega_{n}s + \omega_{n}^{2} = 0, \tag{5.36}$$

where

$$\begin{aligned}\omega_{n} &= \sqrt{\hat{\omega}_{e}^{2} + 1/L_{d}^{2} \cdot (m/k)^{2}}, \\ \xi &= \frac{m/k}{\sqrt{L_{d}^{2}\hat{\omega}_{e}^{2} + (m/k)^{2}}}.\end{aligned} \tag{5.37}$$

ω_n refers to the natural oscillation angular frequency and ξ the damping ratio.

According to the automatic control theory, the convergence rate of the EEMF observation errors $\tilde{e}_{s\alpha}$ and $\tilde{e}_{s\beta}$ depends on $\xi\omega_n$, i.e., the value of $1/L_d \cdot m/k$. The larger

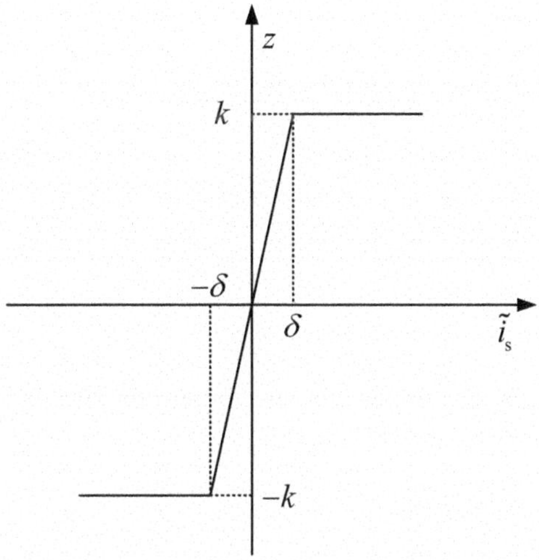

Fig. 5.4 Characteristic curve of saturation function

the value of $\xi\omega_n$ is, the faster the convergence of the EEMF observation error will be. For $\xi \in (0.4, 0.8)$, the overshoot of the system is suitable, with a shorter adjustment time and better dynamic performance. In order to diminish the chattering phenomenon inherent in the sliding mode observer, the saturation function is used to replace the traditional sign function, as shown in Fig. 5.4.

$$z_{\alpha,\ \beta} = \begin{cases} k, & \tilde{i}_s \geq \delta \\ k \cdot \tilde{i}_s/\delta, & -\delta < \tilde{i}_s < \delta \\ -k, & \tilde{i}_s \leq -\delta, \end{cases} \tag{5.38}$$

where δ is the boundary layer constant of current error. The saturation function is a quasi-sliding mode observer method, which is essentially the linear control in the boundary layer and a noncontinuous function switching control outside the boundary layer. Therefore, this method could effectively weaken the chattering phenomenon.

5.3.3 Orthogonal Phase-Locked Loop Position Tracker

Normally, the back EMF estimation information can be obtained through the sliding mode observer, and the method of arctangent function calculation is used to obtain the rotor position, that is,

$$\hat{\theta}_e = -\arctan\left(\frac{\hat{e}_\alpha}{\hat{e}_\beta}\right). \tag{5.39}$$

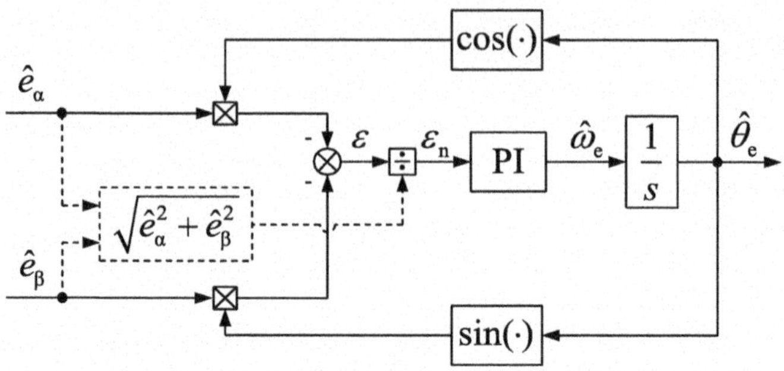

Fig. 5.5 Block diagram of orthogonal phase-locked loop position tracker

However, due to the inherent chattering phenomenon of the sliding mode observer, there are high-frequency harmonics and noise in the EEMF observation, which will inevitably affect the observation performance. Since the arctangent function calculation includes the division operation, the observation error of the rotor position is further amplified when the EEMF observation value approaches to zero. Therefore, as shown in Fig. 5.5, the quadrature phase-locked loop is used to process the EEMF observation information to obtain the rotor position [8]. The rotor position estimation error signal is normalized to simplify the phase-locked loop parameter design. This solution can track the changes in rotor position and speed effectively [11, 12].

As shown in Fig. 5.5, theoretically, the estimated EEMF obtained by the sliding mode observer only contains the fundamental component, and the position error signal after heterodyne processing can be shown as

$$\begin{aligned} \varepsilon &= -\hat{e}_\alpha \cos\hat{\theta}_\mathrm{e} - \hat{e}_\beta \sin\hat{\theta}_\mathrm{e} \\ &= E_\mathrm{ex}\sin\theta_\mathrm{e}\cos\hat{\theta}_\mathrm{e} - E_\mathrm{ex}\cos\theta_\mathrm{e}\sin\hat{\theta}_\mathrm{e} \\ &= E_\mathrm{ex}\sin\left(\theta_\mathrm{e} - \hat{\theta}_\mathrm{e}\right), \end{aligned} \tag{5.40}$$

where E_{ex} is the magnitude of the EEMF, which is proportional to the motor speed ω_e. When $\theta_\mathrm{e} - \hat{\theta}_\mathrm{e} < \frac{\pi}{2}$, (5.40) could be simplified as

$$\varepsilon = E_\mathrm{ex}\left(\theta_\mathrm{e} - \hat{\theta}_\mathrm{e}\right). \tag{5.41}$$

Further, the position error after the normalization of the EEMF can be obtained:

$$\varepsilon_\mathrm{n} = \theta_\mathrm{e} - \hat{\theta}_\mathrm{e}. \tag{5.42}$$

According to the analysis, the simplified block diagram of the quadrature phase-locked loop position tracker can be obtained, as shown in Fig. 5.6.

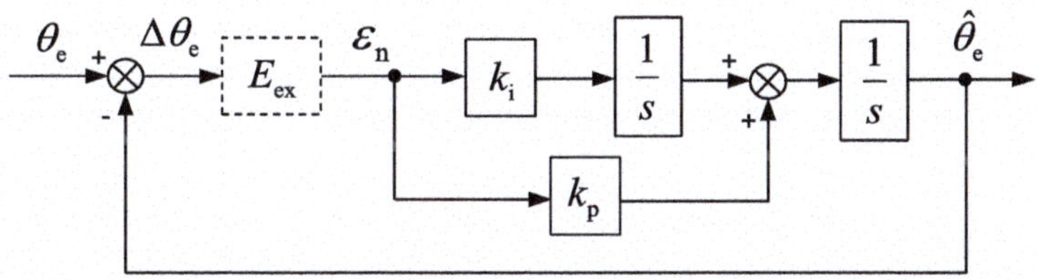

Fig. 5.6 Simplified block diagram of orthogonal phase-locked loop position tracker

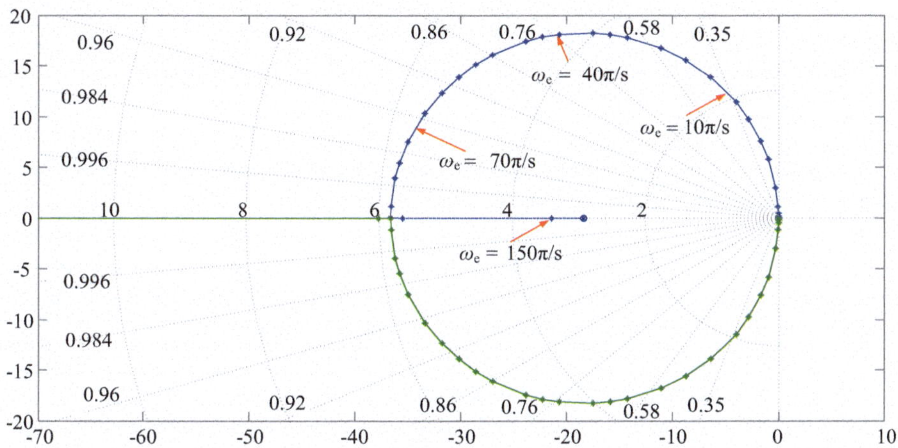

Fig. 5.7 Root-track of phase-locked loop system

At the same time, the closed-loop transfer functions of the quadrature phase-locked loop before and after the EEMF normalization can be obtained:

$$G_{PLL} = \frac{\hat{\theta}_e}{\theta_e} = \frac{E_{ex}k_p s + E_{ex}k_i}{s^2 + E_{ex}k_p s + E_{ex}k_i}, \tag{5.43}$$

$$G_{PLL} = \frac{\hat{\theta}_e}{\theta_e} = \frac{k_p s + k_i}{s^2 + k_p s + k_i}. \tag{5.44}$$

Figure 5.7 shows root trajectory of the orthogonal phase-locked loop with speed as the variable parameter before the EEMF normalization. As shown in the figure, for different frequencies the closed-loop characteristic roots of the system are all located in the left half plane, which means the system is stable. However, the amplitude of the EEMF varies with the speed, which leads to the characteristic root of the closed-loop system varying with the operation frequency. As a result, the system bandwidth is not a fixed value, as shown in (5.45). Before the normalization of the estimated back EMF, the system exhibits time-varying characteristics for the determined PI parameters k_p and k_i. However, in the actual drive system, in order to improve the observation performance, it is often desirable that the rotor position observer has fixed bandwidths at different operation speeds, and the EEMF normalization can effectively solve the problem that observer bandwidth varies with different speeds.

$$\begin{aligned}\omega_b &= \sqrt{\left(E_{ex}k_i + \frac{E_{ex}^2 k_p^2}{2}\right) + \sqrt{\left(E_{ex}k_i + \frac{E_{ex}^2 k_p^2}{2}\right)^2 + E_{ex}^2 k_i^2}} \\ &\approx \sqrt{\left(\omega_e\psi_f k_i + \frac{\omega_e^2\psi_f^2 k_p^2}{2}\right) + \sqrt{\left(\omega_e\psi_f k_i + \frac{\omega_e^2\psi_f^2 k_p^2}{2}\right)^2 + \omega_e^2\psi_f^2 k_i^2}}.\end{aligned} \tag{5.45}$$

The pole assignment method is used to design the quadrature phase-locked loop parameters. Usually, the two poles of the system are placed on the real axis of the left plane, satisfying $k_p = 2\rho$, and $k_i = \rho^2$. When the bandwidth of the phase-locked loop is too narrow, the position observation will generate a large error of DC offset. Conversely, if the bandwidth of PLL is too wide, the influence of HF noise on the position observation performance cannot be effectively suppressed.

Figure 5.8 shows the Bode plot of the system after normalization of the EEMF.

Due to the inherent chattering phenomenon of the sliding mode observer, the EEMF observation usually contains high-frequency noise. To highlight its influence, the EEMF observation can be expressed as follows:

$$\begin{cases}\hat{e}_\alpha = e_\alpha + n_1(t) = -E_f \sin(\omega_e t + \theta_{ei}) + n_1(t) \\ \hat{e}_\beta = e_\beta + n_2(t) = E_f \cos(\omega_e t + \theta_{ei}) + n_2(t),\end{cases} \tag{5.46}$$

where E_f is the amplitude of the EEMF fundamental frequency component, θ_{ei} is the initial phase angle, and $n_1(t)$ and $n_2(t)$ are the sum of the nonideal high-frequency noise.

Accordingly, the output value of the phase-locked loop, i.e., the estimated rotor position, can be expressed as

$$\hat{\theta}_e = \hat{\omega}_e t + \hat{\theta}_{ei}, \tag{5.47}$$

where $\hat{\theta}_{ei}$ refers to the initial phase angle of the rotor position observation.

After EEMF is normalized, the rotor position error signal can be expressed as

$$\varepsilon_n = \frac{1}{\sqrt{\hat{e}_\alpha^2 + \hat{e}_\beta^2}}\left(-\hat{e}_\alpha \cos\hat{\theta}_e - \hat{e}_\beta \sin\hat{\theta}_e\right). \tag{5.48}$$

According to (5.46) and (5.47), the result (5.49) can be obtained as

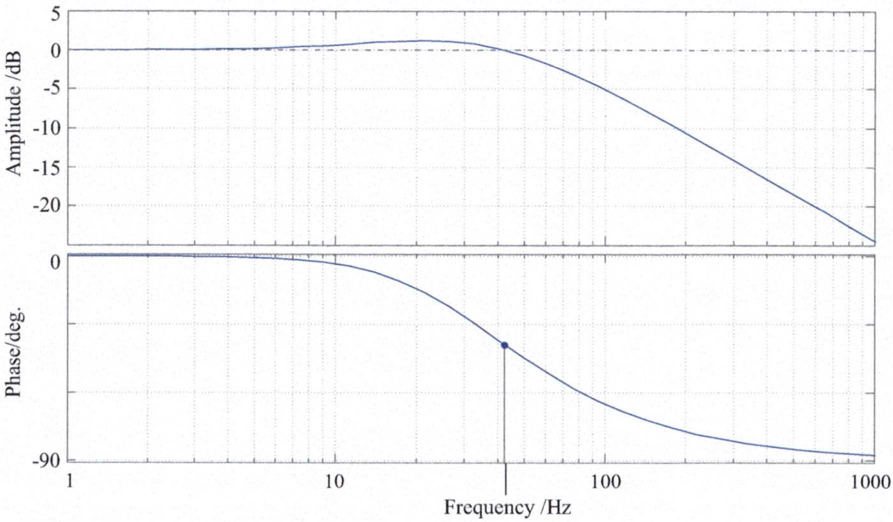

Fig. 5.8 Bode plot of the system after normalization of the EEMF

$$
\begin{aligned}
\varepsilon_{\mathrm{n}} = & \frac{1}{\sqrt{\hat{e}_{\alpha}^{2}+\hat{e}_{\beta}^{2}}} E_{\mathrm{f}}\left(\sin\left((\omega_{\mathrm{e}}-\hat{\omega}_{\mathrm{e}})t+(\theta_{\mathrm{ei}}-\hat{\theta}_{\mathrm{ei}})\right)\right) \\
& -\frac{1}{\sqrt{\hat{e}_{\alpha}^{2}+\hat{e}_{\beta}^{2}}}\left(n_{1}(t)\cos\left(\hat{\omega}_{\mathrm{e}}t+\hat{\theta}_{\mathrm{ei}}\right)+n_{2}(t)\sin\left(\hat{\omega}_{\mathrm{e}}t+\hat{\theta}_{\mathrm{ei}}\right)\right).
\end{aligned}
\tag{5.49}
$$

Equation (5.49) shows that the output of the PLL phase detector contains a difference frequency term and two HF terms, and the rotor position information can be obtained by processing the difference frequency term. Normally, the output of the common PLL phase detector also contains the second harmonic term. However, the second harmonic term can be cancelled out in the phase-locked loop. Therefore, the bandwidth of PLL can be designed to be relatively wide, which can improve system dynamic performance and simplify the design of PLL parameters. In addition, due to the higher frequency, the reasonable design of the phase-locked loop bandwidth can filter out the two HF terms effectively.

5.3.4 Experimental Results of Sliding Mode Observer

Figure 5.9 shows the block diagram of the position-sensorless vector control system based on the extended back-EMF model sliding mode observer. In order to verify the effectiveness of the extended back-EMF model-based sliding mode observer and the quadrature PLL position tracker, the experiment has been carried out on a 2.2 kW

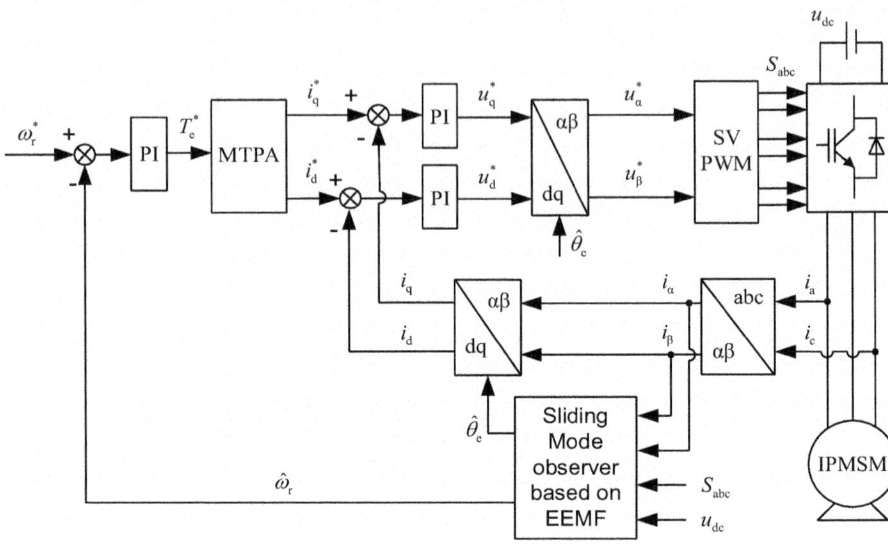

Fig. 5.9 Block diagram of position sensorless vector control system based on EEMF model sliding mode observer

IPMSM drive platform. Figure 5.10 shows the experimental results at 500 r/min with 50% rated load.

Figure 5.10 shows the comparison between the estimated current and the actual current at the α-axis. Figure 5.11 shows the waveform of the estimated current at the α-β axes. Figures 5.10a and 5.11a are the experimental results obtained by using the full-order sliding mode observer. Figures 5.10b and 5.11b are the results of using the traditional reduced-order sliding mode observer. It can be observered from the experiment results that the estimated current in the α-β axis system can follow the actual value by using the designed sliding mode observer more effectively, which is obtained by using the full-order sliding mode observer compared with the traditional second-order observer. The current estimation results are smoother and the Lissajous figure is closer to a circle with less harmonics.

Figure 5.12 shows the experiment comparison of the EEMF estimation and the a-phase current. Figure 5.12a is the experiment results obtained by the full-order sliding mode observer, and Fig. 5.12b is the experiment results obtained by the traditional second-order sliding mode observer. The experiment results show that the EEMF observation value and the phase current curve are smoother, and the harmonic component are significantly reduced after using the full-order sliding mode observer. Therefore, the rotor position estimation performance can be effectively improved.

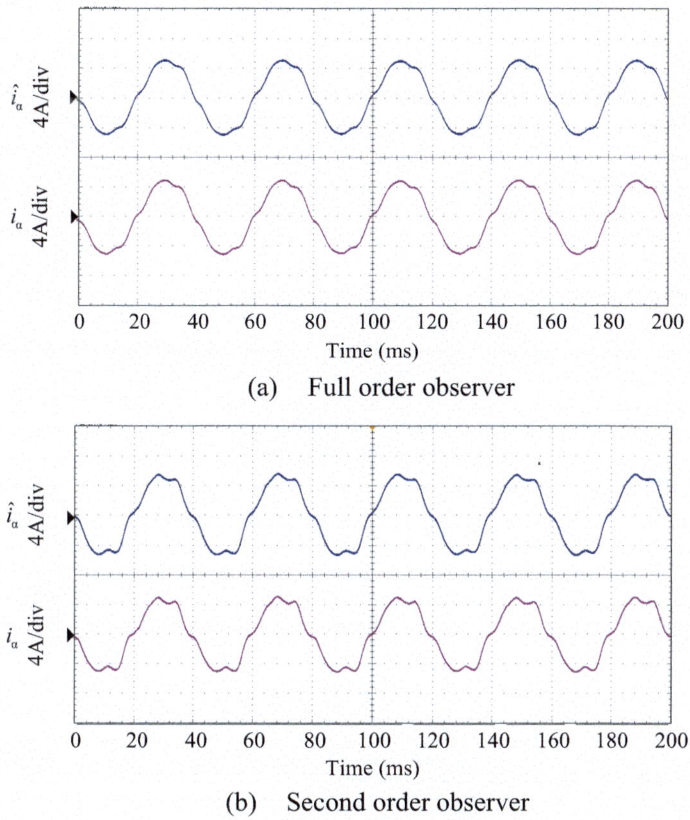

Fig. 5.10 Comparison of the estimated and the actual α-axis currents

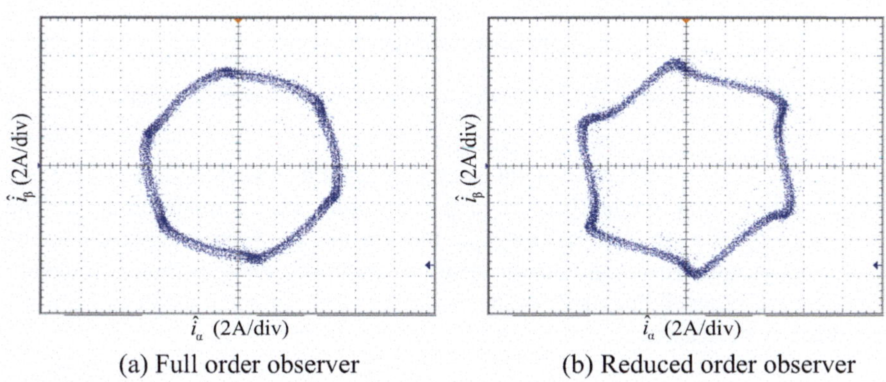

Fig. 5.11 Lissajous waveforms of the estimated α-β axis currents

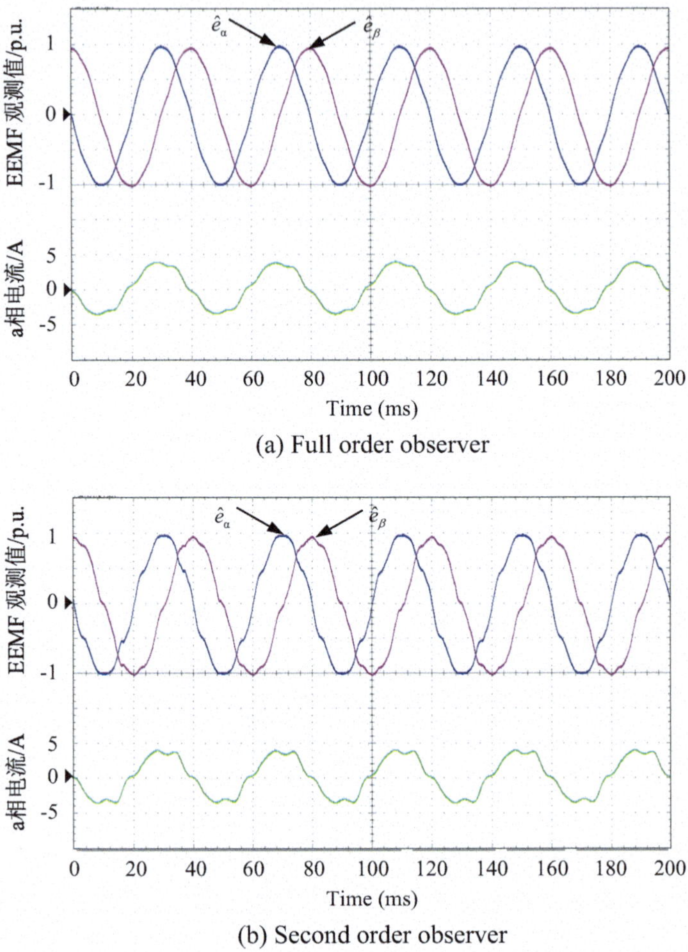

Fig. 5.12 Experimental comparison of the EEMF estimation and the *a*-phase current

Figure 5.13 shows the experiment results obtained by switching from traditional second-order sliding mode observer to the full-order sliding mode observer. The switching time is set to 100 ms. The experiment results show that compared with the traditional second-order sliding mode observer, estimated rotor position is smoother using the full-order sliding mode observer and the *sixth*-harmonic fluctuation error is significantly decreased.

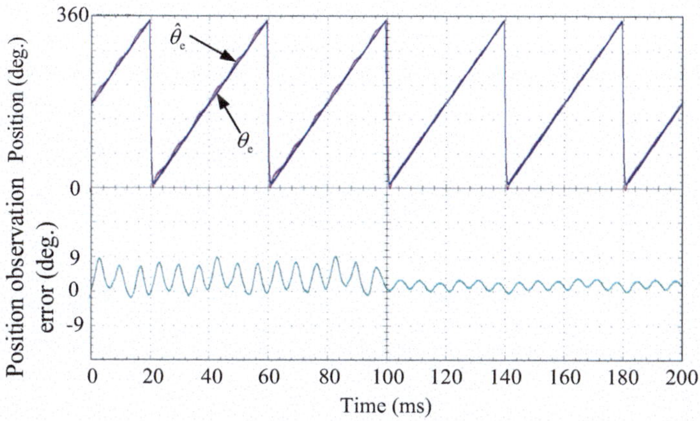

Fig. 5.13 Experimental results of switching from traditional second-order sliding mode observer to full-order sliding mode observer

5.4 Hybrid Sensorless Control Strategy in Whole Speed Range

The position estimation strategy based on signal injection method is only applicable to the zero-speed and low-speed range, while the model method is only applicable to the medium-speed and high-speed range. In order to make the PMSM realize the position sensorless control in the all speed domain, it is necessary to research the switching strategy of the two kinds of position estimation scheme.

Take the sensorless control method based on the HF square wave injection proposed in Chap. 3 and the estimation scheme based on the EEMF-based sliding mode observer proposed in this chapter as examples. In both schemes, the rotor position information can be obtained by using arctangent calculation as follows:

$$\theta_{el} = 0.5 \cdot \tan^{-1}[\sin(2\theta_e + \theta_m)/\cos(2\theta_e + \theta_m)] - 0.5\theta_m, \tag{5.50}$$

$$\theta_{eh} = \tan^{-1}\frac{\hat{x}_\alpha^1}{\hat{x}_\beta^1}, \tag{5.51}$$

where θ_{el} and θ_{eh} are the rotor positions obtained at low speed and high speed by using the different estimation methods, respectively.

In order to combine the two methods together, the estimated rotor position information should be used according to different operating speeds. As mentioned above, the arctangent calculation introduces additional noise, which should be filtered out by an observer. In order to reduce the number of observers, this chapter uses the hybrid control strategy.

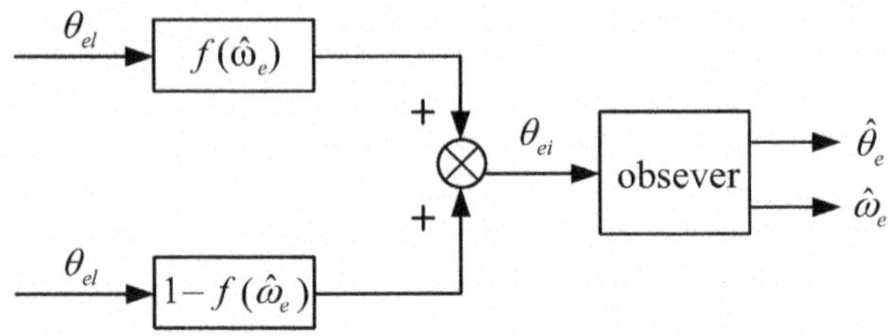

Fig. 5.14 Block diagram of rotor position combination fusion strategy

As shown in Fig. 5.14, at the zero-speed and low-speed range, the rotor position θ_{el} estimated by the signal injection method is used as the input θ_{ei} of the observer to perform position signal filtering and speed estimation. Define ω_1 and ω_2 as the switching points for the injection method and the model-based method, respectively. When the motor speed rises and enters into the switching area $\omega_1 < \omega_e < \omega_2$, the rotor position θ_{ei}, input of the observer, is the linear weighting of θ_{el} and θ_{eh} with respect to the operation speed. When the speed continues to rise and exceeds the switching region, the signal input to the observer is only θ_{eh}. The weighted rotor position can be expressed as

$$\theta_{ei} = \theta_{el} f(\hat{\omega}_e) + \theta_{eh}(1 - f(\hat{\omega}_e)), \tag{5.52}$$

where

$$f(\hat{\omega}_e) = \begin{cases} 1, & |\hat{\omega}_e| \le \omega_1 \\ \dfrac{\omega_2 - \hat{\omega}_e}{\omega_2 - \omega_1}, & \omega_1 < |\hat{\omega}_e| < \omega_2 \\ 0, & |\hat{\omega}_e| \ge \omega_2 \end{cases}. \tag{5.53}$$

The selection of the switching speeds ω_1 and ω_2 is related to the estimation performance of the signal injection method and the model method. In order to ensure the higher accuracy of position estimation for two methods in the switching interval, the switching speeds ω_1 and ω_2 are usually taken as 10% and 20% the rated speed, respectively.

By calculating the rotor position estimated by the two different kinds of methods through the speed-based switching function, the position sensorless control in the all-speed range can be realized.

Figure 5.15 shows the experimental results of the position sensorless control from 0 r/min to rated speed (1000 r/min) at full load. When the speed is less than 10% of the rated speed (i.e., 100 r/min), the position estimation algorithm uses the HF square wave injection method proposed in Chap. 3. When the speed is greater than 20% of the rated speed (i.e., 200 r/min), the position estimation algorithm adopts the sliding mode EEMF observer proposed in this chapter. When the operation speed is at 10–20% of the rated speed, the switching strategy of the two position estimation methods introduced above is adopted.

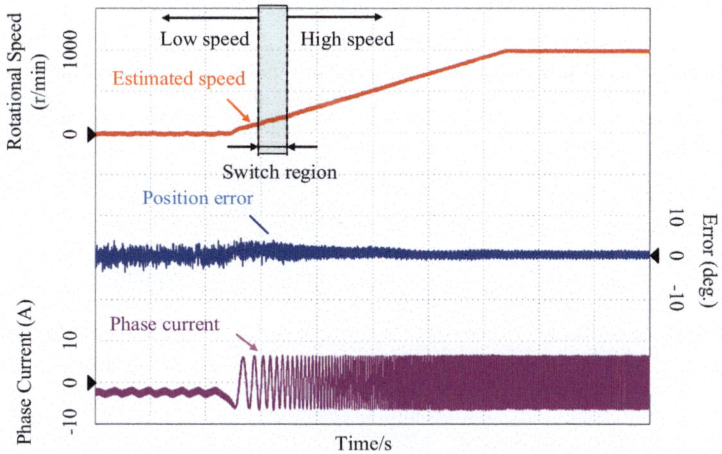

Fig. 5.15 Experiment results of hybrid sensorless control in full speed range

The experimental results show that the proposed hybrid sensorless control strategy can combine the low-speed and the high-speed position estimation methods effectively, which proves the effectiveness of sensorless operation of PMSM drives in the wide-speed range.

5.5 Summary

This chapter introduces the model-based methods to achieve sensorless control scheme for IPMSM drives. In this chapter, a full-order sliding mode observer is proposed based on the stator current and the EEMF as state variables. And the detailed derivation and the simplified sliding mode gain design method are given. The quadrature phase-locked loop can obtain the rotor position information from the sliding mode observer. The EEMF normalization can ensure the fixed bandwidth of the quadrature PLL when the motor speed change. Furthermore, in order to realize the sensorless all-speed range, the hybrid sensorless control strategy is designed combined with the signal injection method and the model-based method.

References

1. W. Qiao, X. Yang, and X. Gong. Wind speed and rotor position Sensorless control for direct-drive PMG wind turbines. IEEE Trans. Ind. Appl. **48**(1), 3–11 (January/February 2012)
2. V. Smidl, and X. Peroutka. Advantages of square-root extended Kalman filter for Sensorless control of AC drives. IEEE Trans. Ind. Electron. **59**(11), 4189–4196 (November 2012)

3. G. D. Andreescu, C. I. Pitic, F. Blaabjerg, and I. Boldea. Combined flux observer with signal injection enhancement for wide speed range Sensorless direct torque control of IPMSM drives. IEEE Trans. Energy Convers. **23**(2), 393–402 (June 2008)
4. G. Foo, and M. F. Rahman. Sensorless direct torque and flux-controlled IPM synchronous motor drive at very low speed without signal injection. IEEE Trans. Ind. Electron. **57**(1), 395–403 (January 2008)
5. Boldea I, Paicu M C, Andreescu G-D. Active flux concept for motion-sensorless unified ac drives. IEEE Trans. Power Electron. **23**(5), 2612–2618 (September 2008)
6. Boldea I, Paicu M C, Andreescu G-D, et al. 'Active flux' DTFC-SVM sensorless control of IPMSM. IEEE Trans. Energy Convers. **24**(2), 314–322 (May 2009)
7. Vaclavek P, Blaha P, Herman I. AC drive observability analysis. IEEE Trans. Ind. Electron. **60** (8), 3047–3059 (August 2013)
8. G. Wang, Z. Li, G. Zhang, Y. yu, and D. Xu, Quadrature PLL-based high-order sliding-mode observer for IPMSM sensorless control with online MTPA control strategy. IEEE Trans. Energy Convers. **28**(1), 214–224 (March 2013)
9. Wang G, Yang R, Xu D. DSP-based control of sensorless IPMSM drives for wide-speed-range operation. IEEE Trans. Ind. Electron. **60**(2), 720–727 (February 2013)
10. Z. Chen, M. Tomita, S. Doki, and S. Okuma, An extended electromotive force model for sensorless control of interior permanent-magnet synchronous motors. IEEE Trans. Ind. Electron. **50**(2), 288–295 (April 2003)
11. L. Harnefors, and H. P. Nee. A general algorithm for speed and position estimation of AC motors. IEEE Trans. Ind. Electron. **47**(1), 77–83 (February 2000)
12. I. Carugati, P. Donato, S. Maestri, D. Carrica, and M. Benedetti. Frequency adaptive PLL for polluted single-phase grids. IEEE Trans. Power Electron. **27**(5), 2396–2404 (May 2012)

Chapter 6
Position Estimation Error Ripple Elimination for Model-Based Method

The position sensorless control methods can be divided into two groups, the back-EMF-based methods [1–8] and the HF injection methods. According to Chap. 5, fundamental excitation methods based on EMF or flux-linkage model are frequently used for medium- and high-speed sensorless operation in industry applications. The second-order position observer, the full-order position observer, and the seamless hybrid control are proposed. Other fundamental excitation methods can be roughly categorized as the state observer, the Kalman filter, the disturbance observer, the MRAS observer, the sliding-mode observer (SMO), and the artificial intelligence-based estimation method [1]. However, there are some common shortcomings among these EMF-based methods.

Since the amplitude of EMF which contains rotor position information is proportional to the rotor speed, the EMF-based methods become invalid at low and zero speeds. More than this, the model-based methods rely on the accuracy of model and parameters. Due to flux spatial harmonics and the inverter nonlinearities effects, there exists substantial harmonic error in the position estimation, resulting in the degradation of the vector control performance. This chapter proposes some solutions focused on the position estimation error ripple elimination for model-based methods.

6.1 Causes of the Position Estimation Error Ripple

The model-based methods rely on the accuracy of the observer model. Especially, it is more critical to sensorless control of PMSM drives whose parameter variation and flux spatial harmonics are significant. Due to the flux spatial harmonics and the inverter nonlinearities, there exists substantial harmonic error in the position estimation. Generally, the position estimation error is comprised of a DC drift and a harmonic ripple, resulting in the degradation of the operation performance of the PMSM drives.

G. Wang et al., *Position Sensorless Control Techniques for Permanent Magnet Synchronous Machine Drives*, https://doi.org/10.1007/978-981-15-0050-3_6

6.1.1 Analysis of the Harmonic Components in Extended EMF Estimation

The voltage model of IPMSM in the α-β SRF can be expressed as (5.15) and E_{ex} is the extended EMF amplitude [1]:

$$E_{ex} = \omega_e\left[\left(L_d - L_q\right)i_d + \lambda_{PM}\right] - \left(L_d - L_q\right)\left(di_q/dt\right), \tag{6.1}$$

where λ_{PM} is the permanent magnet flux.

Equation (5.15) can be written in the vector expression:

$$\boldsymbol{u}_{\alpha\beta} = \left(R + pL_d\right)\mathbf{i}_{\alpha\beta} - j\omega_e\left(L_d - L_q\right)\mathbf{i}_{\alpha\beta} + \mathbf{e}_{\alpha\beta}, \tag{6.2}$$

where $\boldsymbol{i}_{\alpha\beta} = i_\alpha + ji_\beta$,$\boldsymbol{u}_{\alpha\beta} = u_\alpha + ju_\beta$, $\boldsymbol{e}_{\alpha\beta} = e_\alpha + je_\beta$, and j denotes the imaginary unit.

Taking the Laplace transform on (6.2), the result can be obtained:

$$\boldsymbol{e}_{\alpha\beta} = \boldsymbol{u}_{\alpha\beta} + j\omega_e\left(L_d - L_q\right)\boldsymbol{i}_{\alpha\beta} - \left(R + L_d s\right)\boldsymbol{i}_{\alpha\beta}. \tag{6.3}$$

By adopting the state filter method [2], the back-EMF can be calculated as follows:

$$\hat{\boldsymbol{e}}_{\alpha\beta} = \frac{\left(k_p s + k_i\right)\left(L_d s + R\right)}{L_d s^2 + \left(R + k_p\right)s + k_i}\left(\frac{\boldsymbol{u}^*_{\alpha\beta} + j\hat{\omega}_e\left(L_d - L_q\right)\boldsymbol{i}_{\alpha\beta}}{L_d s + R} - \boldsymbol{i}_{\alpha\beta}\right), \tag{6.4}$$

where "^" denotes the estimated value, $\boldsymbol{u}^*_{\alpha\beta}$ denotes the reference voltage vector, and k_p and k_i are the proportional and integral gains of the back-EMF estimator.

The gains of the back-EMF estimator are calculated by the pole-zero cancellation method as follows:

$$k_p = L_d\omega_c, \quad k_i = R\omega_c, \tag{6.5}$$

where ω_c is the bandwidth of the back-EMF estimator. By substituting (6.5) into (6.4), the following equation can be obtained:

$$\frac{\hat{\boldsymbol{e}}_{\alpha\beta}}{\boldsymbol{e}_{\alpha\beta}} = \frac{\omega_c}{s + \omega_c}. \tag{6.6}$$

So, it is clear that the back-EMF estimator has the same characteristics with the first-order LPF.

A voltage difference between the reference and the actual values is produced due to the dead-time effect, which results in the stator current distortion. In addition, the flux spatial harmonics of IPMSM can make the extended back-EMF distorted [3–7]. Both the two cases result in the ($6k\pm1$)th harmonic components in the stator currents, which can be expressed as

$$\begin{aligned} i_x =& I_1 \sin\left(\omega_e t + \theta_1 + i\cdot 2\pi/3\right) + I_{6k-1}\sin\left[-(6k-1)\omega_e t + \theta_{6k-1} + i\cdot 2\pi/3\right] \\ &+ I_{6k+1}\sin\left[(6k+1)\omega_e t + \theta_{6k+1} + i\cdot 2\pi/3\right] \\ =& I_1 \sin\left(\omega_e t + \theta_1 + i\cdot 2\pi/3\right) + \sum_{k=1}^{n} I_{6k\pm1}\sin\left[\pm(6k\pm1)\omega_e t + \theta_{6k\pm1} + i\cdot 2\pi/3\right], \end{aligned} \tag{6.7}$$

where the subscript x represents a, b, or c phase, of which corresponding i is 0,1, or 2. I_1, $I_{6k\text{-}1}$, and I_{6k+1} are the amplitudes of the fundamental and ($6k$-1)th and ($6k$+1)th stator current harmonics, respectively. θ_1, $\theta_{6k\text{-}1}$, and θ_{6k+1} are the initial phases of the fundamental and ($6k$-1)th and ($6k$+1)th stator current harmonics, respectively.

Using the Clarke and Park transformation, the stator currents in the d-q rotating reference frame can be expressed as

$$i_d = I_1 \sin\theta_1 + \sum_{k=0}^{n} I_{6k\pm1}\sin\left(\pm 6k\omega_e t + \theta_{6k\pm1}\right), \tag{6.8}$$

$$i_q = -I_1 \cos\theta_1 - \sum_{k=0}^{n} I_{6k\pm1}\cos\left(\pm 6k\omega_e t + \theta_{6k\pm1}\right). \tag{6.9}$$

According to (6.8) and (6.9), the ($6k$-1)th and ($6k$+1)th harmonics in the α-β axes correspond to the ($6k$)th harmonics in the d-q axes. Substituting (6.8) and (6.9) into (6.1), and only considering the ($6k$-1)th and the ($6k$+1)th harmonics, the extended back-EMF in the α-β axes can be obtained:

$$\begin{aligned} e_\alpha =& -E_{ex}\sin\omega_e t \\ =& (L_d - L_q)\omega_e\Big\{-\left(I_1\sin\theta_1 + \frac{\lambda_{PM}}{L_d - L_q}\right)\sin\omega_e t \\ &\mp \sum_{k=0}^{n}\frac{6k\mp1}{2} I_{6k\pm1}\left[\cos\left(\pm(6k\pm1)\omega_e t + \theta_{6k\pm1}\right) - \cos\left(\pm(6k\mp1)\omega_e t + \theta_{6k\pm1}\right)\right]\Big\}, \end{aligned} \tag{6.10}$$

$$\begin{aligned} e_\beta =& E_{ex}\cos\omega_e t \\ =& (L_d - L_q)\omega_e\{\left(I_1\sin\theta_1 + \frac{\lambda_{PM}}{L_d - L_q}\right)\cos\omega_e t \\ & \mp\sum_{k=0}^{n}\frac{6k\mp 1}{2}I_{6k\pm 1}[\sin(\pm(6k\pm 1)\omega_e t + \theta_{6k\pm 1}) + \sin(\pm(6k\mp 1)\omega_e t + \theta_{6k\pm 1})]\}. \end{aligned} \tag{6.11}$$

In practical applications, the fifth and the seventh harmonics take the dominant ingredient. From (6.6), it is known that the back-EMF estimator produces phase delay between the estimated and the actual values. On the other hand, it can suppress high-order harmonics to some extent. The phase delay and the harmonic suppression are related to the bandwidth ω_c and the operating frequency ω_e (as shown in Fig. 6.1). It can be seen that for certain ω_e, larger ω_c corresponds to smaller phase delay of the fundamental EMF estimates but corresponds to weaker capability of suppressing harmonics. In order to obtain smaller phase delay, the bandwidth ω_c should be a relatively large value, resulting in harmonic components in the back-EMF estimates.

6.1.2 Position Estimation Error Due to the Harmonics in the Estimated Back-EMF

From the above analysis, it can be noted that the back-EMF estimate contains (6k±1)th harmonics. Under this condition, the voltage model of IPMSM can be written as follows:

$$\begin{bmatrix} u_\alpha \\ u_\beta \end{bmatrix} = \begin{pmatrix} R + pL_d & \omega_e L_d - L_q) \\ -\omega_e(L_d - L_q) & R + pL_d \end{pmatrix}\begin{bmatrix} i_\alpha \\ i_\beta \end{bmatrix} + \boldsymbol{e}_f + \boldsymbol{e}_h, \tag{6.12}$$

where $\boldsymbol{e}_h$ and $\boldsymbol{e}_f$ denote the (6k±1)th harmonics and the fundamental component in the back-EMF estimate, which can be expressed as

$$\boldsymbol{e}_f = \begin{bmatrix} e_{f\alpha} \\ e_{f\beta} \end{bmatrix} = \begin{bmatrix} -e_1\sin(\omega_e t + \theta_{e1}) \\ e_1\cos(\omega_e t + \theta_{e1}) \end{bmatrix}, \tag{6.13}$$

$$\begin{aligned} \boldsymbol{e}_h = \begin{bmatrix} e_{h\alpha} \\ e_{h\beta} \end{bmatrix} =& \begin{bmatrix} -e_{6k-1}\sin\left(-(6k-1)\omega_e t + \theta_{e(6k-1)}\right) \\ e_{6k-1}\cos\left(-(6k-1)\omega_e t + \theta_{e(6k-1)}\right) \end{bmatrix} \\ & + \begin{bmatrix} -e_{6k+1}\sin\left((6k+1)\omega_e t + \theta_{e(6k+1)}\right) \\ e_{6k+1}\cos\left((6k+1)\omega_e t + \theta_{e(6k+1)}\right) \end{bmatrix}, \end{aligned} \tag{6.14}$$

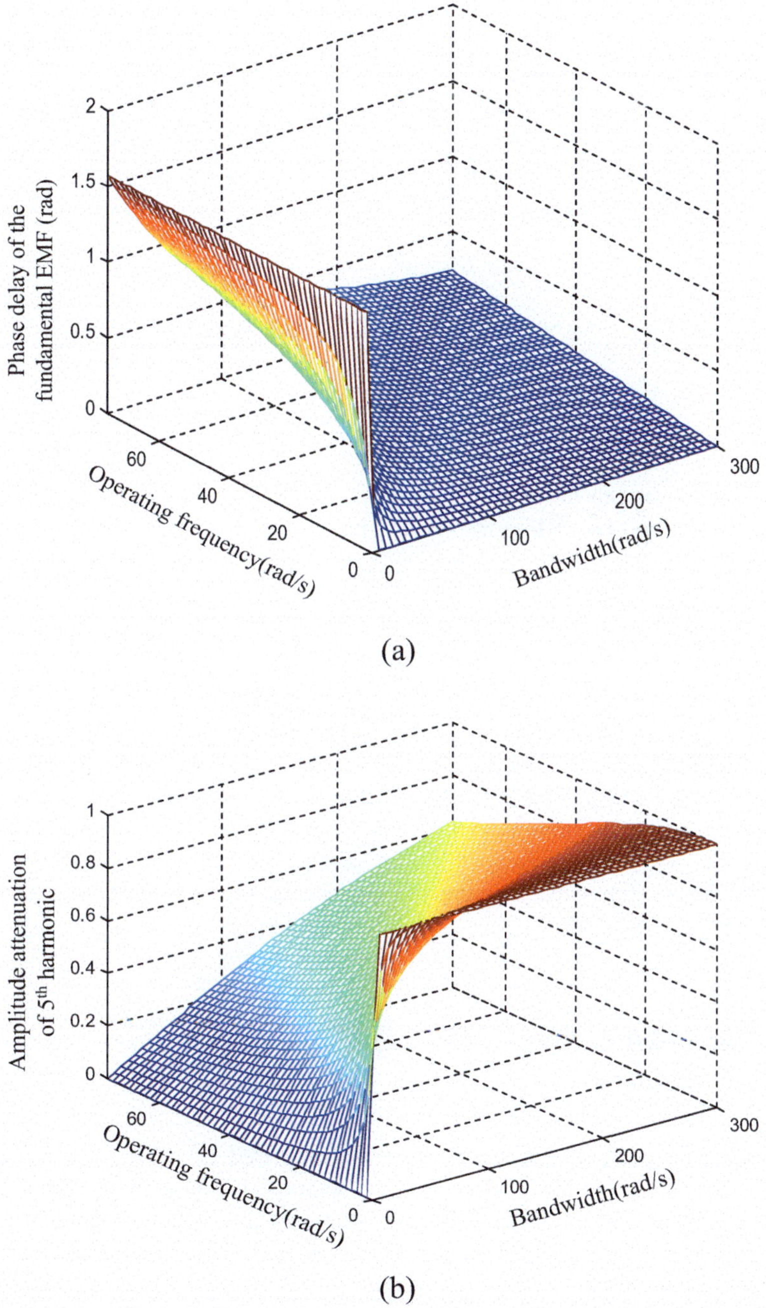

Fig. 6.1 Phase delay and harmonic suppression related to the bandwidth ω_c and operating frequency ω_e. (**a**) Phase delay of the fundamental EMF. (**b**) Amplitude suppression of the fifth harmonic component

where $e_{6k\text{-}1}$ and e_{6k+1} represent the amplitudes of the (6*k*-1)th and the (6*k*+1)th harmonic components and $\theta_{e(6k\text{-}1)}$ and $\theta_{e\ (6k+1)}$ represent the corresponding initial phases, respectively.

The phase difference obtained by the heterodyne method can be given as follows:

$$\begin{aligned}\varepsilon_\theta =& -e_\alpha \cos\hat{\theta}_e - e_\beta \sin\hat{\theta}_e \\ =& -\left(e_{f\alpha} + e_{h\alpha}\right)\cos\hat{\theta}_e - \left(e_{f\beta} + e_{h\beta}\right)\sin\hat{\theta}_e \\ =& e_1 \sin\left[(\omega_e - \hat{\omega}_e)t + \theta_{e1} - \hat{\theta}_{ei}\right] \\ &+ \sum_{k=1}^{n} e_{6k-1}\sin\left[(-(6k-1)\omega_e - \hat{\omega}_e)t + \theta_{e(6k-1)} - \hat{\theta}_{ei}\right] \\ &+ \sum_{k=1}^{n} e_{6k+1}\sin\left[((6k+1)\omega_e - \hat{\omega}_e)t + \theta_{e(6k+1)} - \hat{\theta}_{ei}\right].\end{aligned} \quad (6.15)$$

From (6.15), it can be seen that the phase difference contains one difference-frequency term and two HF terms. When the position estimator converges, (6.15) can be simplified as

$$\varepsilon_\theta = e_1 \sin\left(\theta_{e1} - \hat{\theta}_{ei}\right) + \sum_{k=1}^{n} e_{6k}\sin\left(6k\omega_e t + \theta_{e(6k)}\right), \quad (6.16)$$

where e_{6k} represents the amplitude of the equivalent $6k^{\text{th}}$ back-EMF harmonics and $\theta_{e(6k)}$ represents the initial phase. The $6k^{\text{th}}$ back-EMF harmonics will display as additional errors which exist in the phase difference. As a result, the rotor position obtained from the quadrature PLL whose input signal is the phase difference contains the $6k^{\text{th}}$ harmonics. So the dominant harmonics in the estimated back-EMF should be eliminated in order to enhance the position estimation accuracy without the harmonic error.

6.1.3 Methods to Reduce Harmonic Ripple in Position Estimation

In general, the position estimation error is comprised of a DC drift [9] and a harmonic ripple in fundamental excitation methods. Because of the influence of inverter nonlinearities and flux spatial harmonics, the (6*k*±1)th harmonics appear in EMF estimates, resulting in (6*k*)th harmonic ripples in the estimated rotor position and speed. To diminish the DC drift resulting from parameter uncertainties, online parameter identification is always required.

In vector-controlled PMSM drives, the knowledge of rotor position should be used to achieve coordinate transformation and decoupling control. So the harmonics existing in position estimate may lead to inaccuracy in coordinate transformation and failure in complete decoupling control, which not only causes harmonic contents in d-q axis currents but also gives rise to torque ripples. As a result, the (6k)th harmonic ripple exists in the estimated position and speed, which deteriorates the performance of the sensorless IPMSM drives. Additionally, the harmonics will also bring about unnecessary energy consumption.

Other researches on the low-speed sensorless control revealed that the low-speed limit of back-EMF method primarily depends on the noise on the estimated speed signal from the inverter harmonics and secondarily from the back-EMF harmonics that have not been decoupled. Since the inverter-generated harmonics became a higher percentage of back-EMF as the rotor speed decreased, the noise on the estimated speed signal increased with decreasing rotor speed. All these effects will result in performance degradation of IPMSM drives.

Some methods were developed to reduce the harmonic error in the position estimation. For example, the difference between the reference voltage and the actual voltage of the inverter can be modeled and compensated. An online parameter-identification technique can be applied in order to reduce the error of position estimation. More than this, a fuzzy mathematical model considering the dq-axis cross coupling magnetic saturation according to the self-commission is also revealed. However, it is difficult for both the inverter nonlinearity compensation and the mathematical model considering spatial harmonics to completely eliminate the sixth harmonic error in the position estimate for sensorless IPMSM drives.

In order to improve the rotor position estimation performance of sensorless IPMSM drives, the harmonic ripples in EMF estimates should be suppressed. The adaptive linear neural (ADALINE) network-based filter has been proven to be an effective harmonic detection approach, and it provides good accuracy and convergence speed in tracking and compensating harmonic contents due to easy-to-implement structure and characteristic of supervised learning mechanism [6]. The ADALINE filtering can make a system operate in an optimal state through continuously self-tuning the filter weights on the basis of signal characteristics and operating environment. Least mean square (LMS) and recursive least square (RLS) algorithms are usually used to update the ADALINE filter weights.

The second-order generalized integrator (SOGI) has been applied in power grid and renewable energy for its capability of detecting the selected frequency harmonic [10]. The SOGI filter performs well in signal processing for distorted and asymmetric systems. Researchers proposed a synchronization system based on an adaptive filter implemented by means of a SOGI, which was self-tuned to the grid frequency, thanks to the action of a frequency-locked loop (FLL). The system is able to perform an excellent estimation of the instantaneous symmetrical components of the grid voltage under unbalanced and distorted grid condition. All the methods using SOGI to detect the harmonic components and obtain the fundamental frequency component in grid have the advantages of frequency adaptive performance, low computational burden, and relatively high filtering capability. In view of these characteristics, the SOGI algorithm can be adopted in sensorless IPMSM drives to extract the harmonic components in the EMF estimates to eliminate the harmonic error in the position estimation.

To diminish the position estimation error with harmonic fluctuations for position sensorless IPMSM drives, a multiple adaptive vector filter (AVF) cross-feedback network (CFN) is proposed for model-based position observer [7]. A full-order Luenberger observer is utilized to obtain back-EMF information. The harmonic fluctuations in the position estimation error caused by inverter nonlinearities and flux spatial harmonics are analyzed. Accordingly, a newly proposed multiple-AVF CFN with easy-to-implement structure is adopted to detect and compensate the back-EMF harmonic distortions and hence to eliminate the position error harmonic fluctuations. Consequently, the position estimation accuracy can be enhanced, which improves the position sensorless control performance.

6.2 Conventional Elimination Scheme for Position Estimation Error Ripple

6.2.1 Position Estimation Based on Extended EMF Sliding-Mode Observer

Figure 6.2 shows the extended EMF model-based sensorless IPMSM drive system.

The SMO is designed using the extended EMF model of IPMSM [1]. As shown in Fig. 6.3, a SMO based on the extended EMF model is used to estimate the rotor position.

SMO provides fast convergence and low sensitivity to parameter variations. It can be expressed as follows:

$$\dot{\hat{\boldsymbol{i}}} = \boldsymbol{A}\hat{\boldsymbol{i}} + \boldsymbol{B}(\boldsymbol{u} - \mathbf{z}), \tag{6.17}$$

where $\boldsymbol{A} = \begin{bmatrix} -R/L_d & \hat{\omega}_e(L_q - L_d)/L_d \\ -\hat{\omega}_e(L_q - L_d)/L_d & -R/L_d \end{bmatrix}$, $\boldsymbol{B} = \begin{bmatrix} 1/L_d & 0 \\ 0 & 1/L_d \end{bmatrix}$, $\hat{\mathbf{i}} = \begin{bmatrix} \hat{i}_\alpha \\ \hat{i}_\beta \end{bmatrix}$, $\mathbf{u} = \begin{bmatrix} u_\alpha \\ u_\beta \end{bmatrix}$, $\hat{\omega}_e$ is the estimated electrical speed, z is the equivalent sliding-mode control function using a saturation function defined as follows:

$$\mathbf{z} = \begin{cases} -k, & \text{for } \left(\hat{\mathbf{i}} - \mathbf{i}\right) \leq -\delta \\ k/\delta \cdot \left(\hat{i} - i\right), & \text{for } -\delta < \left(\hat{\mathbf{i}} - \mathbf{i}\right) < \delta, \\ k, & \text{for } \left(\hat{\mathbf{i}} - \mathbf{i}\right) \geq \delta \end{cases} \tag{6.18}$$

where k is the sliding-mode gain, δ is the boundary layer, and $\mathbf{S} = \hat{\mathbf{i}} - \mathbf{i} = \mathbf{0}$ is selected to be the sliding-mode surface. It can be proved that if k is positive and greater than the maximum value of the extended EMF, namely, $k > |\mathbf{e}|_{\max}$, the

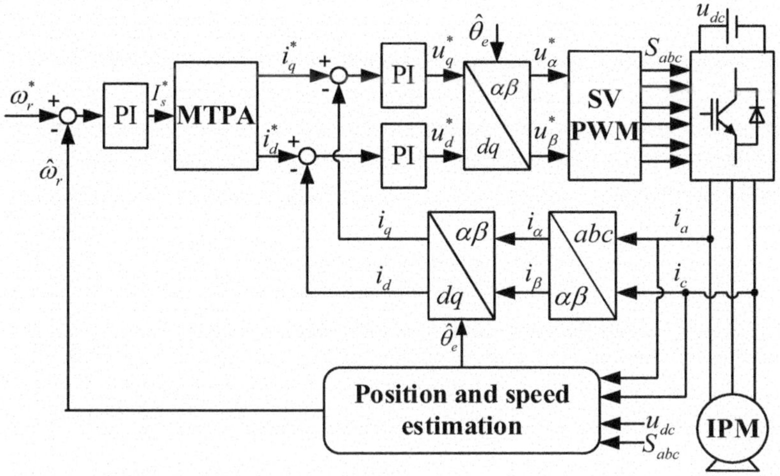

Fig. 6.2 EMF model-based sensorless IPMSM drive

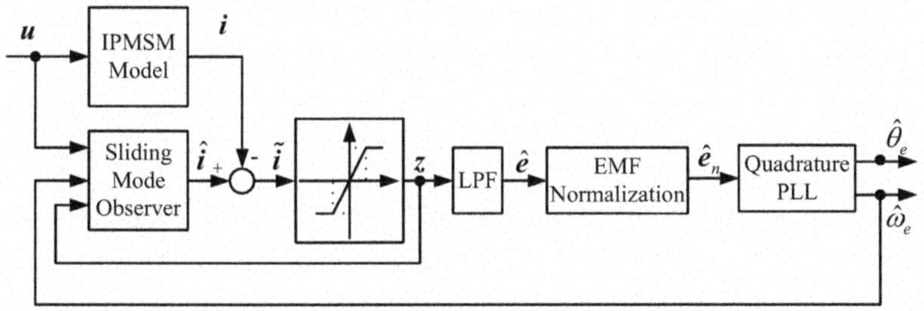

Fig. 6.3 Position observer adopting the proposed ADALINE-based filter with a quadrature PLL

observer can be asymptotically stable [11]. It means that the state can converge to $S=0$ in a finite time. So, the relation between EMF and the control function is

$$\mathbf{e} = \mathbf{z}. \tag{6.19}$$

6.2.2 *Quadrature PLL for Rotor Position Extraction*

Conventionally, the rotor position estimate can be calculated directly from EMF estimates through an arc-tangent function:

$$\hat{\theta}_e = -\tan^{-1}\left(\hat{e}_\alpha / \hat{e}_\beta\right) \tag{6.20}$$

where $\hat{e}_\alpha$ and $\hat{e}_\beta$ are the α-β axis components of the extended EMF estimates.

However, the existence of noise and harmonics may influence the position estimation accuracy. Especially during EMF crossing zero, an obvious estimation error may occur using arc-tangent calculation due to the noise. To improve the position estimation for mitigation of adverse influence, a software quadrature PLL is used to extract rotor position information according to the estimated EMF from SMO. The rotor position estimate obtained from the quadrature PLL can be expressed as follows:

$$\hat{\theta}_e = (1/s)(k_i/s + k_p)(-\hat{e}_\alpha \cos\hat{\theta}_e - \hat{e}_\beta \sin\hat{\theta}_e), \tag{6.21}$$

where k_p is the proportional gain and k_i is the integral gain.

Since the mechanical time constant is far greater than the electrical one, the rotor speed ω_e can be considered as a constant during a current sampling period. The transfer function of the quadrature PLL tracking observer without EMF normalization can be expressed as

$$G_{PLL} = \frac{\hat{\theta}_e}{\theta_e} = \frac{E_{ex}k_p s + E_{ex}k_i}{s^2 + E_{ex}k_p s + E_{ex}k_i}, \tag{6.22}$$

where E_{ex} is the extended EMF amplitude. Figure 6.4 shows the frequency analysis of PLL with different EMF amplitudes. Note that due to the variations of EMF amplitudes with different speeds, the bandwidth of the PLL system changes with the operating frequency, which deteriorates the performance and increases parameter design complexity. Therefore, the normalization of EMF for PLL tracking estimator is adopted. The transfer function of the quadrature PLL tracking observer with EMF normalization can be expressed as

$$G_{PLL} = \frac{\hat{\theta}_e}{\theta_e} = \frac{k_p s + k_i}{s^2 + k_p s + k_i}. \tag{6.23}$$

With normalization, the bandwidth of the PLL system is constant with the variations of EMF amplitudes at different speeds, which makes the quadrature PLL suitable for the whole operating range.

In order to highlight the effect of high-order harmonics and noises, the estimated extended EMF considering nonideal contents can be defined as follows:

$$\begin{cases} \hat{e}_\alpha = -E_f \sin(\omega_e t + \theta_{ei}) - E_{6k\pm1} \sin((6k \pm 1)\omega_e t + \theta_{6k\pm1}) + n_1(t) \\ \hat{e}_\beta = E_f \cos(\omega_e t + \theta_{ei}) + E_{6k\pm1} \cos((6k \pm 1)\omega_e t + \theta_{6k\pm1}) + n_2(t), \end{cases} \tag{6.24}$$

where E_f and $E_{6k\pm1}$ represent the amplitude of fundamental and high-order harmonic contents, θ_{ei} and $\theta_{6k\pm1}$ represent the corresponding initial phase angle, and $n_1(t)$ and $n_2(t)$ are the total noise contents.

Correspondingly, $\hat{\theta}_e$ the estimated rotor position which is the output of the quadrature PLL can be expressed as

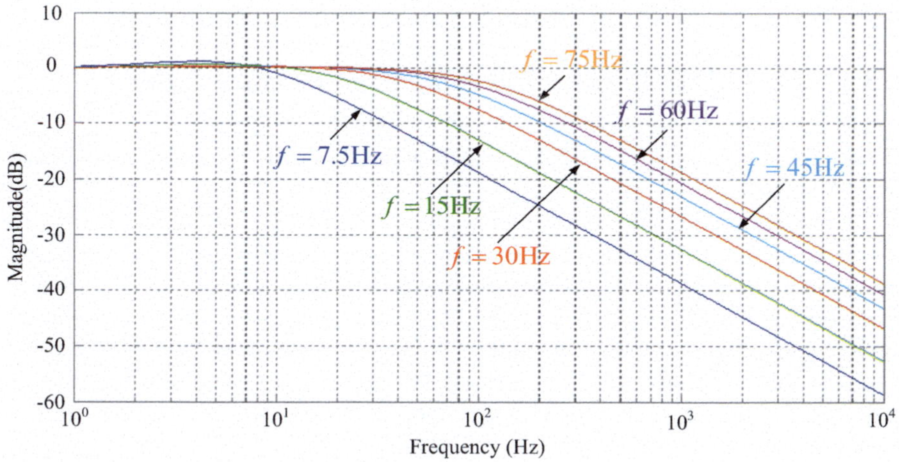

Fig. 6.4 Frequency analysis of PLL without normalization

$$\hat{\theta}_e = \hat{\omega}_e t + \hat{\theta}_{ei}, \tag{6.25}$$

where $\hat{\theta}_{ei}$ is the initial phase angle of the estimation value.

After normalization for extended EMF, the equivalent position error signal can be expressed as

$$\varepsilon_f = \frac{1}{\sqrt{\hat{e}_\alpha^2 + \hat{e}_\beta^2}} \left(-\hat{e}_\alpha \cos \hat{\theta}_e - \hat{e}_\beta \sin \hat{\theta}_e\right). \tag{6.26}$$

Through the analysis above, ε_f can be obtained as

$$\begin{aligned} \varepsilon_f = & \frac{1}{\sqrt{\hat{e}_\alpha^2 + \hat{e}_\beta^2}} \left\{E_f \sin\left((\omega_e - \hat{\omega}_e)t + \left(\theta_{ei} - \hat{\theta}_{ei}\right)\right)\right. \\ & + E_{6k\pm1} \sin\left(((6k \pm 1)\omega_e - \hat{\omega}_e)t + \left(\theta_{6k\pm1} - \hat{\theta}_{ei}\right)\right) \\ & \left. - n_1(t) \cos\left(\hat{\omega}_e t + \hat{\theta}_{ei}\right) - n_2(t) \sin\left(\hat{\omega}_e t + \hat{\theta}_{ei}\right)\right\}. \end{aligned} \tag{6.27}$$

Note that the output of the phase detector consists of one difference frequency term, one high frequency harmonic term, and two high frequency noise terms, and the first term is used to obtain rotor position information. Besides, it can be seen that the (6*k*)th harmonic error appears resulting from (6*k*±1)th harmonic distortions of the extended EMF estimate when the position observer converges. The two high-frequency noise terms can be easily suppressed by the loop filter in PLL. However, since the low bandwidth of the loop filter may bring about amplitude attenuation and phase lag, it is difficult to determine a suitable cutoff frequency of the quadrature PLL in reaching a tradeoff between elimination of the (6*k*)th harmonic error and risk of influencing the fundamental content.

6.3 ADALINE-Based Position Error Ripple Elimination

6.3.1 ADALINE-Based Filter Adopted in Position Estimator

Figure 6.5 shows the proposed ADALINE network-based PLL position estimator using SMO [6].

Because of the influence of inverter nonlinearities and flux spatial harmonics, the $(6k\pm1)$th harmonics appear in EMF estimates, resulting in the $(6k)$th harmonic ripples in the estimated rotor position and speed, which deteriorates the performance of sensorless IPMSM drives [2]. In the $(6k)$th harmonic ripples, the sixth harmonic content takes the dominant role due to its bigger amplitude by FFT spectrum analysis. Since high-frequency noises can be filtered out by the quadrature PLL, considering high-order harmonics only, the voltage model can be described as (6.12).

The harmonic contents can be modeled as a variable amplitude sinusoidal waveform shown as (6.28). The shape of $e_{h\alpha\beta}$ depends on permanent magnet flux distribution in a machine along with speed and load conditions:

$$e_{h\alpha\beta} = E_{6k\pm1}\sin\left((6k \pm 1)\omega_e t + \theta_{6k\pm1}\right). \tag{6.28}$$

Figure 6.6 shows the position comparison and estimation error waveforms at 500 r/min with 50% rated load. Obviously, there's a sixth harmonic ripple in the rotor position estimate whose amplitude can reach up to 9°.

The ADALINE network has been proven to be an effective harmonic detection approach and provides good accuracy and convergence speed in tracking harmonic contents due to its simple structure and characteristic of supervised learning mechanism. The block diagram of the ADALINE-based filter is shown in Fig. 6.7. Note that the ADALINE-based filter is a multi-input, single-output, and single-layer linear neural element that can be trained online based on the input signals, the target response, and the weight updating adaptive algorithm. Where, $\hat{e}_{n\alpha\beta}$, the arbitrary component of the extended EMF acts as the primary input by estimating with normalization. $\boldsymbol{X}(n)$ represents the reference input vector, which can be obtained through taking cosine and sine terms of the rotor position estimate outputted from the

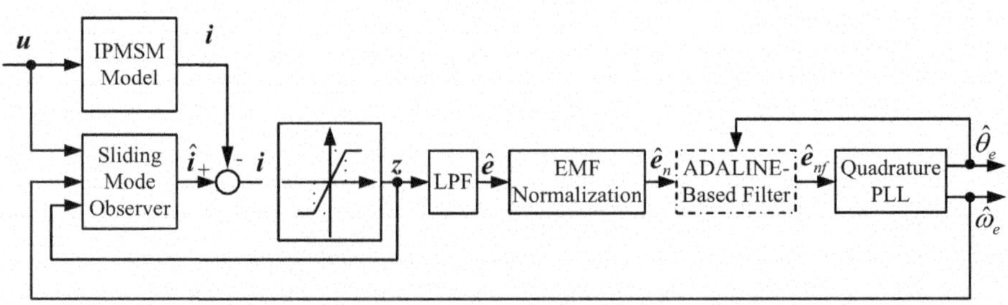

Fig. 6.5 Position observer adopting the proposed ADALINE-based filter with a quadrature PLL

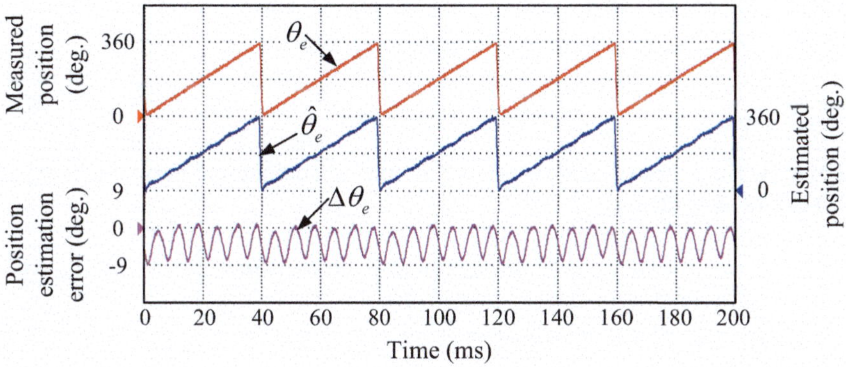

Fig. 6.6 Harmonic error exists in the position estimate

quadrature PLL. $\boldsymbol{w}(n)$ represents the adjustable weight vector. The weighted versions of the reference are summed to form the filter's output $\hat{e}_{h\alpha\beta}$; for any reference input x, $\hat{e}_{h\alpha\beta}$ can be calculated as

$$\hat{e}_{h\alpha\beta} = \mathbf{X}^T(n)\mathbf{w}(n) = \sum_{k=1} x_k(n)w_k(n). \tag{6.29}$$

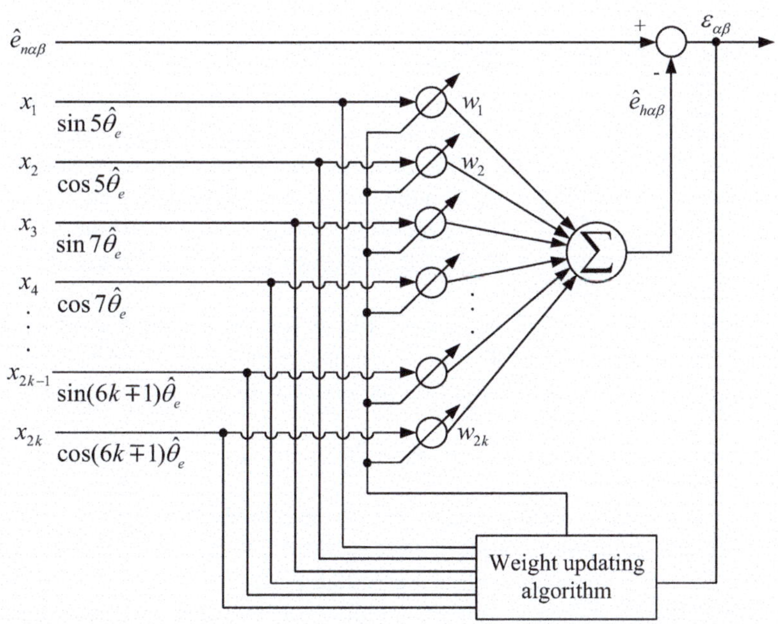

Fig. 6.7 Block diagram of the ADALINE-based filter

The desired fundamental content can be obtained directly from the error signal $\varepsilon_{\alpha\beta}$ given by subtracting $\hat{e}_{h\alpha\beta}$ from $\hat{e}_{n\alpha\beta}$, described as follows:

$$\varepsilon_{\alpha\beta} = \hat{e}_{n\alpha\beta} - \hat{e}_{h\alpha\beta}. \tag{6.30}$$

To allow the proposed estimation techniques to compensate harmonic contents, (6.28) can be rewritten as

$$e_{h\alpha\beta} = A(t)(\sin\omega_h t\cos\theta_h + \cos\omega_h t\sin\theta_h), \tag{6.31}$$

and in the vector form described as

$$\begin{aligned} e_{h\alpha\beta} =& \begin{bmatrix} \sin\omega_h t & \cos\omega_h t \end{bmatrix} \begin{bmatrix} A(t)\cos\theta_h \\ A(t)\sin\theta_h \end{bmatrix} \\ =& \mathbf{X}^T(n)\mathbf{w}(n), \end{aligned} \tag{6.32}$$

where $A(t)$ is the amplitude of the high-order harmonic content $e_{h\alpha\beta}$.

Through the analysis above, it can be seen that when the adaptive filter converges, the weight vector tends to be equal to the discrete Fourier coefficient of $e_{h\alpha\beta}$. This model can be used to detect and compensate EMF high-order harmonics caused by both inverter nonlinearities and flux spatial harmonics.

6.3.2 *Online Training Algorithms of Filter Weights*

To allow the ADALINE network-based filter to detect and compensate the selective harmonic ripple in the position estimation error, the adaptive filtering algorithm is necessary to update the filter weights online according to the harmonic characteristics of the estimated EMF. The LMS algorithm and the RLS algorithm are widely used in various applications of ADALINE network due to easy-to-implement structure and computational simplicity.

B. Widrow proposed LMS algorithm, which has been widely used in signal processing and adaptive control due to its simplicity to design and robustness. The weight vector trained with LMS algorithm is given as follows:

$$\mathbf{w}(n+1) = \mathbf{w}(n) + 2\mu\cdot\mathbf{X}(n)\cdot\varepsilon_{\alpha\beta}, \tag{6.33}$$

where μ is the learning rate parameter, which determines convergence speed and steady-state misadjustment. In order to increase the convergence speed, the adaptive filter is usually made underdamped, namely, that $\mu < \omega_h$.

For simplicity in real-time implementation for practical position sensorless IPMSM drives, the ADALINE network-based filter with single weight is adopted,

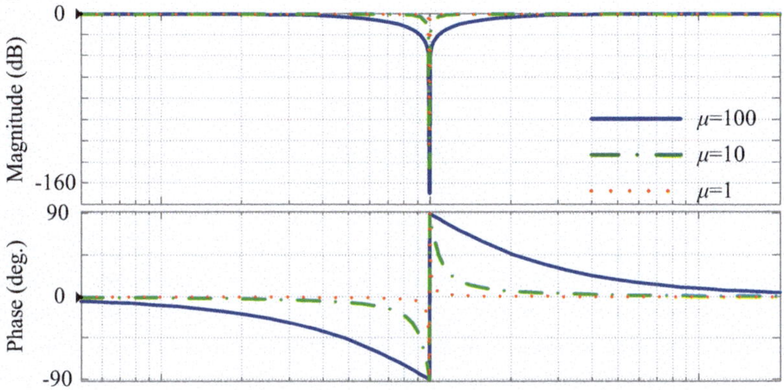

Fig. 6.8 Frequency response of the ADALINE-based filter trained LMS algorithm

and the frequency response of such a filter trained by LMS algorithm with different learning rate parameters is shown in Fig. 6.8. The bandwidth of the ADALINE-based filter trained with LMS algorithm is 2 μ. As μ gets larger, the system responds faster, but the steady-state misadjustment gets bigger. Therefore, the convergence speed has to be traded off with the steady state misadjustment. Though the LMS training algorithm is widely used due to its simplicity to implement and low computational burden, it suffers from a relatively slow convergence speed, which limits its applications.

In order to increase the convergence speed of the ADALINE-based filter, the RLS algorithm can be adopted [12]. The formula of RLS algorithm is expressed as follows:

$$\mathbf{w}(n+1) = \mathbf{w}(n) + \mathbf{k}(n) \cdot \varepsilon_{\alpha\beta}, \tag{6.34}$$

where the gain vector $\boldsymbol{k}(n)$ is given as

$$\mathbf{k}(n) = \frac{\mathbf{P}(n-1)\mathbf{X}(n)}{\lambda + \mathbf{X}^T(n)\mathbf{P}(n-1)\mathbf{X}(n)} \tag{6.35}$$

and the inverse of the autocorrelation matrix $\boldsymbol{P}(n)$ is given as

$$\mathbf{P}(n) = \frac{\mathbf{P}(n-1) - \mathbf{k}(n)\mathbf{X}^T(n)\mathbf{P}(n-1)}{\lambda}, \tag{6.36}$$

where λ is the forgetting factor, which is a positive constant close to but less than 1, namely, $0<\lambda<1$. When $\lambda = 1$, the ordinary method of least squares can be made. The introduction of λ is to ensure that the data in the distant past are forgotten and more emphases are put on the recent samples.

In practical implementations, the initial value of $\boldsymbol{w}(n)$ and $\boldsymbol{P}(n)$ are chosen as follows:

$$\begin{cases} \mathbf{P}(0) = \sigma \mathbf{I} \\ \mathbf{w}(0) = 0\mathbf{I}, \end{cases} \tag{6.37}$$

where $\boldsymbol{I}$ is the unit matrix and σ is a positive constant. The initial value of $\boldsymbol{P}(n)$ contributes to convergence of ADALINE-based filter. For low σ values, the response is overdamped, and for high σ values, the response is underdamped. In this application, σ is chosen as 0.001 to get rid of overshoot. The convergence speed of the ADALINE-based filter trained with RLS algorithm is directly related to λ, and the bandwidth is inversely related to λ. In addition, when λ is very close to 1, the RLS algorithm achieves low misadjustment and good stability, but its tracking capability is reduced. A smaller value of λ improves the tracking capability, but the misadjustment is increased, and the stability of RLS algorithm can be affected. In general, the forgetting factor λ can be set to $0.95 < \lambda < 1$ to avoid undesired divergence, and in this application, λ is set to $\lambda = 0.9996$.

6.3.3 Experimental Results on ADALINE-Based Filter

Figure 6.9 shows the vector control scheme of sensorless IPMSM drive using the proposed ADALINE filter-based PLL position observer. SMO based on the extended EMF model is used to obtain rotor position information. The proposed position observer has been verified for a 2.2 kW IPMSM sensorless vector control drive with a DSP. The experimental setup is shown in Fig. 6.10. The rated parameters of IPMSM are listed as follows: 380 V, 5.0A, 14 N·m, 75 Hz, 1500 r/min, $R = 2.75\ \Omega$, $L_d = 45$ mH, $L_q = 60$ mH, $\psi_f = 0.48$ Wb. DSP-TMS320F2808 is adopted to execute the whole sensorless control algorithm. An IM is mechanically coupled with IPMSM to apply the load torque, and the PWM switching frequency of the inverter is 10 kHz. An absolute encoder (ECN1113) is used to obtain the actual position that is solely used for comparison but not for control purposes. For LMS training algorithm, the learning rate parameter μ is 1. And for RLS training algorithm, the forgetting factor λ is 0.9996, $\boldsymbol{P}(0) = \boldsymbol{0}$, and $\boldsymbol{w}(0) = \boldsymbol{0}$. Performance of the ADALINE-based filter trained with both LMS and RLS algorithms for harmonic tracking and compensation of extended EMF estimate is evaluated both at steady and transient operations. All the following experimental results were achieved with the IPMSM drive operating in sensorless control. Since the sixth harmonic ripple takes the dominant role in the position estimation error due to its bigger amplitude, only the fifth and seventh harmonic contents of EMF estimates are tracked and compensated in the experiments.

Figure 6.11 shows the experimental comparison of the estimated extended EMF with normalization and FFT spectrum analysis before and after using the

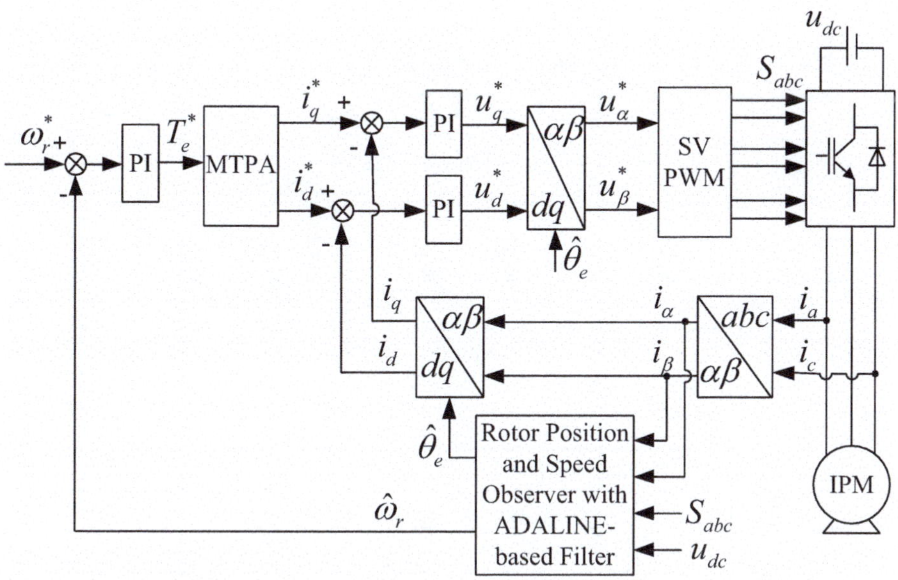

Fig. 6.9 EMF model-based sensorless IPMSM drive

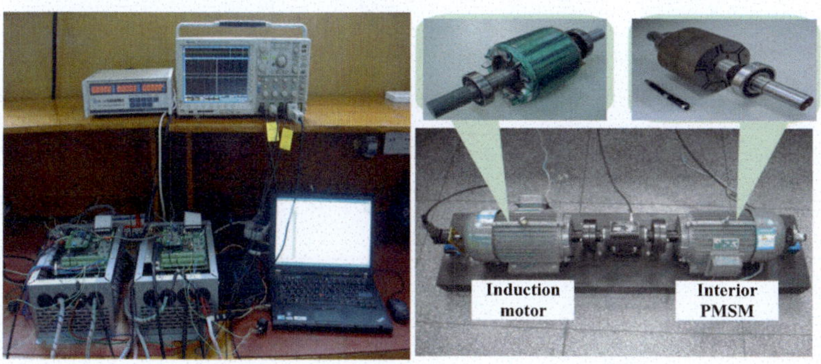

(a) (b)

Fig. 6.10 2.2 kW sensorless IPMSM drive based on DSP. (**a**) Drivers with DC bus connection. (**b**) Motor platform

ADALINE-based filter trained with LMS and RLS algorithms at 600 r/min and 50% rated load. Tenfold magnified FFT analysis of the eleventh and thirteenth harmonics in the vertical axis is depicted in the green dotted area. From the experimental results, it can be seen obviously that the estimated EMF waveforms get more sinusoidal, and the fifth and seventh harmonics are effectively eliminated using the ADALINE-based filter trained with LMS and RLS algorithms, but the eleventh and thirteenth harmonics with small amplitudes still remain since they are not compensated in the test.

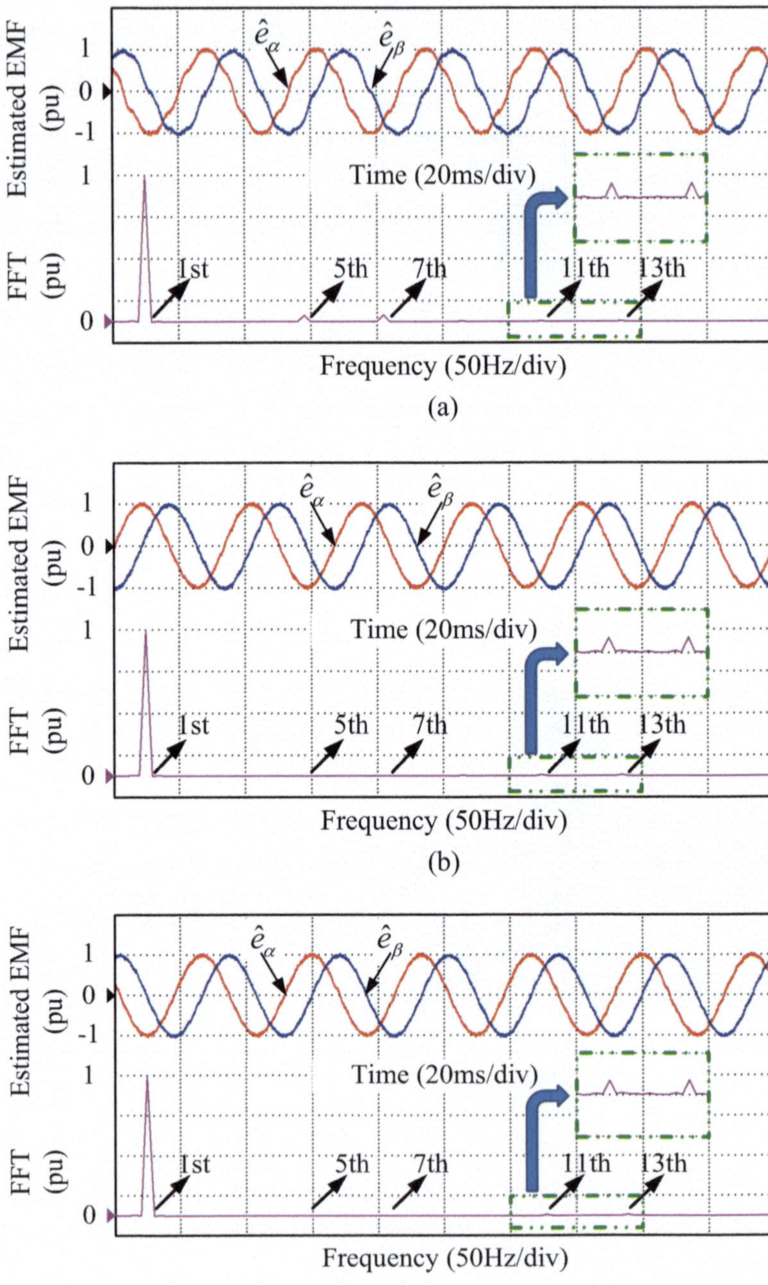

Fig. 6.11 Experimental comparison of estimated EMF waveforms before and after using ADALINE-based filter at 600 r/min with 50% rated load. (**a**) Without adaptive filter. (**b**) Using ADALINE-based filter trained with LMS algorithm. (**c**) Using ADALINE-based filter trained with RLS algorithm

Figure 6.12 shows the comparison of the estimated rotor position, the position estimation error, and the harmonic estimate waveforms with and without an ADALINE-based filter trained by LMS and RLS algorithms at 500 r/min and 50% rated load. Since the ADALINE-based filter is disabled, the EMF harmonic estimate cannot be detected in Fig. 6.12a. Note that the estimated position gets relatively smooth and the sixth harmonic content can be eliminated with the ADALINE-based filter trained with both LMS and RLS algorithms in the steady state. The maximum harmonic ripple of the position estimation error is reduced to 2° from 9° with the ADALINE-based filter. Since the eleventh and thirteenth harmonics are not compensated, there still is a twelfth harmonic ripple with smaller amplitude in the estimated rotor position.

Since the inverter nonlinearities are more obvious in low-speed range, the signal-to-noise ratio of the EMF amplitude decreases as speed decreases. Therefore, the experimental tests for low-speed operation under 50% rated load have also been conducted as shown in Fig. 6.13. As can be seen, the lowest speed with the conventional SMO can operate down to 120 r/min which corresponds to 8% rated speed. And the estimated position is distorted with an obvious sixth harmonic ripple up to 18°. In contrast, with the proposed ADALINE-based filter trained with LMS and RLS algorithms, the method can still be survival at 75 r/min which corresponds to 5% rated speed. The estimated rotor position is in good agreement with the measured position with a 12th position estimation error ripple. And it can be concluded that the proposed ADALINE-base filter is effective in improving the accuracy of the position estimation at low speed.

Figure 6.14 shows the comparison of the estimated rotor position and the harmonic estimate waveforms before and after using the ADALINE-based filter trained with LMS and RLS algorithms at 500 r/min and 50% rated load, and the ADALINE-based filter is enabled at 4 s. From the experimental results, it can be concluded that the ADALINE-based filter trained with both LMS and RLS algorithms can effectively reduce the position estimation error, but RLS algorithm converges faster than LMS algorithm. The execution times of the sensorless control using the ADALINE-based filter trained with LMS algorithm and RLS algorithm are 19.3 μs and 21.5 μs, respectively. The results offer a guideline on selecting the training algorithms for the ADALINE-based filter. The performance of fast convergence makes RLS algorithm attractive in variable frequency applications, and the simple structure makes LMS algorithm suitable for low computational burden applications.

Figure 6.15 illustrates the convergence of position estimation error under various λ using the ADALINE-based filter trained with RLS algorithm at 600 r/min under 50% rated load. As can be seen, smaller λ results in faster error convergence but gives rise to a larger position estimation error up to 16 deg., which decreases the position estimation accuracy. However, larger λ results in slower error convergence but guarantees better position estimation accuracy. Furthermore, the position estimation error comparison under various λ {0.95, 0.96, 0.97, 0.98, 0.99, 0.9996} at different speed operating conditions with 50% rated load is given in Fig. 6.16, and an identical conclusion can be made from the experimental results.

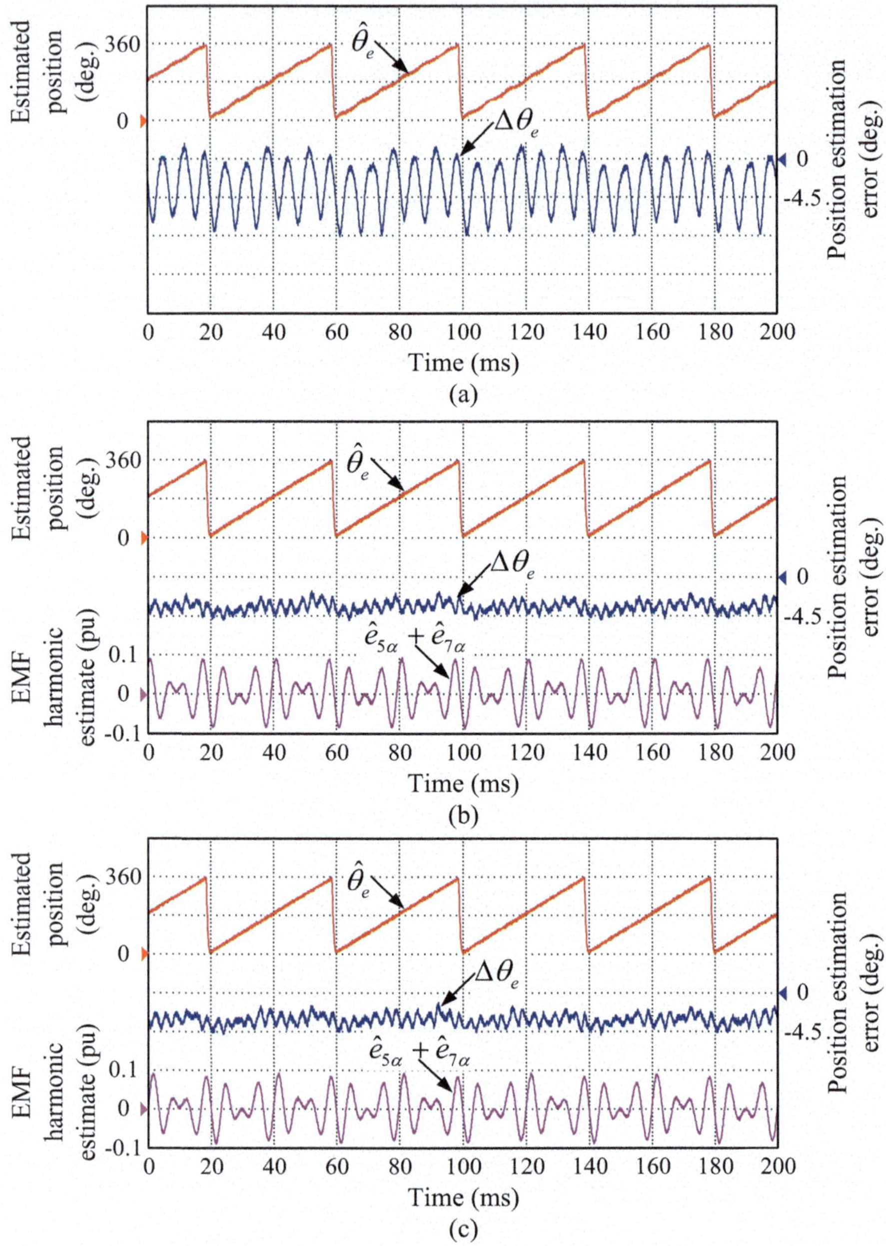

Fig. 6.12 Experimental comparison of position estimation error and harmonic estimates at 500 r/min with 50% rated load. (**a**) Without ADALINE-based filter. (**b**) Using ADALINE-based filter trained with LMS algorithm. (**c**) Using ADALINE-based filter trained with RLS algorithm

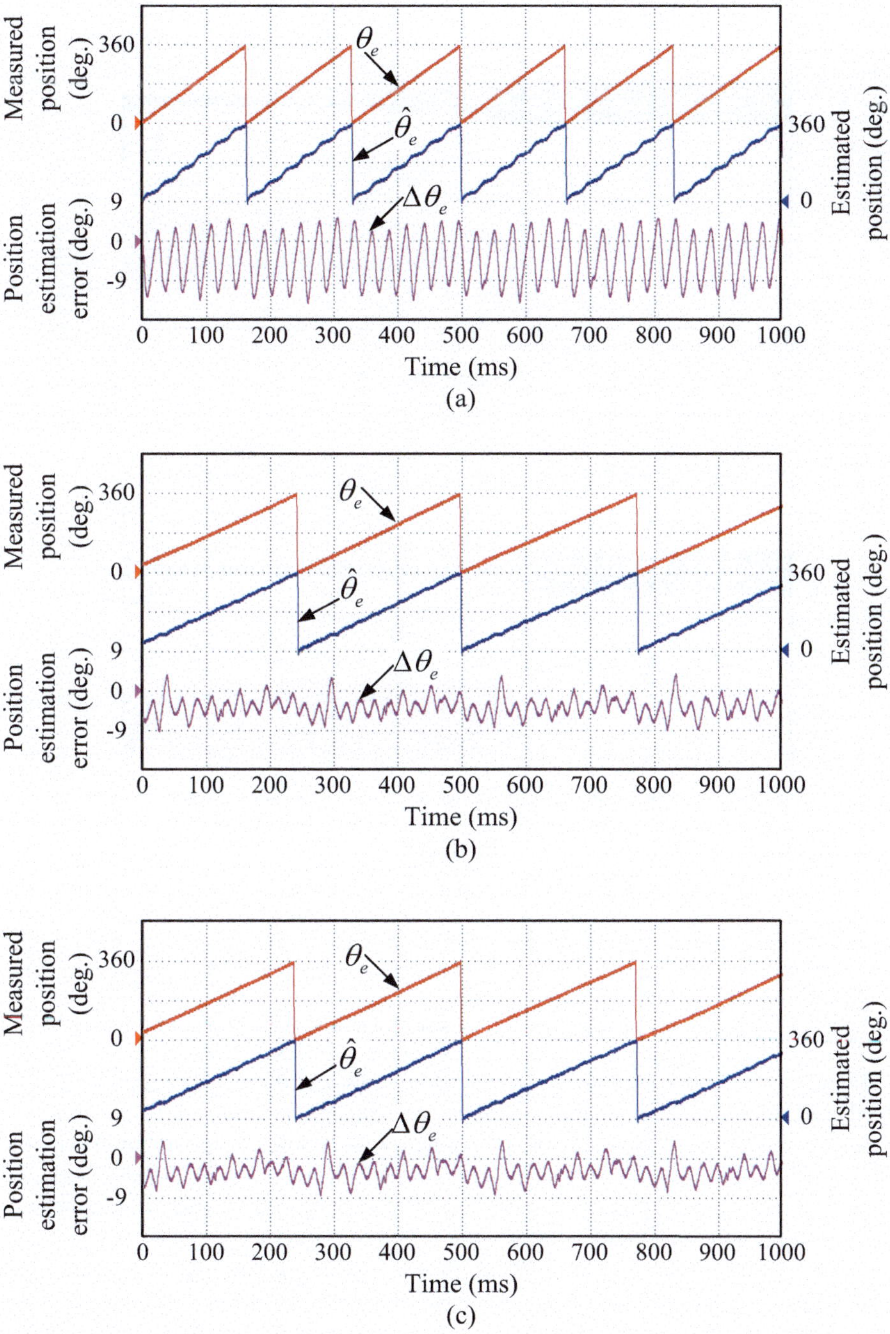

Fig. 6.13 Experimental comparison of low-speed operation with 50% rated load. (**a**) Without ADALINE-based filter, 120 r/min. (**b**) Using ADALINE-based filter trained with LMS algorithm, 75 r/min. (**c**) Using ADALINE-based filter trained with RLS algorithm, 75 r/min

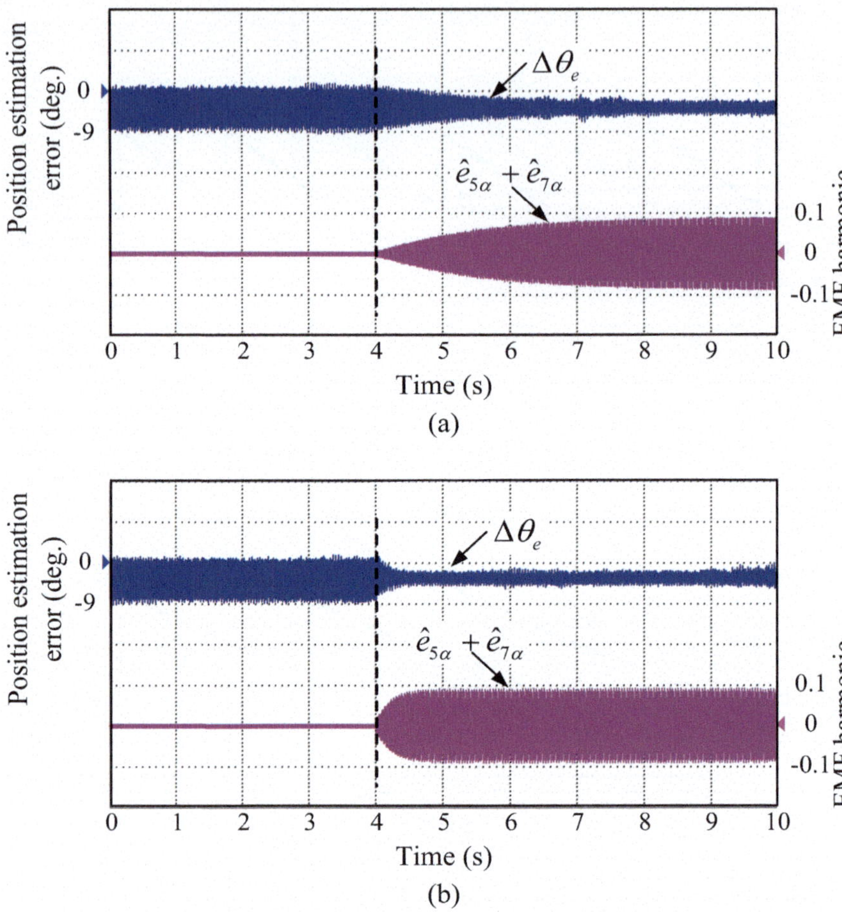

Fig. 6.14 Experimental comparison of position estimation error and harmonic estimates at 500 r/min with 50% rated load. (**a**) Using ADALINE-based filter trained with LMS algorithm. (**b**) Using ADALINE-based filter trained with RLS algorithm

Figure 6.17 shows the control performance comparison at 600 r/min with 75% rated load. In each figure, *d*-*q* axis current, electromagnet torque, and speed estimation error are given, respectively. From the experimental results, note that the ripples of i_d, i_q, T_e, and $\Delta\omega_e$ (the difference between the reference speed and the feedback speed) are reduced obviously after the ADALINE-based filter process compared to the conventional sensorless-based process (i_d, from 2 to 0.4A; i_q, from 1.5 to 0.3A; T_e, from 2 to 0.5 N·m; $\Delta\omega_e$, from 12 to 3 r/min). It can be concluded that since the harmonics existing in position estimate can be reduced through the ADALINE-based filter, the accuracy in coordinate transformation and complete decoupling

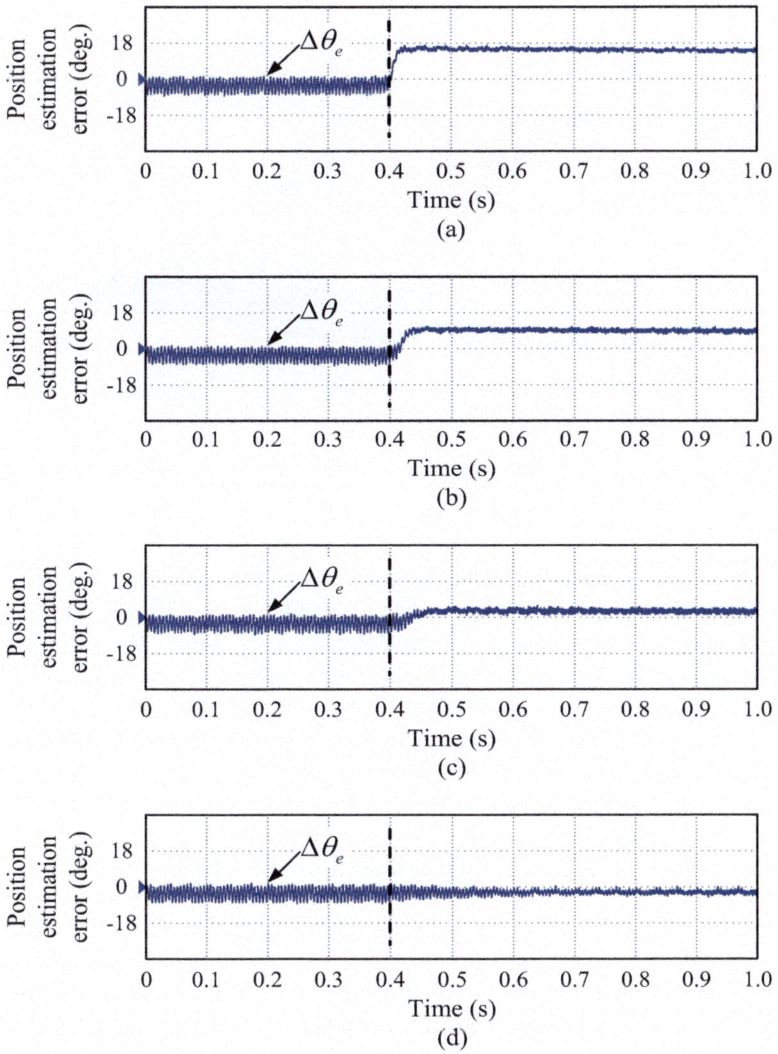

Fig. 6.15 Convergence of position estimation error under various λ. (a) λ = 0.97. (b) λ = 0.98. (c) λ = 0.99. (d) λ = 0.9996

control can be guaranteed. However, the *d*-*q* axis current ripples can be reduced by the ADALINE-based filter, which diminishes the torque and speed ripples.

Figure 6.18 shows the speed and the position estimation error comparison during continuous speed variation at the speed slew rate of 300 r/min/s with 25% rated load over the whole speed range in which the extended EMF-based SMO is applied. From top to bottom, Fig. 6.18a corresponds to the results without the ADALINE-based

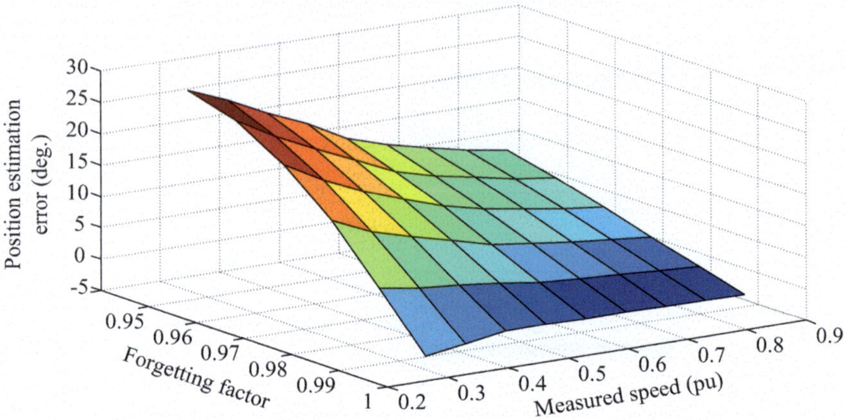

Fig. 6.16 Position estimation error comparison under various λ at different speed operating conditions

filter; Fig. 6.18c, e correspond to the results with the ADALINE-based filter trained with LMS and RLS algorithms; Fig. 6.18b, d, and f are zoomed-in position estimation error with the acceleration transient of Fig. 6.18a, c, and e. IPMSM accelerates from 150 r/min (10% rated speed) to 1500 r/min (rated speed) and then decelerates to 150 r/min. From the experimental results, note that the position estimation ripple error can be reduced to some degree using the ADALINE-based filter trained with LMS and RLS algorithms over the whole acceleration and deceleration process during continuous speed variation, and the RLS algorithm outperforms the LMS algorithm in the sense of the harmonic tracking performance.

Figure 6.19 shows the speed and the position estimation error comparison during a step load disturbance from 25% to 100% rated load at 750 r/min. From the experimental results, note that the harmonic ripple in the position estimation error can be reduced obviously during the whole operation with step load disturbance, and the proposed ADALINE-based filter trained with LMS and RLS algorithms is effective for sensorless control. Compared with LMS algorithm, the RLS algorithm offers a better dynamic response with low ripples under load disturbance.

In order to highlight the position ripple error suppressing effectiveness with the ADALINE-based filter, Fig. 6.20 gives the peak-to-peak value measurement comparison of position estimation error ripple at different operating conditions. From the experimental results, it can be noted that the peak-to-peak value of position estimation error ripple obtained by the conventional SMO can reach up to 8.3°, and the one obtained by the ADALINE-based filter trained with LMS and RLS algorithms can be reduced to 2.2°. Therefore, it can be further concluded that the ADALINE-based filter trained with LMS and RLS algorithms can suppress the position estimation error ripple effectively and offer low sensitivity to harmonic amplitude variations with different operating conditions.

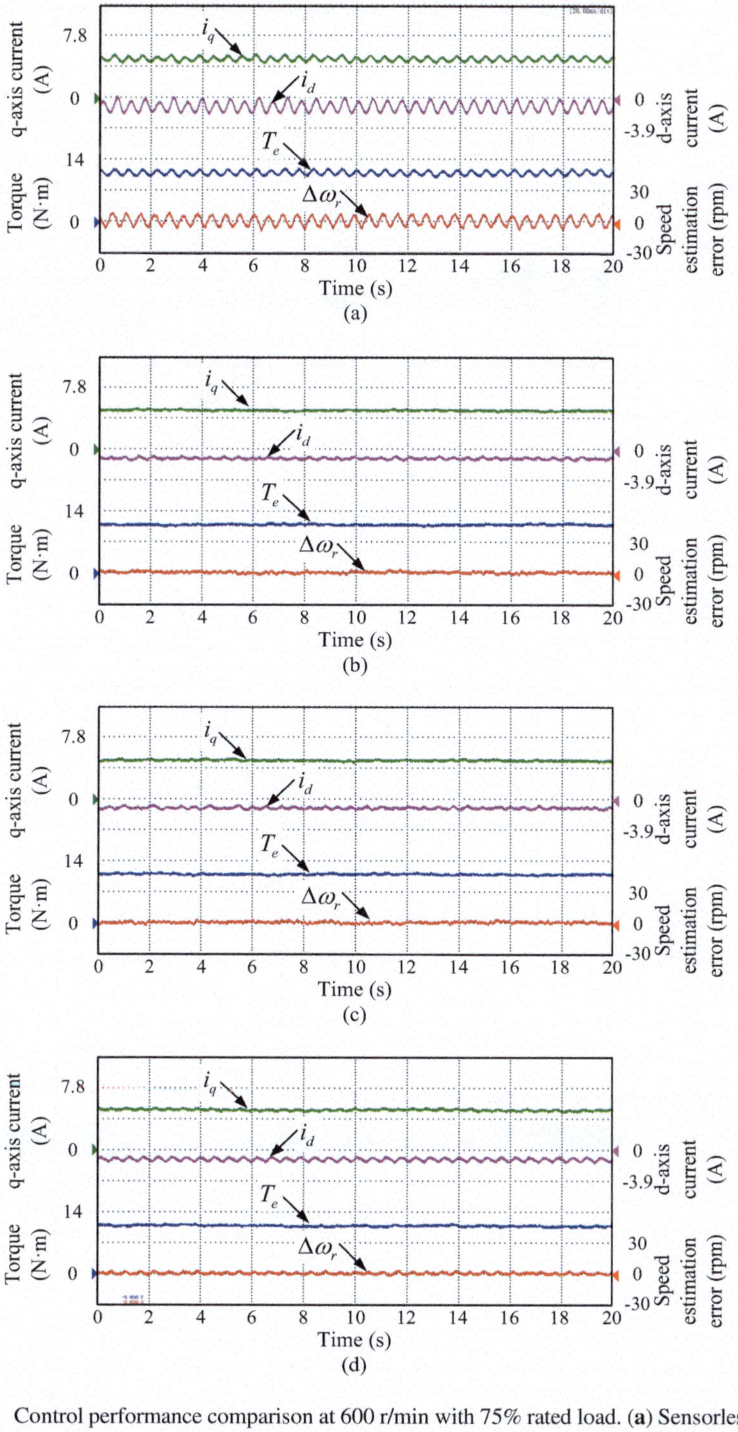

Fig. 6.17 Control performance comparison at 600 r/min with 75% rated load. (**a**) Sensorless-based results without ADALINE-based filter. (**b**) Sensorless-based results using ADALINE-based filter trained with LMS algorithm. (**c**) Sensorless-based results using ADALINE-based filter trained with RLS algorithm. (**d**) Sensor-based results

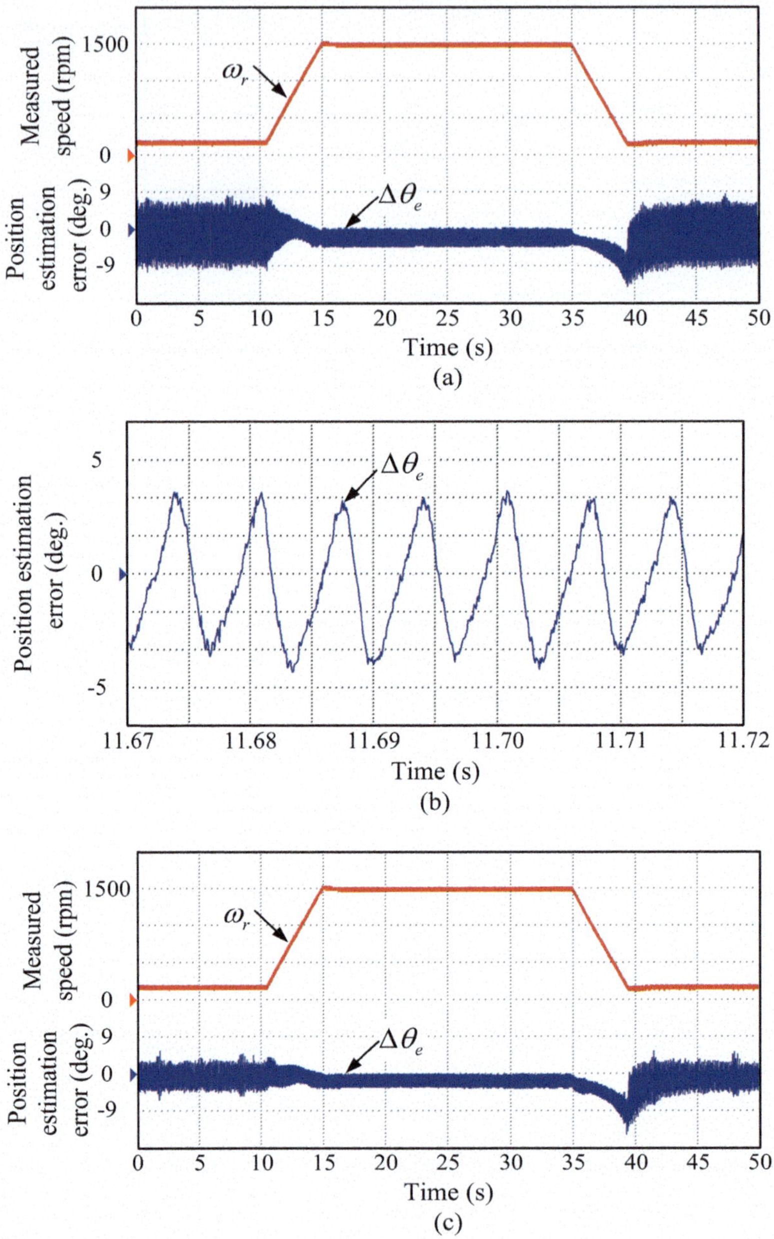

Fig. 6.18 Experimental comparison of speed and position estimation errors during continuous speed variation with 25% rated load. (**a**) Without ADALINE-based filter. (**b**) Zoomed-in position estimation error of (**a**). (**c**) Using ADALINE-based filter trained with LMS algorithm. (**d**) Zoomed-in position estimation error of (**c**). (**e**) Using ADALINE-based filter trained with RLS algorithm. (**f**) Zoomed-in position estimation error of (**e**)

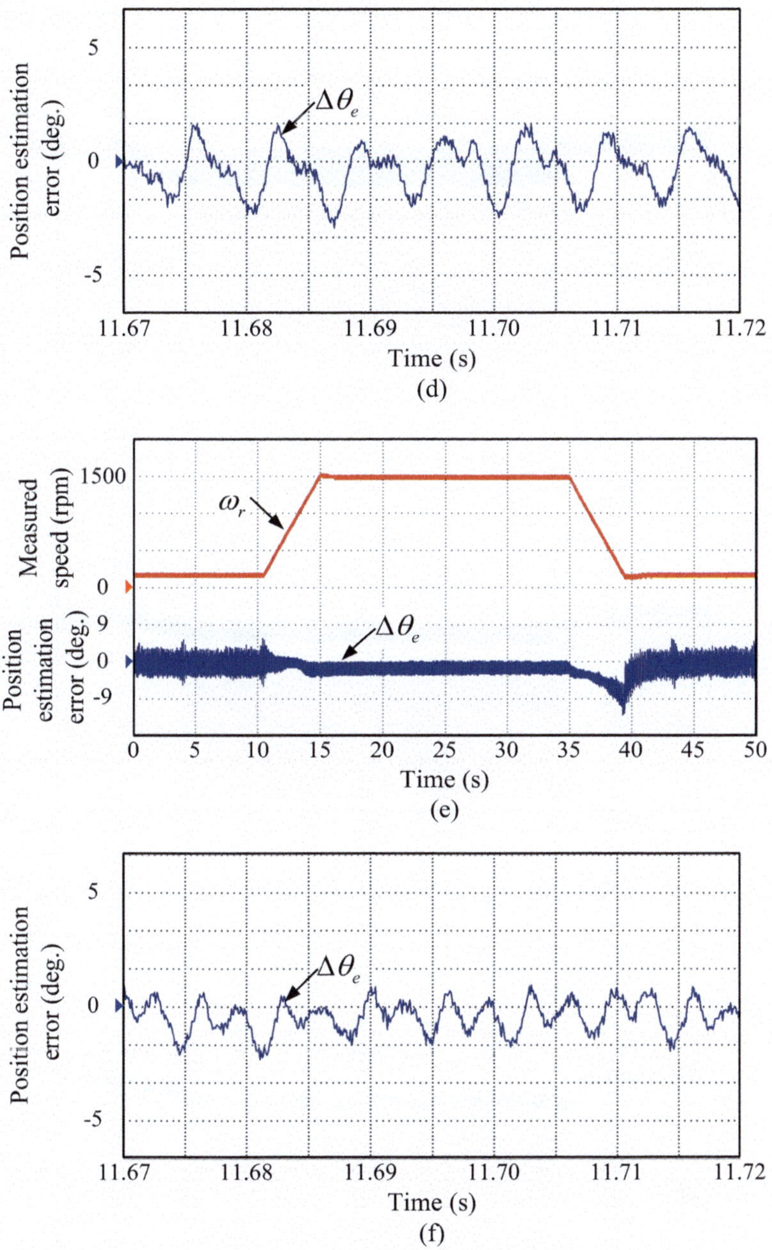

Fig. 6.18 (continued)

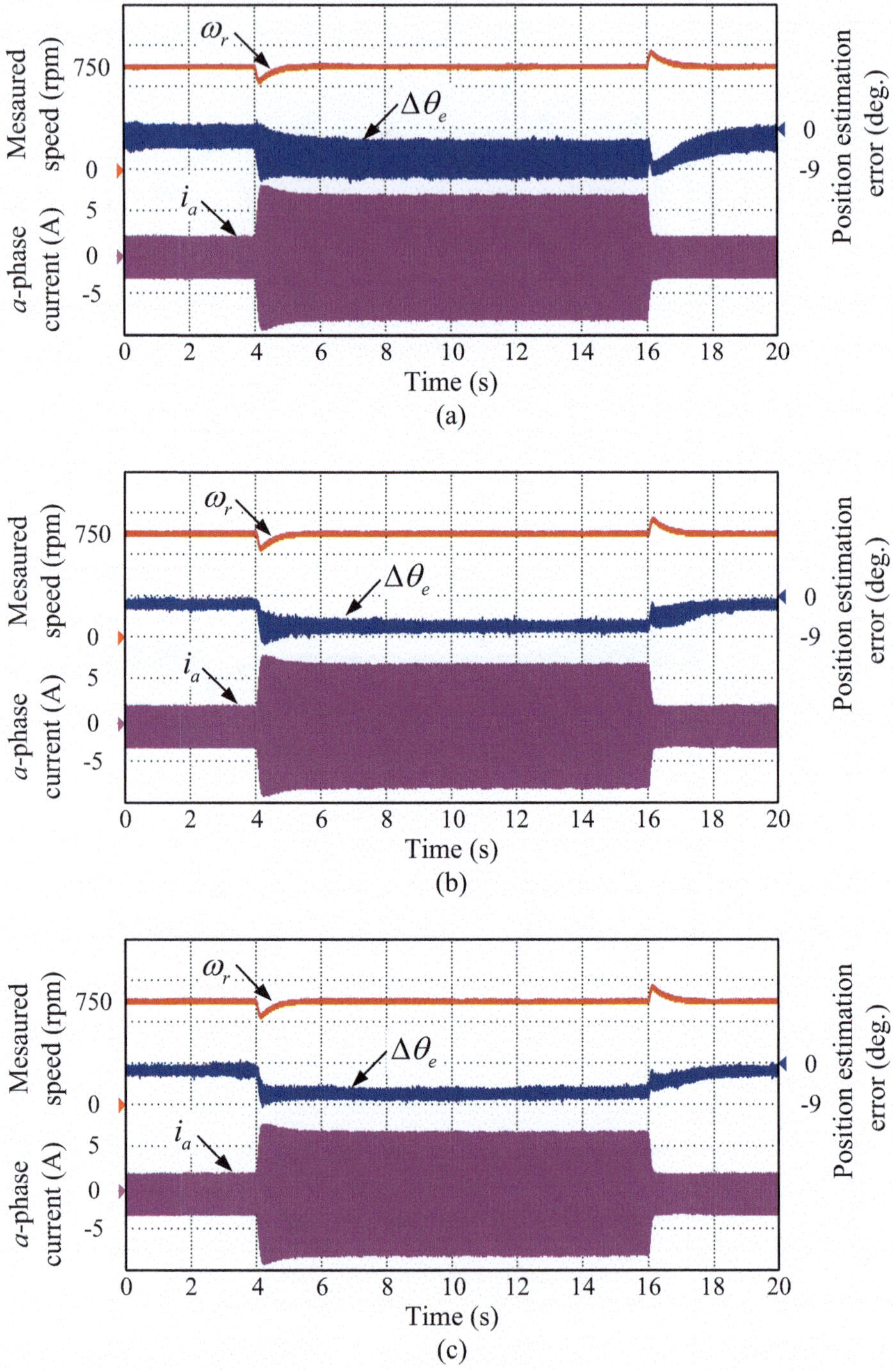

Fig. 6.19 Experimental comparison of speed, position estimation error, and a-phase current with load step disturbance at 750 r/min. (**a**) Without ADALINE-based filter. (**b**) Using ADALINE-based filter trained with LMS algorithm. (**c**) Using ADALINE-based filter trained with RLS algorithm

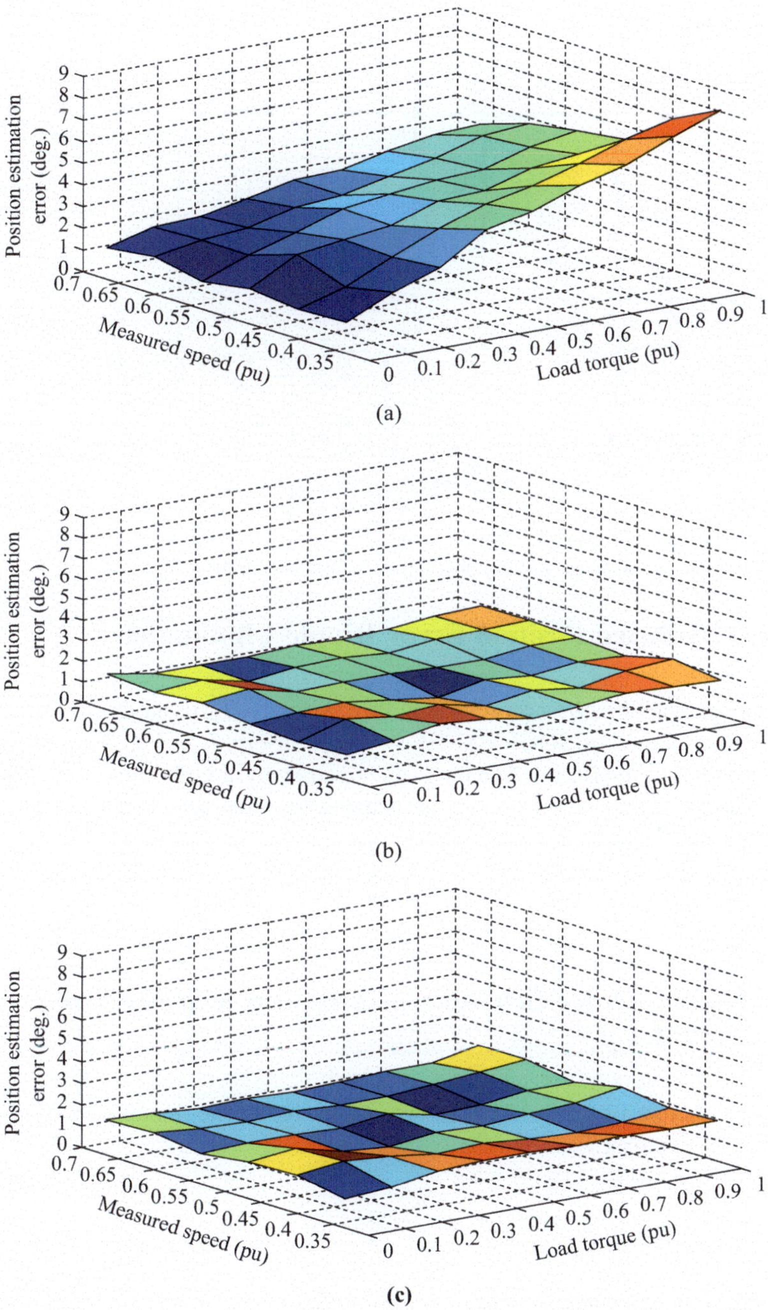

Fig. 6.20 Peak-to-peak value measurement comparison of the position ripple error. (**a**) Without ADALINE-based filter. (**b**) Using ADALINE-based filter trained with LMS algorithm. (**c**) Using ADALINE-based filter trained with RLS algorithm

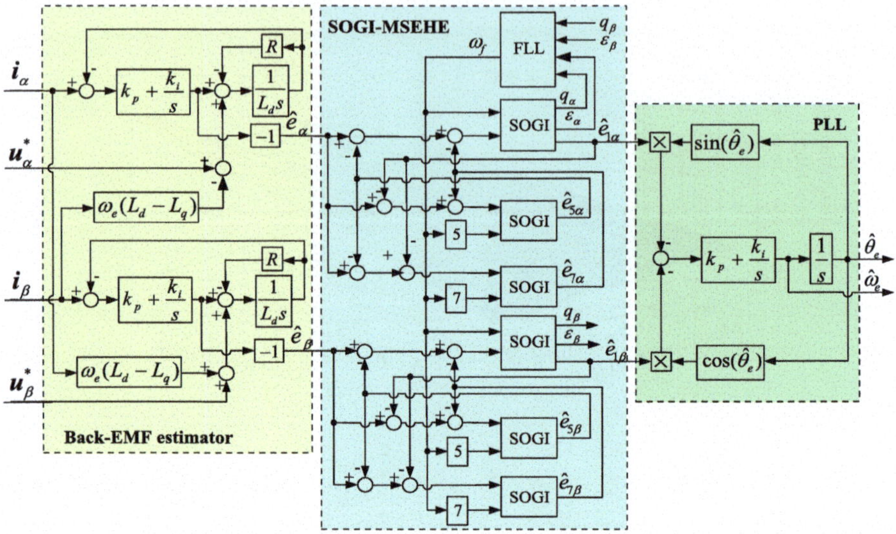

Fig. 6.21 Proposed enhanced position observer with SOGI-MSEHE

6.4 SOGI-Based Position Error Ripple Elimination

The proposed position observer with a multiple selective EMF harmonic elimination adaptive filter named as SOGI-MSEHE is shown in Fig. 6.21 [3]. After estimating the extended back-EMF, the SOGIs work together by a CFN which is frequency adaptive due to the use of FLL. It can achieve multiple selective EMF harmonic elimination before the PLL-based position calculation. As a result, the corresponding position estimation harmonic error can be eliminated.

6.4.1 Second-Order Generalized Integrator with Frequency Lock Loop

The basic structure of the SOGI is shown in Fig. 6.22. The transfer function of the SOGI can be expressed as follows:

$$SOGI(s) = \frac{u_o}{k\varepsilon}(s) = \frac{\omega_f s}{s^2 + {\omega_f}^2}, \tag{6.38}$$

where ω_f is the resonance frequency of the SOGI.

The two output signals of the SOGI in Fig. 6.22, i.e., v_o and q, are defined by the following transfer functions:

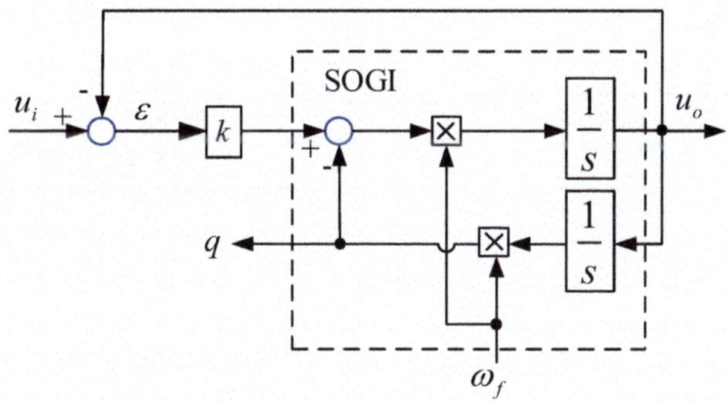

Fig. 6.22 Basic structure of the SOGI

$$D(s) = \frac{u_o}{v_i}(s) = \frac{k\omega_f s}{s^2 + k\omega_f s + \omega_f{}^2}, \tag{6.39}$$

$$Q(s) = \frac{q}{v_i}(s) = \frac{k\omega_f{}^2}{s^2 + k\omega_f s + \omega_f{}^2}. \tag{6.40}$$

$D(s)$ can be regarded as a BPF whose bandwidth is exclusively set by the gain k and is independent of the resonance frequency ω_f. The center frequency of the bandwidth is the resonance frequency of SOGI. The smaller gain k corresponds to the smaller bandwidth and the better selectivity in frequency but a longer settling time. $Q(s)$ can be regarded as a LPF, of which output signal q is always 90° lagged from the v_o output.

From (6.39) and (6.40), it is clear that the frequency of the output signal equals ω_f. Only when the frequency of the input signal is equal to ω_f, the two outputs have the same amplitude that is equal to that of the input signal. So when the input of the SOGI is the extended EMF estimate, in order to extract the fundamental component of the extended EMF, a FLL is adopted to adapt ω_f. The diagram of FLL adopting the α-β axis EMF estimates to tune the resonance frequency of SOGI is shown in Fig. 6.23. The frequency adaptation of the FLL can be simplified as a first-order frequency adaptation loop whose transfer function is given by

$$\frac{\overline{\omega}_f}{\omega} = \frac{\Gamma}{s + \Gamma}, \tag{6.41}$$

where ω is the fundamental frequency of input signal, $\overline{\omega}_f$ is the average value of output frequency, and Γ is the gain of the FLL.

Using the frequency adaptation of the FLL, it can be deduced that for a given sinusoidal input signal $u = U\sin(\omega t + \phi)$, the output of the SOGI can be given as follows:

$$\bar{y} = \begin{bmatrix} v_o \\ q \end{bmatrix} = U\begin{bmatrix} \sin(\omega t + \phi) \\ -\cos(\omega t + \phi) \end{bmatrix}. \tag{6.42}$$

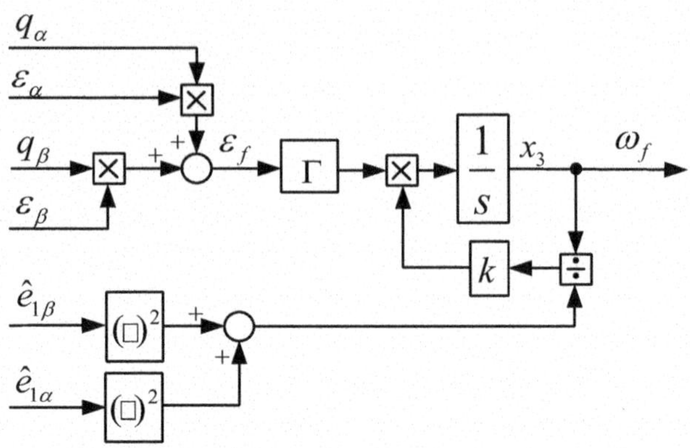

Fig. 6.23 Diagram of FLL adopting the α-β axis EMF estimates to tune the resonance frequency of SOGI

Therefore, by using the FLL, the SOGI is able to track the extended back-EMF with variable frequency.

6.4.2 *Multiple Selective EMF Harmonic Elimination Using SOGIs with a CFN*

Due to the BPF characteristic of $D(s)$, even in the presence of harmonics in the estimated EMF, it can reject the high-order harmonics. However, low-order harmonics, such as the fifth or the seventh, are very close to the fundamental frequency, and they will still lead to some distortion on the EMF detected by the single SOGI shown in Fig. 6.21.

From the previous analysis, the gain k of SOGI cannot be too small in order to obtain a comparatively small phase delay between the estimated and actual values of the EMF. However, a relative larger gain k will cause the low-order harmonic components difficult to be filtered out by the single SOGI. In order to ensure the dynamic performance, a CFN consisting of multiple SOGIs is proposed as shown in Fig. 6.21. The scheme called the SOGI-MSEHE is presented to eliminate the multiple low-order harmonics. The cross-feedback network is an effective solution to accurately extract the fundamental components of the back-EMF estimates, even under very extreme distortion condition by working in a collaborative way and tuning at different operating frequencies. The SOGI-MSEHE can achieve elimination of selective multiple EMF harmonics, only if the resonance frequency is equal to the corresponding EMF harmonic frequency. Because the fifth and seventh harmonics take the dominant roles in the total harmonics of the back-EMF estimate, the SOGI-MSEHE contains three SOGIs for each phase to extract the fifth and seventh harmonic components and a FLL to realize frequency adaptation.

From Fig. 6.21, it can be obtained that the relationships between the output signals $\hat{e}_{1\alpha}$, $\hat{e}_{5\alpha}$, $\hat{e}_{7\alpha}$, $\hat{e}_{1\beta}$, $\hat{e}_{5\beta}$, $\hat{e}_{7\beta}$ and the input signals $\hat{e}_\alpha$, $\hat{e}_\beta$ can be expressed as follows:

$$\hat{e}_{1m} = D_1(s)(\hat{e}_m - \hat{e}_{5m} - \hat{e}_{7m}), \tag{6.43}$$

$$\hat{e}_{5m} = D_5(s)(\hat{e}_m - \hat{e}_{1m} - \hat{e}_{7m}), \tag{6.44}$$

$$\hat{e}_{7m} = D_7(s)(\hat{e}_m - \hat{e}_{1m} - \hat{e}_{5m}), \tag{6.45}$$

where the subscript m represents α- or β-axis, the version of $D_1(s)$, $D_5(s)$, and $D_7(s)$ is the same as that of the transfer function shown in (6.39) in which the resonance frequencies are ω_f, $5\omega_f$, and $7\omega_f$, respectively. So, these transfer functions can be represented by a general expression $D_n(s)$ as follows:

$$D_n(s) = \frac{u_o}{u_i}(s) = \frac{k(n \cdot \omega_f)s}{s^2 + k(n \cdot \omega_f)s + (n \cdot \omega_f)^2}, \tag{6.46}$$

where n represents 1, 5, or 7.

According to (6.43)–(6.46), the transfer functions from $\hat{e}_\alpha$ to $\hat{e}_{1\alpha}$ and from $\hat{e}_\beta$ to $\hat{e}_{1\beta}$ can be obtained:

$$F(s) = D_1 \frac{1 - D_5 - D_7 + D_5 D_7}{1 - D_5 D_7 - D_1 D_5 - D_1 D_7 + 2 D_1 D_5 D_7} = \frac{\hat{e}_{1\alpha}}{\hat{e}_\alpha} = \frac{\hat{e}_{1\beta}}{\hat{e}_\beta}. \tag{6.47}$$

Taking the operating frequency of the IPMSM $\omega_e = 100$ rad/s as an example, the Bode diagram of $F(s)$ is shown as Fig. 6.24. The dash curve in Fig. 6.24 represents the frequency response using the single SOGI (as shown in Fig. 6.22). And the solid curve represents the frequency response using the SOGI-MSEHE (as shown in

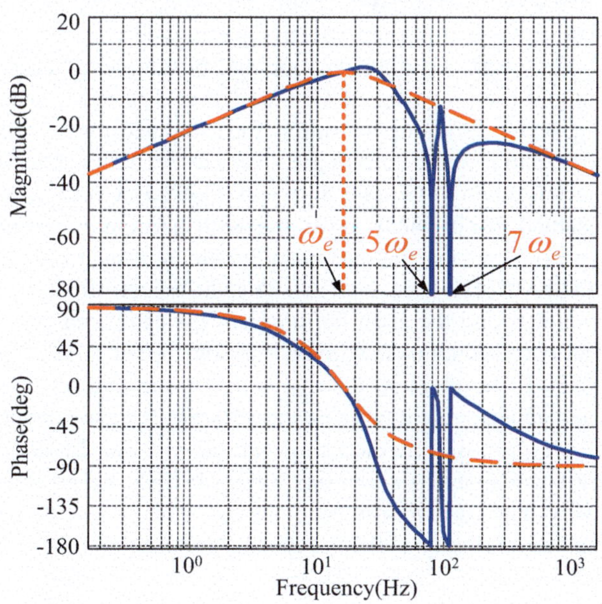

Fig. 6.24 Bode diagram of the EMF estimation characteristic F(s)

Fig. 6.21). It should be noted that the latter scheme can eliminate the fifth and seventh harmonics effectively. Therefore, the obtained back-EMF estimates from the SOGI-MSEHE are clean sinusoidal signal, assuring the accuracy enhancement of the rotor position estimate for sensorless IPMSM drive.

6.4.3 Experimental Results on SOGI-MSEHE Method

For demonstrating the effectiveness of the proposed method, experiments are conducted at a 2.2 kW IPMSM sensorless vector control platform. The whole sensorless control algorithm of the proposed method is realized by a 32-bit DSP-TMS320F2808. The switching frequency of the PWM in the inverter is set to 10 kHz. The gain of the SOGI extracting the fundamental frequency is $k = \sqrt{2}$, and the gains of the corresponding extracting the fifth and seventh harmonics are $k/5 = \sqrt{2}/5, k/7 = \sqrt{2}/7$, respectively. The gain of the FLL is $\Gamma = 50$. An absolute encoder (ECN1113) is adopted to acquire the actual motor position which is for comparison rather than control. All of the following experimental waveforms are sampled under the IPMSM operating under sensorless control.

The waveforms of the extended EMF estimates obtained with and without the SOGI-MSEHE at 500 r/min and 75% rated load are shown in Fig. 6.25. From top to bottom, the estimated EMF in the α-β axes and the corresponding FFT analysis are given, respectively. The experimental result without the SOGI- MSEHE is shown in Fig. 6.25a, and the result with the SOGI-MSEHE is shown in Fig. 6.25b for comparison. Based on the experimental waveforms, it can be seen that the estimated EMF waveforms become more sinusoidal, and the fifth and seventh harmonics in the back-EMF estimate are effectively removed with the SOGI-MSEHE.

The estimated rotor position and position estimation error waveforms with and without the proposed SOGI-MSEHE at 500 r/min and rated load are displayed in Fig. 6.26a. From top to bottom, the estimated rotor position, the position estimation error, and a-phase current are given, respectively. The SOGI-MSEHE is disabled in the left side of the dash line and enabled at 0.25 s. When the SOGI-MSEHE is disabled, it obviously exists the sixth harmonic component in the estimated rotor position, causing the maximum harmonic error to reach 0.06π. After the proposed SOGI-MSEHE is enabled at 0.25 s, the waveform of the estimated position becomes smoother and the sixth harmonic in the estimated position is removed effectively. The corresponding maximum harmonic error is reduced to 0.01π with the SOGI-MSEHE. In addition, the fast convergence rate of the proposed SOGI-MSEHE is validated because it takes no more than 0.015 s to converge. In order to see the effect of the proposed method more intuitively, the Lissajous waveforms of the stator current with and without the SOGI-MSEHE are given in Fig. 6.26b, and c, respectively. The current orbit becomes closer to a round by using the SOGI-MSEHE. So, the torque ripples will be reduced effectively.

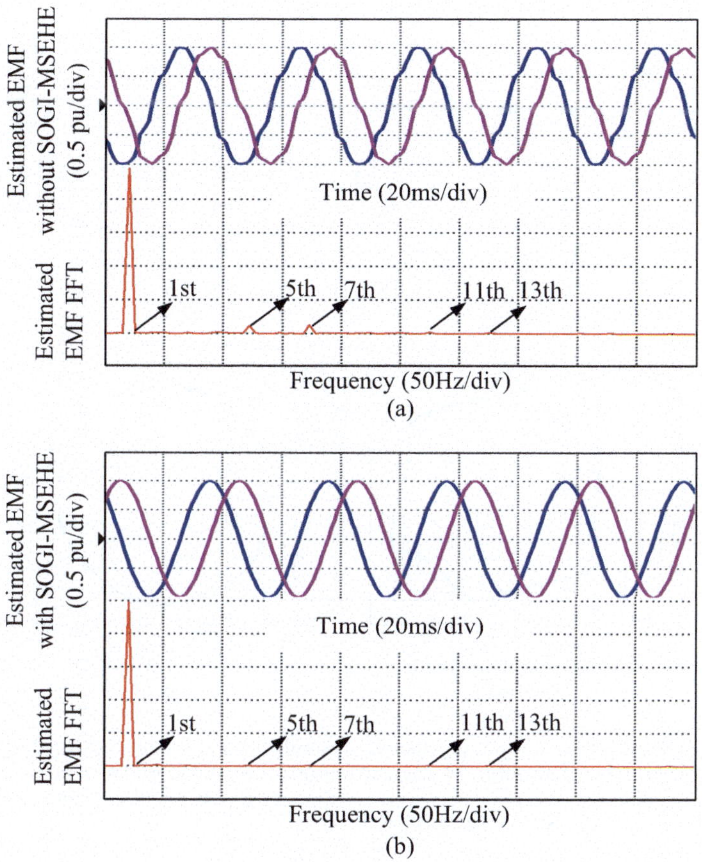

Fig. 6.25 Experimental waveforms of estimated back-EMF at 500 r/min and 75% rated load. (**a**) Without the SOGI-MSEHE. (**b**) With the SOGI-MSEHE

The waveforms of the α-axis extended back-EMF estimate with and without SOGI-MSEHE and the fifth and seventh harmonic components extracted by the proposed method with SOGI-MSEHE at 600 r/min and 50% rated load are displayed in Fig. 6.27. From top to bottom, the extended back-EMF estimate $\hat{e}'_{1\alpha}$ outputted directly from the EMF estimator, the extracted fifth and seventh harmonic components, and the extended back-EMF estimate $\hat{e}_{1\alpha}$ from the SOGI-MSEHE are given, respectively.

The experimental results at the speed acceleration and deceleration with and without the SOGI-MSEHE are shown in Fig. 6.28, where the rotor speed estimate, the speed estimate error, and the position estimate error waveforms are given from top to bottom, respectively. Figure 6.28a displays the result without the SOGI-MSEHE, and Fig. 6.28b displays the result with the SOGI-MSEHE. The IPMSM

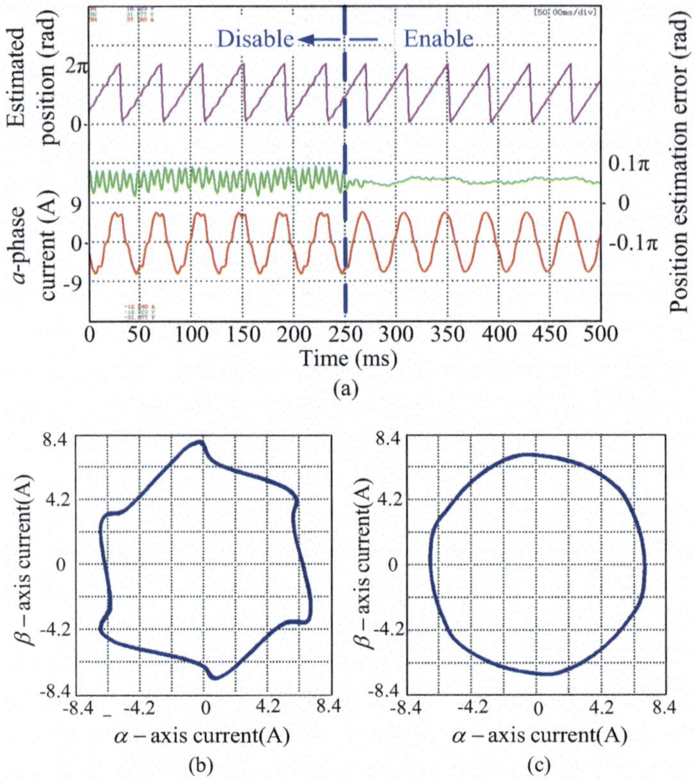

Fig. 6.26 Experimental waveforms of sensorless control with and without SOGI-MSEHE at 500 r/min and rated load. (**a**) Waveforms of estimated position, position estimation error, and a-phase current. (**b**) Lissajous waveform of stator current without the SOGI-MSEHE. (**c**) Lissajous waveform of stator current with the SOGI-MSEHE

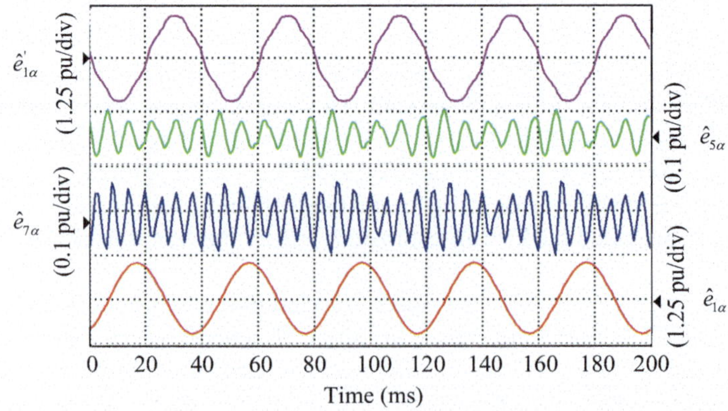

Fig. 6.27 Experimental waveforms of α-axis EMF estimate inputted into the SOGI-MSEHE, the extracted fifth and seventh harmonics components, and the α-axis EMF estimate outputted from the SOGI-MSEHE

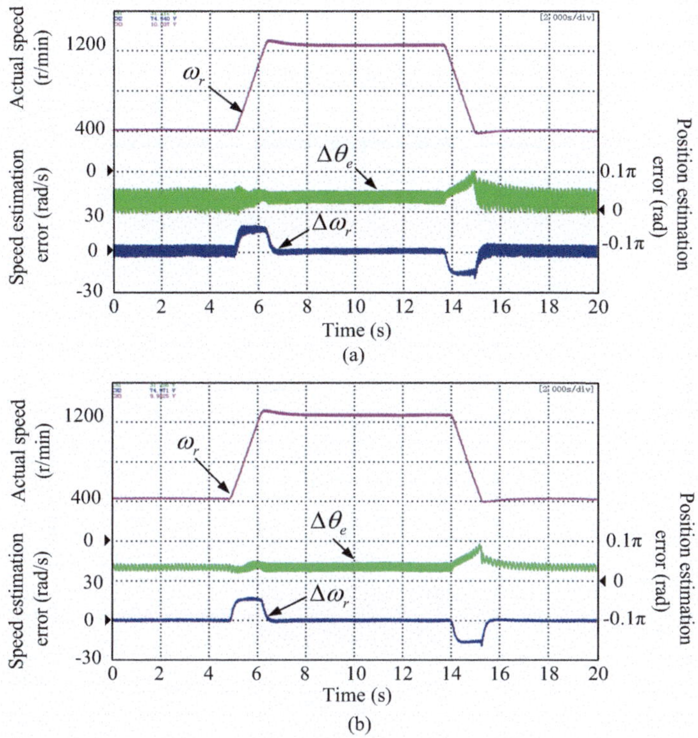

Fig. 6.28 Experimental results of speed acceleration and deceleration with 50% rated load. (**a**) Without the SOGI-MSEHE. (**b**) With the SOGI-MSEHE

accelerates from 400 to 1200 r/min and then decelerates to 400 r/min. From the waveforms, it can be seen that both the speed and position estimation errors are reduced. When the SOGI-MSEHE is disabled, the maximum harmonic errors of the speed and position estimation are 10 r/min and 0.06π, respectively. When the SOGI-MSEHE is enabled, the corresponding maximum harmonic errors are reduced to 2 r/min and 0.02π, respectively.

The experimental results of operating at 600 r/min with step load disturbance from 25% to 100% rated load with and without the SOGI-MSEHE are displayed in Fig. 6.29 The rotor speed, the position estimation error, the speed estimation error, and a-phase current are displayed from top to bottom, respectively. It can be seen that when the SOGI-MSEHE is enabled, the harmonic error of the rotor position estimation is reduced from 0.06 to 0.02π, and the harmonic error of the speed estimation is reduced from 9 to 2 r/min, which validates the effectiveness of the proposed method with the SOGI-MSEHE for sensorless control.

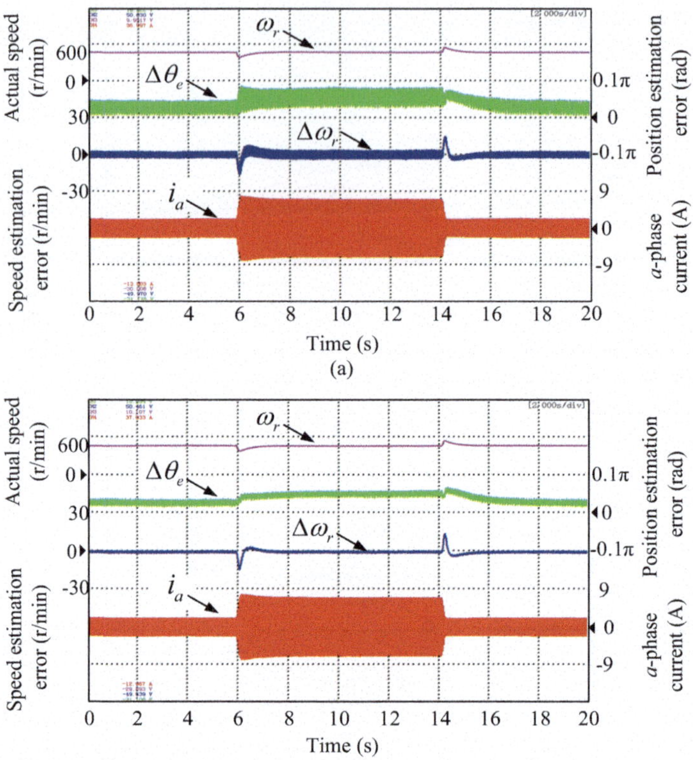

Fig. 6.29 Experimental results of step load disturbance from 25% to 100% rated torque at 600 r/min. (**a**) Without the SOGI-MSEHE. (**b**) With the SOGI-MSEHE

6.5 AVF-Based Position Error Ripple Elimination

6.5.1 Active Flux-Based Back-EMF Observer

The *d*-*q* axis model for IPMSM is presented as follows (3.36). Owing to the active flux or the torque-producing flux concept, salient-pole AC machines can be turned to nonsalient-pole ones. The active flux ψ_{af} is defined as the flux which multiplies the current i_q in the *d*-*q* axis model torque expression (1.56) with

$$\psi_{af} = \psi_f + \left(L_d - L_q\right) i_d. \tag{6.48}$$

Note that the active flux is comprised of the conventional permanent magnet flux linkage and the one due to saliency, and it has the real rotor *d*-axis orientation, which can facilitate the position and speed estimation. The α-β axis model for IPMSM can be expressed as (5.15). e_α and e_β are the α-β axis equivalent back-EMF components presented as

$$\begin{bmatrix} e_\alpha \\ e_\beta \end{bmatrix} = \psi_{af}\omega_e \begin{bmatrix} -\sin\theta_e \\ \cos\theta_e \end{bmatrix}. \tag{6.49}$$

A dynamic equation of the equivalent back-EMF is given as (6.50). Since the mechanical variables vary slowly compared with the electrical variables [13], i.e., $p\omega_e = 0$, (6.50) can be simplified to (6.51) under this assumption.

$$p\begin{bmatrix} e_\alpha \\ e_\beta \end{bmatrix} = \omega_e \begin{bmatrix} -e_\beta \\ e_\alpha \end{bmatrix} + p\omega_e \begin{bmatrix} e_\alpha \\ e_\beta \end{bmatrix}, \tag{6.50}$$

$$p\begin{bmatrix} e_\alpha \\ e_\beta \end{bmatrix} = \omega_e \begin{bmatrix} -e_\beta \\ e_\alpha \end{bmatrix}. \tag{6.51}$$

Therefore, on the basis of the α-β axis machine model (5.15) and (6.51), the full-order Luenberger observer can be used to estimate the equivalent back-EMF as follows:

$$p\begin{bmatrix} \hat{i}_\alpha \\ \hat{i}_\beta \end{bmatrix} = -\frac{1}{L_q}\begin{bmatrix} R & 0 \\ 0 & R \end{bmatrix}\begin{bmatrix} \hat{i}_\alpha \\ \hat{i}_\beta \end{bmatrix} + \frac{1}{L_q}\begin{bmatrix} u_\alpha - \hat{e}_\alpha \\ u_\beta - \hat{e}_\beta \end{bmatrix} + K_1\begin{bmatrix} i_\alpha - \hat{i}_\alpha \\ i_\beta - \hat{i}_\beta \end{bmatrix}, \tag{6.52}$$

$$p\begin{bmatrix} \hat{e}_\alpha \\ \hat{e}_\beta \end{bmatrix} = \hat{\omega}_e \begin{bmatrix} -\hat{e}_\beta \\ \hat{e}_\alpha \end{bmatrix} + K_2\begin{bmatrix} i_\alpha - \hat{i}_\alpha \\ i_\beta - \hat{i}_\beta \end{bmatrix}, \tag{6.53}$$

where $\hat{i}_\alpha$, $\hat{i}_\beta$ and $\hat{e}_\alpha$, $\hat{e}_\beta$ are the α-β axis current and back-EMF estimates and $\hat{\omega}_e$ is the electrical speed estimate. K_1 and K_2 are observer gains.

Then the transfer function between the estimated and the actual back-EMFs can be obtained in the complex variable form as follows:

$$\frac{\hat{\mathbf{E}}_{\alpha\beta}}{\mathbf{E}_{\alpha\beta}} = \frac{-K_2}{(s - j\hat{\omega}_e)(sL_q + R_s + K_1L_q) - K_2}. \tag{6.54}$$

As can be seen, the full-order Luenberger back-EMF observer has the characteristics of second-order LPF.

To obtain the rotor position, the normalized PLL presented in [8] is adopted as a substitute for the conventional arc-tangent calculation, thus enhancing the position estimation performance. The block diagram of the full-order Luenberger back-EMF observer with normalized PLL is shown in Fig. 6.30. In practical test setup, the normalized PLL is tuned to achieve a bandwidth equal to 150 Hz with the gains k_p and k_i set to 1885 and 888,264, respectively. The frequency response of the normalized PLL is shown in Fig. 6.31. For comparison, the frequency response of the mechanical model-based PLL is also given with the same PLL gains. As can be seen, with normalization, the bandwidth of the PLL system can keep constant with the variations of back-EMF amplitudes at different speeds, which makes the PLL suitable for whole speed range.

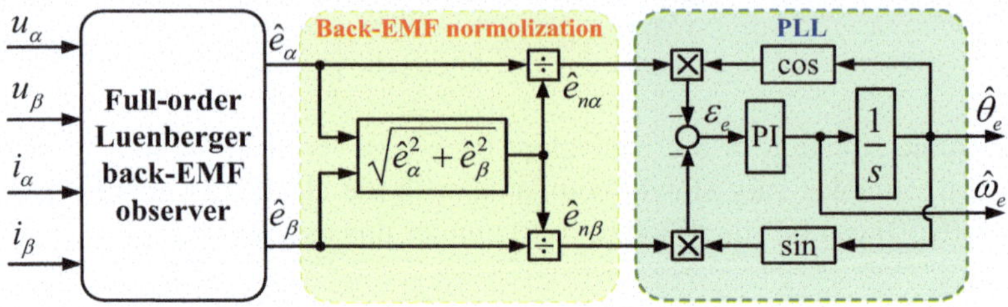

Fig. 6.30 Full-order Luenberger back-EMF observer with normalized PLL

6.5.2 *Proposed Multiple-AVF CFN for Position Error Harmonic Fluctuation Elimination*

To eliminate the position error harmonic fluctuations, and hence improve the position estimation accuracy for sensorless IPMSM drives, a multiple-AVF CFN strategy is proposed [7]. In the proposed approach, the multiple AVFs can work together inside a CFN, achieving back-EMF harmonic decoupling and compensating.

1. AVF

AVF has been a good candidate for harmonic detection and provides promising characteristics of positive and negative frequency selectivity. The block diagrams of AVF in both scalar and vector forms are shown in Fig. 6.32.

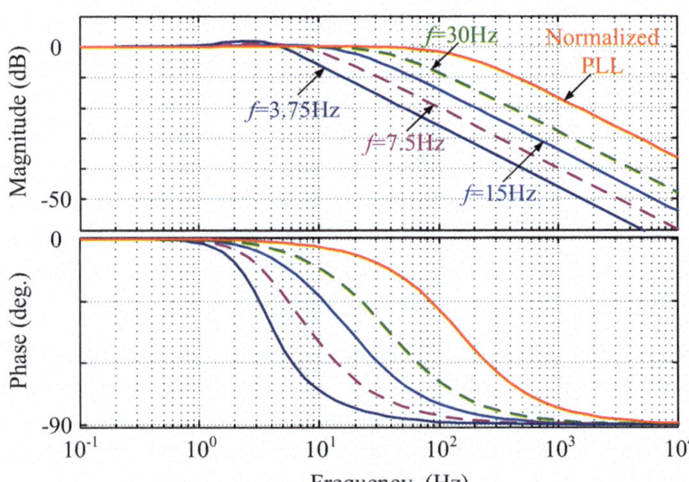

Fig. 6.31 Frequency response comparisons of the normalized PLL and the mechanical model-based one

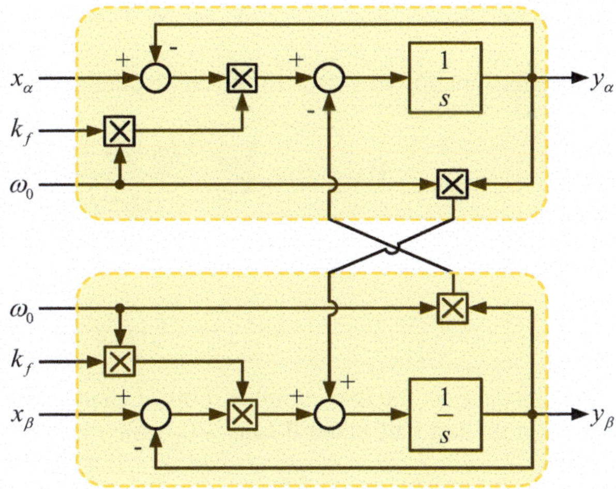

(a)

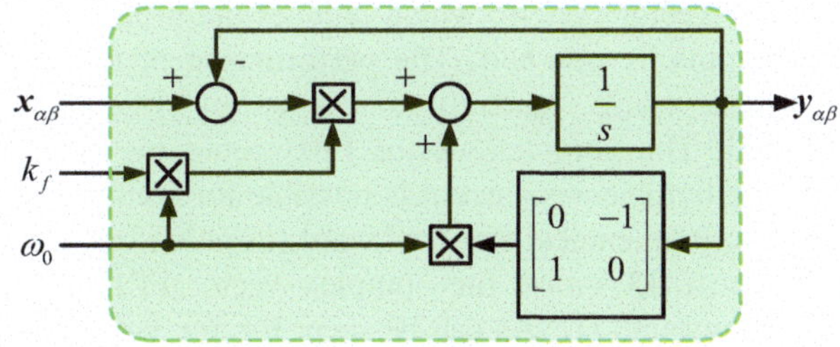

(b)

Fig. 6.32 Block diagram of AVF. (**a**) Scalar form. (**b**) Vector form

From Fig. 6.32, the transfer function in scalar form of AVF is depicted by

$$\begin{bmatrix} y_\alpha \\ y_\beta \end{bmatrix} = \frac{k_f \omega_0}{\left(s + k_f \omega_0\right)^2 + \omega_0^2} \begin{bmatrix} s + k_f \omega_0 & -\omega_0 \\ -\omega_0 & s + k_f \omega_0 \end{bmatrix} \begin{bmatrix} y_\alpha \\ y_\beta \end{bmatrix}, \tag{6.55}$$

where $\mathbf{x}_{\alpha\beta} = \left[x_\alpha, x_\beta\right]^{\mathrm{T}}$ and $\mathbf{y}_{\alpha\beta} = \left[y_\alpha, y_\beta\right]^{\mathrm{T}}$ denote the input vector and the desired output vector, ω_0 is the resonance frequency, and k_f is the gain of AVF.

Then, through the complex vector concept, introducing

$$\mathbf{X}_{\alpha\beta} = x_\alpha + j x_\beta, \tag{6.56}$$

$$\mathbf{Y}_{\alpha\beta} = y_\alpha + jy_\beta, \tag{6.57}$$

the transfer function in vector form can be obtained as

$$\mathbf{H}_{AVF}(s) = \frac{\mathbf{Y}_{\alpha\beta}}{\mathbf{X}_{\alpha\beta}} = \frac{k_f\omega_0(s + k_f\omega_0 + j\omega_0)}{(s + k_f\omega_0)^2 + \omega_0^2} \tag{6.58}$$

with the characteristic equation roots

$$s_{1,\ 2} = -k_f\omega_0 \pm j\omega_0. \tag{6.59}$$

As can be seen, the two complex roots of characteristic equation could always be located in the left half plane if $k_f\omega_0 > 0$, which guarantees the stability of the AVF system.

The use of complex vector concept simplifies the model of AVF from a multiple-input and multiple-output system to an equivalent single-input and single-output complex vector system, which reduces the order and the number of system inputs and outputs by one-half. The performance of this complex vector model can be evaluated using generalized forms of frequency-response function (FRF) [14, 15]. The complex vector FRF could be exhibited with both positive and negative frequencies since it is possible for complex vectors to rotate both forward (positive frequencies) and backward (negative frequencies).

Figure 6.33 shows the complex vector FRF of AVF under different k_f with ω_0 setting to 30 Hz. As can be seen, for any value of k_f, AVF has a unity gain and a zero phase shift at the frequency ω_0; and for any other frequency, the input signal can be attenuated. The characteristics guarantee the extraction of the desired frequency content. Besides, it can be noted that the filter is more selective with smaller k_f, but the bandwidth becomes narrower and vice versa. This conclusion could also be made from the characteristic equation of AVF. In practical applications, a tradeoff between harmonic selectivity and convergence speed should be taken into consideration. Moreover, it can be noticed that AVF has the unique feature of both frequency and polarity-selective property, which facilitates the extraction of fundamental contents in back-EMF with distorted harmonics for sensorless PMSM drives.

6.5.3 Multiple-AVF CFN

As has been noted, AVF has unity gain and zero phase shift only for the tuned frequency. For other frequencies, the input signal could be attenuated. It can be observed from Fig. 6.32 that attenuation for other frequency contents depends on the selected gain k_f. Accordingly, with regard to the estimated back-EMF with several frequency contents for IPMSM drives, the value of k_f should be decreased to offer

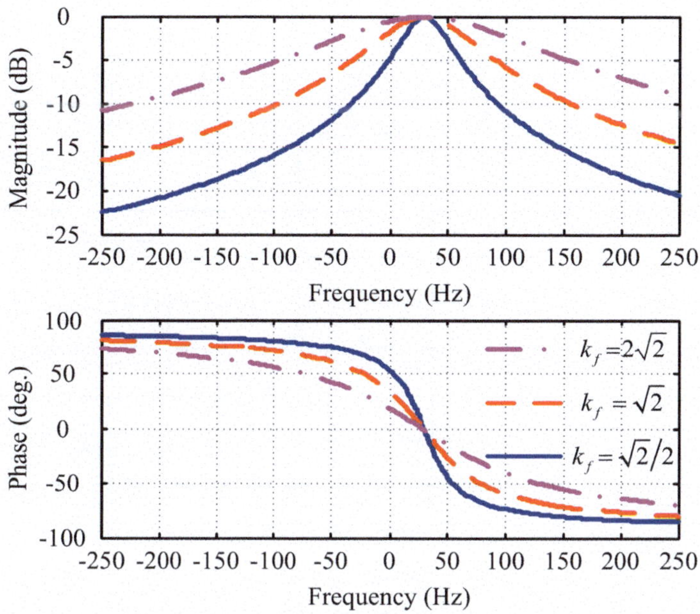

Fig. 6.33 Complex vector FRF of AVF with different k_f

desired attenuation. As a counterpart, lower values of k_f imply longer settling time, which might lead to substantial distortion on the fundamental content and hence degrade the performance of the system. Therefore, the CFN is utilized, in which the multiple-AVFs tuned for separate frequency contents are arranged in parallel working in a collaborative way. This CFN can be understood as a set of selective and adaptive filters and mainly consists of subtracting the undesired frequency contents at the input of each AVF. Only taking the fifth and seventh harmonic contents with bigger amplitudes into account, the proposed multiple-AVF CFN shown in Fig. 6.34 is used to extract the fundamental content in the estimated back-EMF.

As shown in Fig. 6.34, the CFN consists of three AVFs, where $\boldsymbol{H}_n(s)$ is a customized version of the transfer function shown in (6.58), where the center frequency is given by $n\omega_0$, and n {1, −5, 7} is the fundamental and harmonic order for the AVF$_n$ block, i.e.,

$$\hat{\mathbf{e}}_{\alpha\beta n1} = \mathbf{H}_1(s)\left(\hat{\mathbf{e}}_{\alpha\beta n} - \hat{\mathbf{e}}_{\alpha\beta n-5} - \hat{\mathbf{e}}_{\alpha\beta n7}\right), \tag{6.60}$$

$$\hat{\mathbf{e}}_{\alpha\beta n-5} = \mathbf{H}_{-5}(s)\left(\hat{\mathbf{e}}_{\alpha\beta n} - \hat{\mathbf{e}}_{\alpha\beta n1} - \hat{\mathbf{e}}_{\alpha\beta n7}\right), \tag{6.61}$$

$$\hat{\mathbf{e}}_{\alpha\beta n7} = \mathbf{H}_7(s)\left(\hat{\mathbf{e}}_{\alpha\beta n} - \hat{\mathbf{e}}_{\alpha\beta n1} - \hat{\mathbf{e}}_{\alpha\beta n-5}\right), \tag{6.62}$$

where $\hat{\mathbf{e}}_{\alpha\beta n1}$, $\hat{\mathbf{e}}_{\alpha\beta n-5}$, and $\hat{\mathbf{e}}_{\alpha\beta n7}$ denote the fundamental, the fifth, and the seventh harmonic back-EMF estimates.

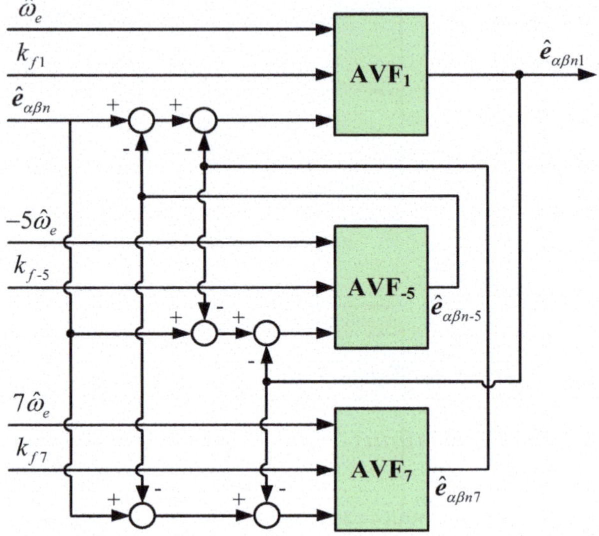

Fig. 6.34 Multiple-AVF CFN

Hence, the transfer function of the proposed multiple-AVF CFN can be given as

$$\mathbf{H}_{CFN}(s) = \mathbf{H}_1 \frac{1 - \mathbf{H}_{-5} - \mathbf{H}_7 + 2\mathbf{H}_{-5}\mathbf{H}_7}{1 - \mathbf{H}_{-5}\mathbf{H}_7 - \mathbf{H}_1\mathbf{H}_{-5} - \mathbf{H}_1\mathbf{H}_7 + 2\mathbf{H}_1\mathbf{H}_{-5}\mathbf{H}_7} \tag{6.63}$$

with

$$\mathbf{H}_n(s) = \frac{k_{fn}(n\omega_0)\left(s + k_{fn}(n\omega_0) + j(n\omega_0)\right)}{\left(s + k_{fn}(n\omega_0)\right)^2 + (n\omega_0)^2}, \tag{6.64}$$

where k_{fn} is the gain of AVF tuned for the selective nth back-EMF harmonic content.

From (6.63) and (6.64), it can be concluded that the CFN structure can be able to extract fundamental and multiple harmonic contents (the fifth and the seventh harmonics in back-EMF estimates) using multiple AVFs connected in parallel. The extracted signals are cross-feedback canceled from the inputs of the multiple AVFs to obtain fundamental and harmonic contents without harmonic distortion.

Through above analysis, it can be concluded that the input signal of each AVF is calculated by subtracting the output of all the rest of AVFs from the original input signal. In this way, after a transient process, the input signal of each AVF is cleaned up from the harmonic contents detected by the rest of AVFs, which will compensate the distortion at its output.

The complex vector FRF of the proposed multiple-AVF CFN is shown in Fig. 6.35 when IPMSM operates at 10 Hz. As can be seen, the multiple-AVF CFN has a unity gain and a zero phase shift at the frequency ω_0; and it has infinite gain at the fifth and seventh harmonics allowing for the removal of the undesired harmonic

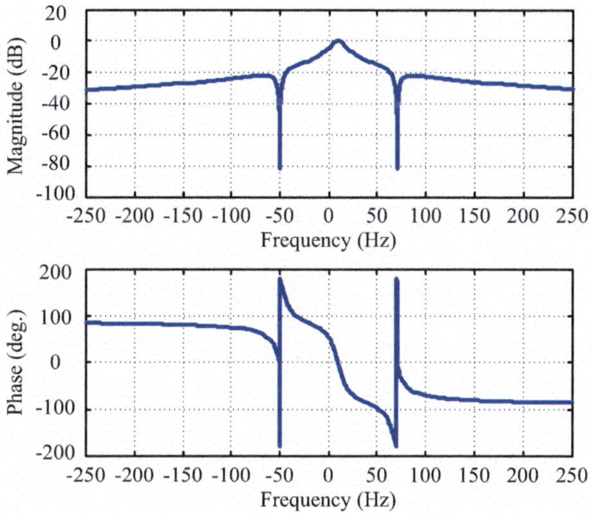

Fig. 6.35 Complex vector FRF of the proposed multiple-AVF CFN

contents. The frequency and polarity-selective characteristics guarantee the extraction of fundamental content in back-EMF and elimination of the undesired $\pm(6k\pm1)$th harmonic distortions for sensorless PMSM drives. Moreover, the computational burden can be reduced one-half for implementation due to the polarity-selective characteristic.

The proposed multiple-AVF CFN is plugged into the full-order Luenberger back-EMF observer with normalized PLL. The normalized estimated back-EMF serves as the input vector, the estimated speed serves as the resonance frequency, and the output back-EMF vector after filtering is utilized to obtain the rotor position and speed estimates through PLL.

6.5.4 Experimental Results on Multiple-AVF CFN Method

Figure 3.18 shows the experimental test setup with a 2.2 kW IPMSM test machine and a mechanically coupled load IM. The IPMSM parameters are shown in Table 3.2. The proposed multiple-AVF CFN for position estimation error harmonic fluctuation elimination is executed using a low-cost ARM chip STM32F103VCT6. All the following experimental results were achieved with the IPMSM operating in sensorless control. An encoder (ECN1113) is utilized to measure the actual position that is solely used for comparison but not for control purposes. Since the sixth harmonic fluctuation takes the dominant role in the position estimation error due to its bigger amplitude, only the fifth and seventh harmonic contents of back-EMF estimates are tracked and compensated in the test.

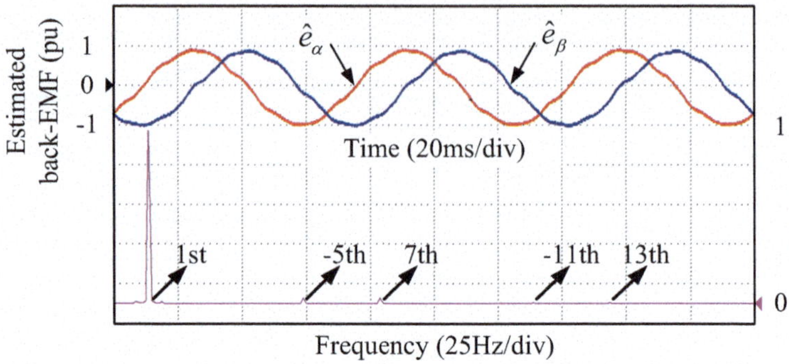

(a)

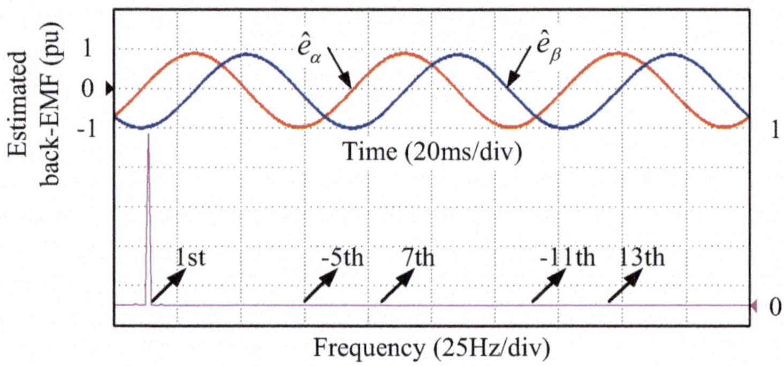

(b)

Fig. 6.36 Back-EMF estimates and FFT analysis. (**a**) Without multiple-AVF CFN. (**b**) With multiple-AVF CFN

Figure 6.36 shows the experimental comparison of the estimated back-EMF with normalization and FFT spectrum analysis before and after using the proposed multiple-AVF CFN at 300 r/min under 75% rated load. From the experimental results, it can be seen obviously that the fifth, seventh, eleventh, and thirteenth harmonic contents in the estimated back-EMF can be observed without utilizing the proposed multiple-AVF CFN, and the fifth and seventh harmonic contents take the critical role for position estimation due to bigger amplitudes. After employing the proposed multiple-AVF CFN, the fifth and seventh harmonic contents can be completely eliminated and hence the back-EMF estimate becomes more sinusoidal, but the eleventh and thirteenth harmonic contents with smaller amplitudes still remain since they are not compensated in the test. Furthermore, the estimated fifth and seventh harmonic contents in α-axis estimated back-EMF are illustrated in Fig. 6.37. As can be seen, the proposed multiple-AVF CFN can effectively extract the fundamental and other harmonic contents.

Since inverter nonlinearities can behave more critically in low-speed range, the signal-to-noise ratio of back-EMF decreases as operating speed decreases. Hence, the comparison tests of estimated position and error in low-speed operation were

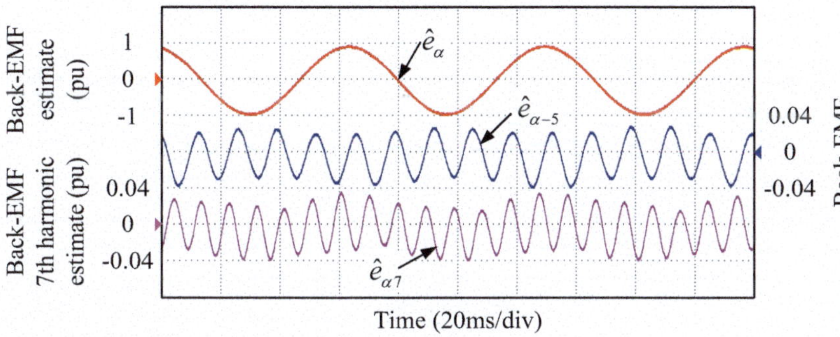

Fig. 6.37 Estimated back-EMF harmonic contents

carried out. The experimental results at 150 r/min (10% rated speed) and 75 r/min (5% rated speed) are shown in Figs. 6.38 and 6.39, respectively. As can be seen, the estimated position is distorted with a remarkable sixth harmonic fluctuation up to 0.025π at 150 r/min and 0.045π at 75 r/min, respectively. In contrast, with the proposed multiple-AVF CFN, the estimated rotor position is in good agreement with the measured position, and the sixth position error harmonic fluctuation is

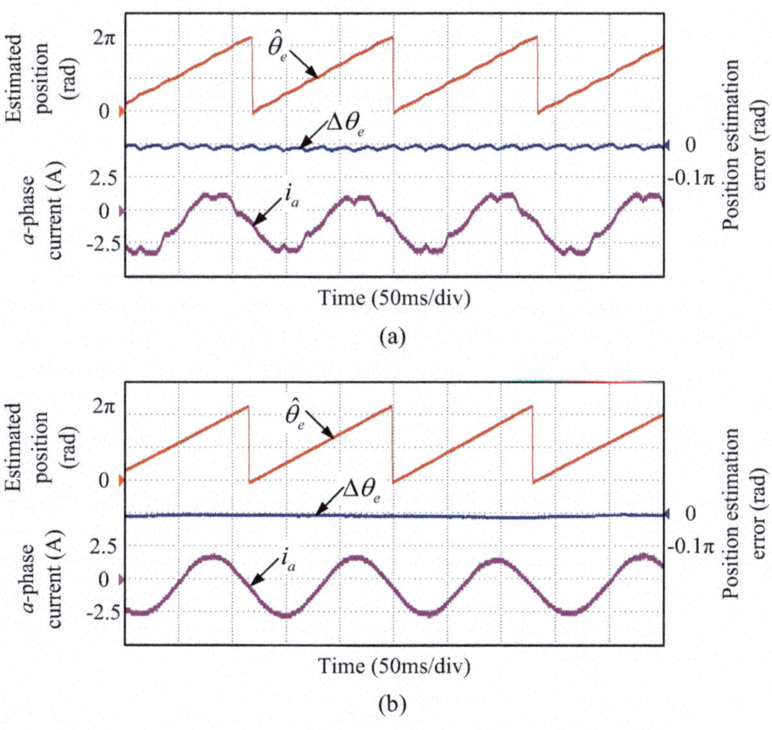

Fig. 6.38 Comparison of estimated rotor position and error at 150 r/min (10% rated speed). (**a**) Without multiple-AVF CFN. (**b**) With multiple-AVF CFN

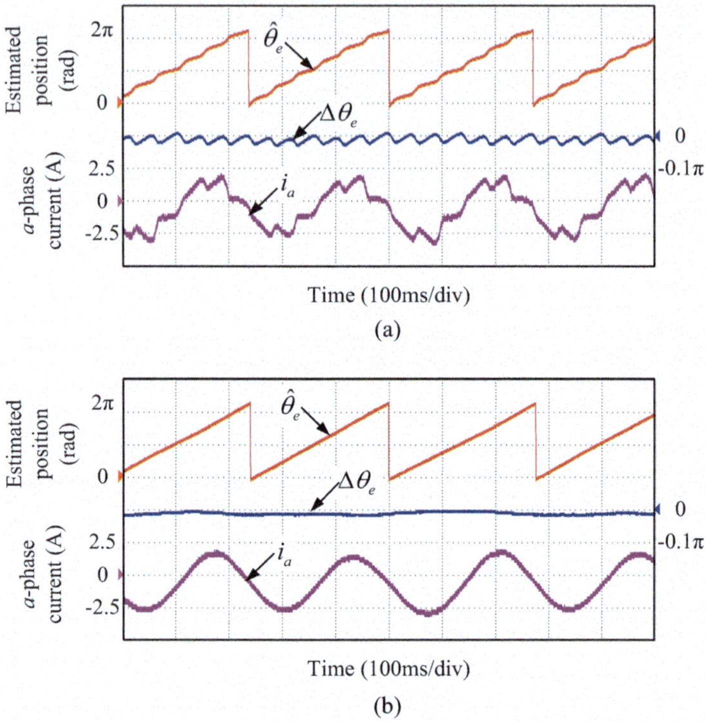

Fig. 6.39 Comparison of estimated rotor position and error at 75 r/min (5% rated speed). (**a**) Without multiple-AVF CFN. (**b**) With multiple-AVF CFN

completely eliminated. Moreover, the phase current becomes more sinusoidal without distortions. Thus, it can be concluded that the proposed multiple-AVF CFN is effective in improving the accuracy of the position estimation at low speed.

To further verify the harmonic fluctuation elimination effectiveness of the proposed multiple-AVF CFN in high-speed operation, the comparison tests of estimated position error at 600 r/min with 50% rated load were carried out as shown in Fig. 6.40. As can be seen, the estimated position is distorted by a sixth harmonic fluctuation with amplitude of 0.015π without compensation. And the proposed multiple-AVF CFN can effectively eliminate the sixth harmonic fluctuation. It is worth noting that the harmonic fluctuation amplitude decreases as the operating speed increases. This result occurs due to the fact the average distortion voltage caused by inverter nonlinearities with the constant control period, when normalized to the fundamental back-EMF magnitude, decreases as the speed increases. The current-loop regulator and the position observer could cancel out partial position error harmonic fluctuations because of the bandwidth limit as operating speed increases.

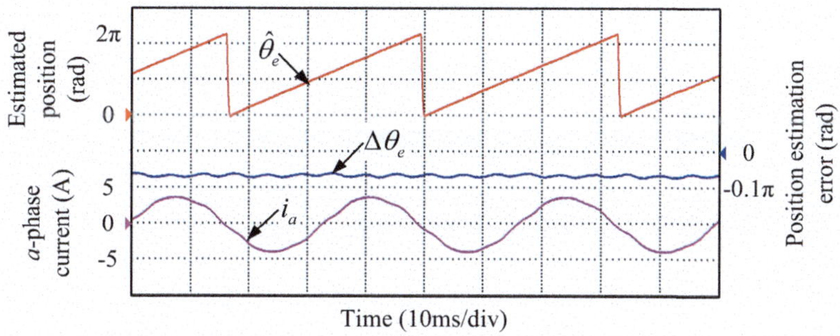

(a)

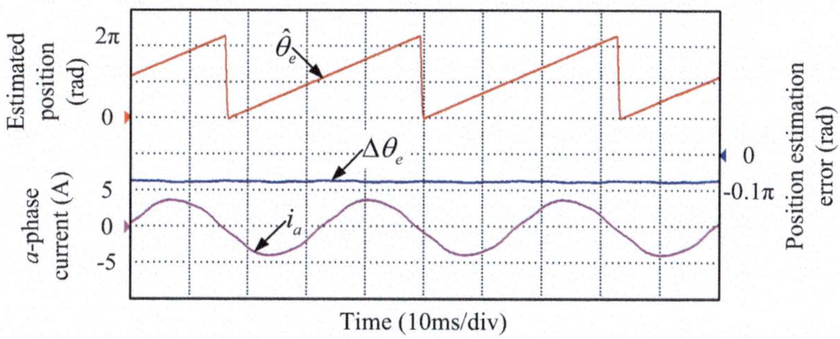

(b)

Fig. 6.40 Comparison of estimated position and error at 600 r/min. (**a**) Without multiple-AVF CFN. (**b**) With multiple-AVF CFN

The proposed multiple-AVF CFN was also tested under parameter variations. Figures 6.41 and 6.42 show the rotor position estimation error when the stator resistance R_s and the q-axis inductance L_q vary from 70% to 130% of their nominal values $\tilde{R}_s$ and $\tilde{L}_q$ (listed in Table 3.2) while the motor is running at 300 r/min with 50% rated load. As can be seen, DC offset occurs under parameter variations. After employing the proposed multiple-AVF CFN, the position error harmonic fluctuations can be effectively attenuated. Furthermore, the proposed multiple-AVF CNF position observer can achieve good tolerance to parameter variations.

Figure 6.43 shows the control performance comparison at 150 r/min. In each figure, d-q axis current and a-phase current are given, respectively. The actual MTPA point in test is verified using the theoretical values of MTPA, depending on the load conditions as shown in Fig. 6.43(c). From the experimental results, it can be seen that with the conventional full-order Luenberger observer, the sixth fluctuation contents of i_d and i_q can reach up to 1 A and 1.5 A, respectively. i_d and i_q become smoother by eliminating the fluctuation contents based on the proposed

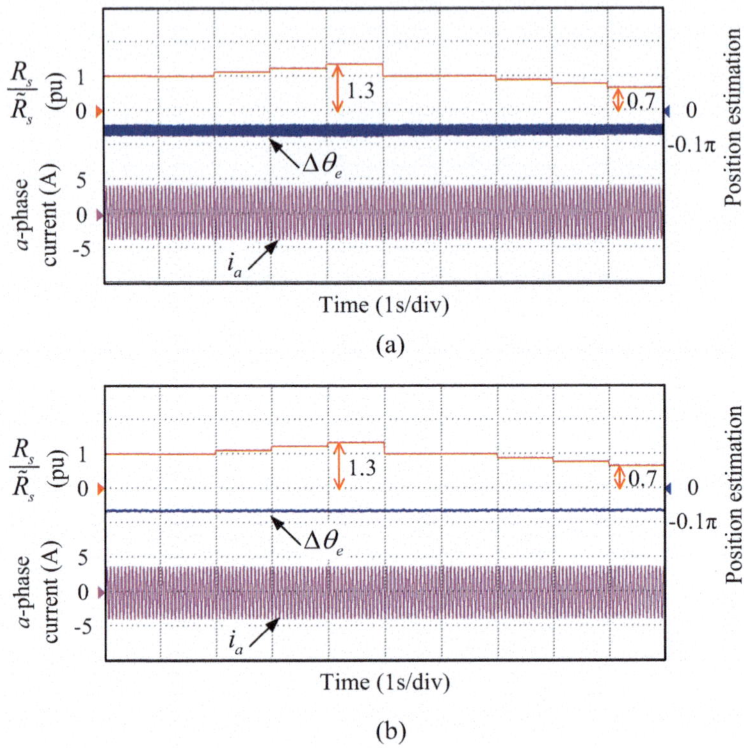

Fig. 6.41 Comparison of estimated position error under stator resistance variations. (**a**) Without multiple-AVF CFN. (**b**) With multiple-AVF CFN

multiple-AVF CFN. It can be concluded that since the fluctuation contents existing in position estimate can be reduced through the proposed multiple-AVF CFN, the accuracy in coordinate transformation and complete decoupling control can be guaranteed. The *d*-*q* axis current fluctuations can be reduced owing to the proposed multiple-AVF CFN, which contributes to the improvement of the sensorless control performance.

Figure 6.44 shows the estimated speed and position error comparison during a step load disturbance from 25% to 100% rated load at 300 r/min. From the experimental results, it can be observed that the harmonic fluctuations in the position estimation error can be attenuated remarkably during the whole operation under step load disturbance with the proposed multiple-AVF CFN.

Figure 6.45 shows the speed and the position estimation error comparison during continuous speed variation at the rate of change of 300 r/min. IPMSM accelerates from 100 to 300 r/min and then decelerates to 100 r/min. From the experimental results, it can be observed that the position error harmonic fluctuations can be

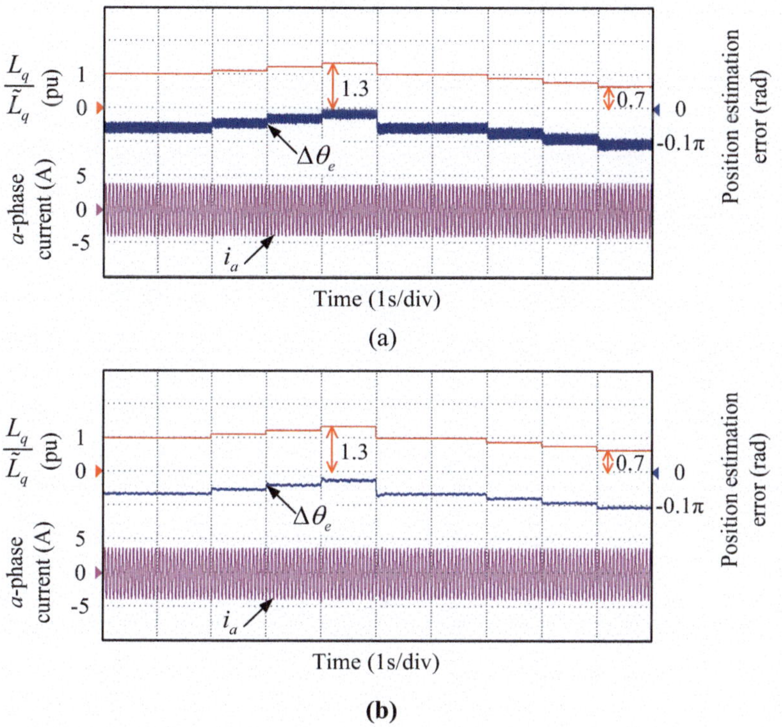

Fig. 6.42 Comparison of estimated position error under q-axis inductance variations. (**a**) Without multiple-AVF CFN. (**b**) With multiple-AVF CFN

reduced to some degree using the proposed multiple-AVF CFN over the whole acceleration and deceleration process during continuous speed variation.

To further evaluate the harmonic fluctuation elimination effectiveness of the proposed multiple-AVF CFN, Fig. 6.46 gives the peak-to-peak value measurement comparison of position and speed error harmonic fluctuations at different operating conditions, respectively. From the experimental results, it can be noted that the peak-to-peak value of position and speed estimation error harmonic fluctuations obtained through the conventional full-order Luenberger observer can reach up to 0.668 rad and 45 r/min, respectively. Hereafter, using the proposed multiple-AVF CFN, the peak-to-peak value of position and speed estimation error harmonic fluctuations can be reduced to 0.06 rad and 8 r/min, respectively. Therefore, it can be further concluded that the proposed multiple-AVF CFN can diminish the position and speed estimation error harmonic fluctuations effectively and achieve low sensitivity to harmonic amplitude variations with different operating conditions.

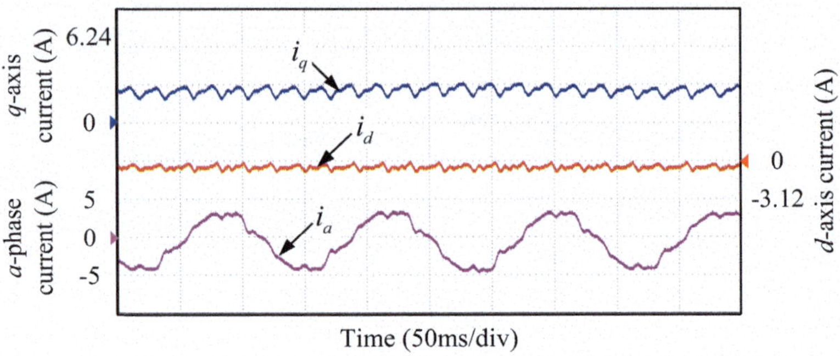

(a)

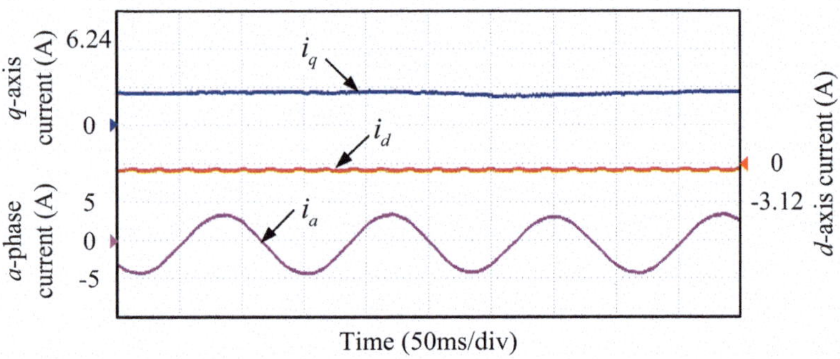

(b)

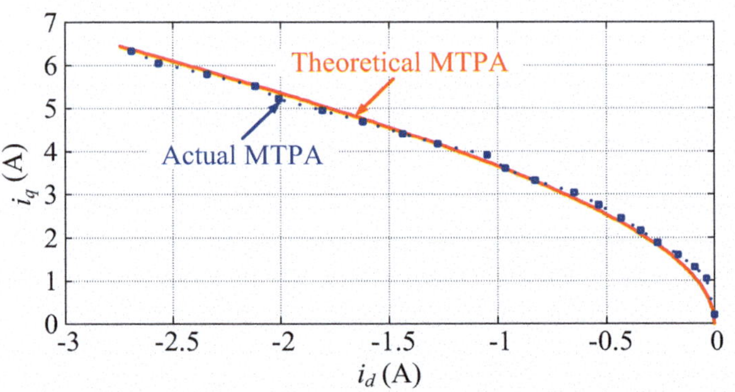

(c)

Fig. 6.43 Sensorless control performance comparison. (**a**) Without multiple-AVF CFN. (**b**) With multiple-AVF CFN. (**c**) MTPA point verification

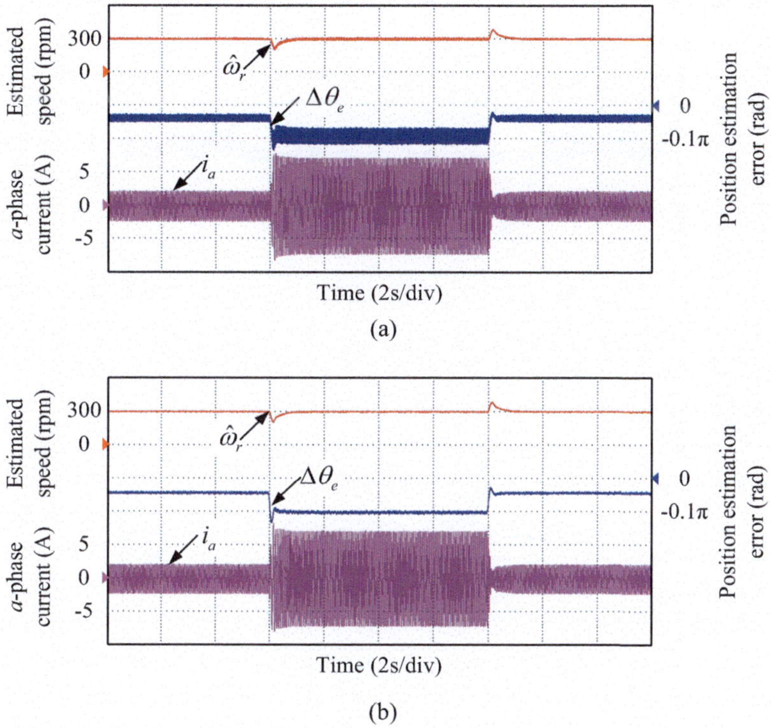

Fig. 6.44 Comparison of load step disturbance response. (**a**) Without multiple-AVF CFN. (**b**) With multiple-AVF CFN

6.6 Summary

This chapter introduces some solutions to eliminate the ripple in the rotor position estimation error for sensorless of IPMSM drives.

ADALINE network-based filter with a quadrature PLL position observer implements elimination to harmonic ripple in the rotor position estimation error. According to the harmonic characteristics of the estimated rotor position, the ADALINE network-based filter is adopted to track and compensate the selective harmonic ripple in the rotor position estimation error through continuously updating the filter weights online.

SOGI-MSEHE is also a suitable solution to eliminate the fifth and seventh harmonic components of the estimated EMF. In the adaptive filter based on SOGI-

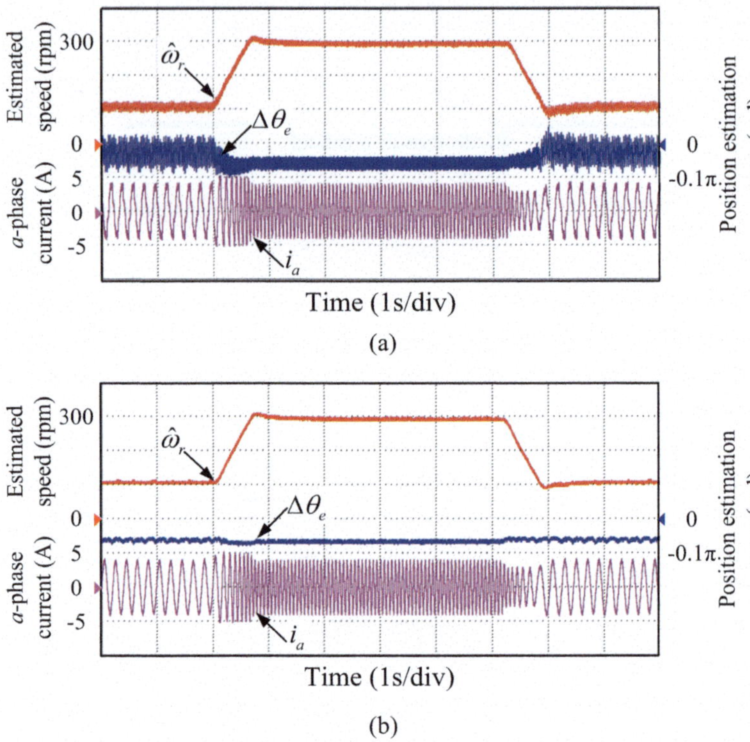

Fig. 6.45 Comparison of speed and position estimation error with speed variation under 50% rated load. (**a**) Without multiple-AVF CFN. (**b**) With multiple-AVF CFN

MSEHE, the enhanced position observer consists of multiple SOGIs tuned at selective harmonic frequency of the estimated EMF. These SOGIs work in a collaborative way by using a CFN, which can decouple the effect of the different harmonics of the extended EMF.

The multiple-AVF CNF is introduced for a model-based full-order Luenberger observer. The multiple-AVF CFN can eliminate the position error harmonic fluctuations induced from inverter nonlinearities and flux spatial harmonics through detecting and compensating the back-EMF harmonics with selective frequency and polarity.

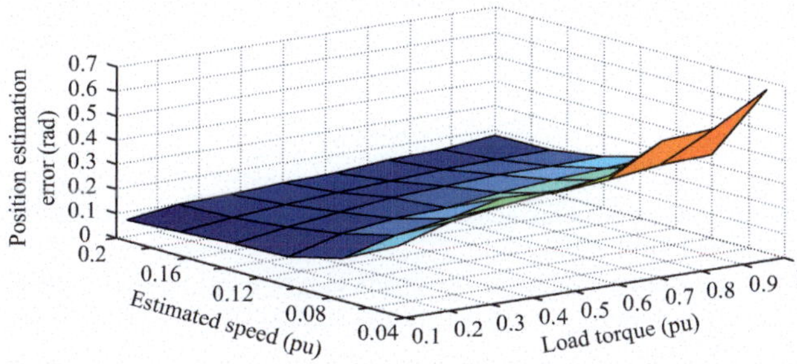

(a)

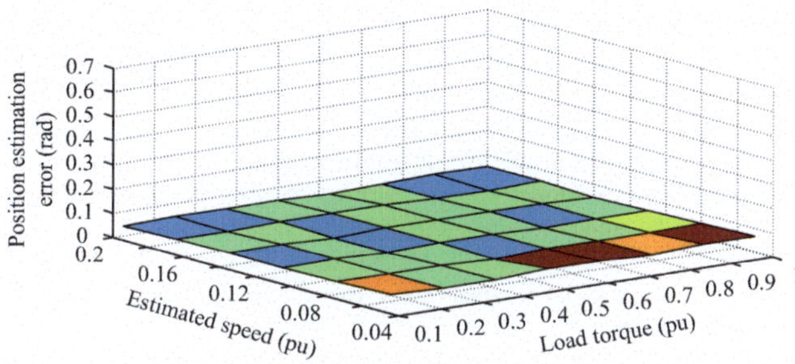

(b)

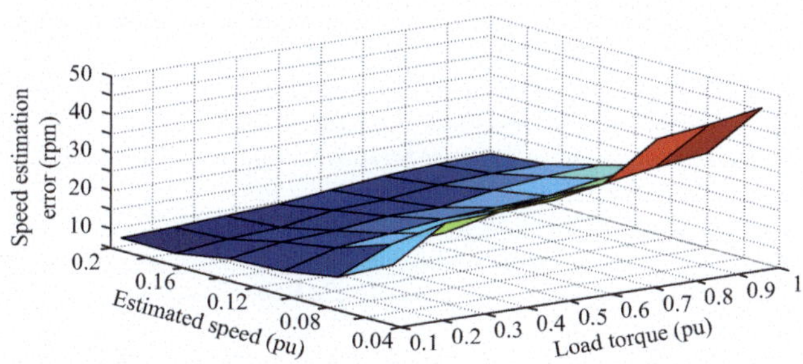

(c)

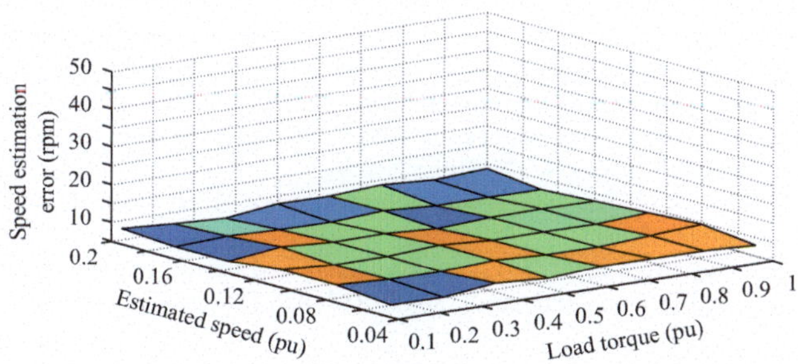

(d)

Fig. 6.46 Peak-to-peak value measurement comparison at different operating conditions. (**a**) Position error harmonic fluctuations without multiple-AVF CFN. (**b**) Position error harmonic fluctuations with multiple-AVF CFN. (**c**) Speed error harmonic fluctuations without multiple-AVF CFN. (d) Speed error harmonic fluctuations with multiple-AVF CFN

References

1. Z. Q. Chen, M. Tomita, S. Doki, S. Okuma, An extended electromotive force model for sensorless control of interior permanent magnet synchronous motors. IEEE Trans. Ind. Electron. **50**(2), 288–295 (April 2003)
2. R. W. Hejny, R. D. Lorenz, Evaluating the practical low-speed limits for back-EMF tracking-based sensorless speed control using drive stiffness as a key metric. IEEE Trans. Ind. Appl. **47** (3), 1337–1343 (May/June 2011)
3. G. Wang, L. Ding, Z. Li, J. Xu, G. Zhang, H. Zhan, R. Ni, D. Xu, Enhanced position observer using second-order generalized integrator for sensorless interior permanent magnet synchronous motor drives. IEEE Trans. Energy Convers. **29**(2), 486–495 (June 2014)
4. G. Wang, H. Zhan, G. Zhang, X. Gui, D. Xu, Adaptive compensation method of position estimation harmonic error for EMF-based observer in sensorless IPMSM drives. IEEE Trans. Power Electron. **29**(6), 3055–3064 (June 2014)
5. G. Wang, T. Li, G. Zhang, X. Gui, D. Xu, Position estimation error reduction using recursive-least-square adaptive filter for model-based sensorless interior permanent-magnet synchronous motor drives. IEEE Trans. Ind. Electron. **61**(9), 5115–5125 (September 2014)
6. G. Zhang, G. Wang, D. Xu, N. Zhao, ADALINE-network-based PLL for position sensorless interior permanent magnet synchronous motor drives. IEEE Trans. Power Electron. **31**(2), 1450–1460 (February 2016)
7. G. Zhang, G. Wang, D. Xu, R. Ni, C. Jia, Multiple-AVF cross-feedback-network-based position error harmonic fluctuation elimination for sensorless IPMSM drives. IEEE Trans. Ind. Electron. **63**(2), 821–831 (February 2016)
8. G. Wang, R. Yang, D. Xu, DSP-based control of sensorless IPMSM drives for wide-speed-range operation. IEEE Trans. Ind. Electron. **60**(2), 720–727 (February 2013)
9. Y. Li, Z.Q. Zhu, D. Howe, C.M. Bingham, Improved rotor position estimation in extended back-EMF based sensorless PM brushless AC drives with magnetic saliency, in *Electric Machines & Drives Conference, 2007. IEMDC'07*, vol. 1, (IEEE International, Antalya, Turkey, 2007), pp. 214–219
10. P. Rodriguez, A. Luna, R. S. Munoz-Aguilar, I. Etxeberria-Otadui, R. Teodorescu, F. Blaabjerg, A stationary reference frame grid synchronization system for three-phase grid-connected power converters under adverse grid conditions. IEEE Trans. Power Electron. **27**(1), 99–112 (January 2012)
11. Z. Qiao, T. Shi, Y. Wang, Y. Yan, C. Xia, X. He, New sliding-mode observer for position sensorless control of permanent-magnet synchronous motor. IEEE Trans. Ind. Electron. **60**(2), 710–719 (Feburary 2013)
12. M. I. Marei, E. F. El-Saadany, M. M. A. Salama, Envelope tracking techniques for flicker mitigation and voltage regulation. IEEE Trans. Power Del. **19**(4), 1854–1861 (October 2004)
13. S. Bolognani, S. Calligaro, and R. Petrella, Design issues and estimation errors analysis of back-EMF-based position and speed observer for SPM synchronous motors. IEEE J. Emerging Sel. Topics Power Electron. **2**(2), 159–170 (June 2014)
14. K. W. Martin, Complex signal processing is not complex. IEEE Trans. Circuits Syst. **51**(9), 1823–1836 (September 2004)
15. F. Briz, M. W. Degner, and R. D. Lorenz, Analysis and design of current regulators using complex vectors. IEEE Trans. Ind. Appl. **36**(3), 817–825 (May/June 2000)

Chapter 7
Low-Frequency Ratio Sensorless Control for High-Speed PMSM Drives

7.1 Applications of High-Speed PMSM

High-speed machines have been widely employed in applications like smaller size of spindle motors, large chiller motors, micro-turbine generators, aircraft generators, compressors, and gas-turbines owing to their high power density and high efficiency.

In conventional electromechanical high-speed drive systems, gear boxes are used to increase the drive speed to be suitable for the applied load. However, high-speed machine drives benefit from the elimination of gear boxes, which enhances system performance such as lower noise, simplified construction and maintenance, increased compactness, and better mechanical stiffness [1].

Squirrel-cage asynchronous motors, solid rotor asynchronous motors, wire-wound synchronous motors, permanent magnet synchronous motors, induction sub-motors, and switched reluctance motors can be used in high-speed applications.

High-speed permanent magnet synchronous motor has good application potential due to its high efficiency and high power density, which covers various power levels.

7.2 Challenges of Low-Frequency Ratio Sensorless Control

Aside from the cost reduction, the applications introduced in the last section are nowadays further constrained with further demands of a low-PWM frequency (f_{PWM}) over the operating fundamental frequency (f_{out}) ratio, $f_{ratio} = f_{PWM} / f_{out}$.

On the one hand, problems emerge in middle- and high-power applications, where, notwithstanding the advances in power device technology, the PWM frequency is limited to values even lower than several hundreds of hertz, which keeps the switching losses of the power semiconductors at a tolerable level. On the other hand, problems have emerged in high-frequency PMSM drives.

G. Wang et al., *Position Sensorless Control Techniques for Permanent Magnet Synchronous Machine Drives*, https://doi.org/10.1007/978-981-15-0050-3_7

High-frequency PMSMs could be divided into two categories: super high-speed machines and high-pole-count machines [2]. Super high-speed machines are used for the high-speed applications like turbo compressors, whose maximum speed could be 60,000 r/min or even higher. High-pole-count machines are used for hybrid (or electric) vehicles, whose number of poles is 20 or higher. In both cases, the maximum speed and, consequently, the maximum operating fundamental frequency could be close to or even higher than 1 kHz. However, the increase of operating fundamental frequency cannot be followed with the increase of the PWM frequency because that would increase switching losses and demand more CPU resource. As a result, in middle- and high-power applications as well as high-frequency PMSM drives, the frequency ratio of the PWM frequency over the fundamental frequency, f_{ratio}, is low, sometimes close to or even smaller than 10. However, due to the effect of time delay inherent in digital control system of position sensorless IPMSM drives, low f_{ratio} is a new source of instability that gains attention and must be analyzed and dealt with. For this reason, a discrete-time rotor position estimation methodology is critical to achieve the satisfactory stability and dynamic performance at low f_{ratio} values.

The specialized discrete-time flux and speed estimation methodology for sensorless controlled IMs operating at low-frequency ratios has been extensively analyzed and investigated. In [3, 4], a Luenberger observer using genetic algorithms was proposed for high-speed sensorless IM drives. The observer matrix values were computed offline at different speeds, which could provide the desired stability and dynamics at a maximum speed 12,000 r/min with the PWM frequency 12 kHz, corresponding to $f_{\text{ratio}} = 30$. In [5], a low-switching-frequency flux observer was developed for IM drives by cross-solving the coupled stator flux linkage and stator current differential equations in the discrete-time domain, which improved the estimation accuracy particularly at low-PWM frequency, and covered f_{out} up to 50 Hz with $f_{\text{PWM}} = 0.5$ kHz, corresponding to $f_{\text{ratio}} = 10$.

The development and implementation of rotor position estimation methodologies for position sensorless PMSM drives with high f_{ratio} have been widely explored in literatures [2, 5–7, 13] mostly using continuous-time domain analysis and transforming into discrete-time domain with Euler or Tustin approximations. Although this approximate discrete-time model works well in many cases, stability problem and performance degradation occur at low f_{ratio} where only a small number of control updates per fundamental cycle are allowed [7–10]. The limitations and the impacts of rotor position estimation at low-frequency ratios are still unexplored areas and only few research institutes were involved on them. A discrete-time quasi-SMO based on Euler approximation was proposed in [11] and evaluated with f_{ratio} down to 13.3, but the performance of such approach degrades with the decreased f_{ratio}.

7.3 Discrete-Time Domain Mathematical Model of PMSM

7.3.1 Continuous-Time Domain Model

This section introduces the continuous-time PMSM machine model for conventional EMF voltage estimation [12]. The extended back EMF model of PMSM in the synchronous frame is given as

$$\boldsymbol{u}_{\mathrm{dq}} = \left(R + pL_d + j\omega_e L_q\right)\boldsymbol{i}_{\mathrm{dq}} + \boldsymbol{e}_{\mathrm{dq}} \tag{7.1}$$

with

$$\mathbf{e}_{dq} = jE_{\mathrm{ex}} = j\left[\left(L_d - L_q\right)\left(\omega_e i_d - p i_q\right) + \omega_e \psi_f\right], \tag{7.2}$$

where $\boldsymbol{u}_{\mathrm{dq}}$ and $\boldsymbol{i}_{\mathrm{dq}}$ are the stator voltage and current vector in d-q axes, R is the stator resistance, L_{d} and L_{q} are the d-q axis inductances, ψ_f is the permanent magnet flux, ω_e is the electrical speed, p is the differential operator, and j is the imaginary unit. The latter term of (7.1), $\boldsymbol{e}_{\mathrm{dq}}$, is defined as an extended back-EMF vector. Transforming (7.1) into the estimated d-q axis frame (also noted as the γ-δ axis frame) yields

$$\mathbf{u}_{\gamma\delta} = \left(R + pL_d + j\omega_e L_q\right)\mathbf{i}_{\gamma\delta} + \mathbf{e}_{\gamma\delta} \tag{7.3}$$

with

$$\mathbf{e}_{\gamma\delta} = jE_{\mathrm{ex}}\left(-\,sin\,\tilde{\theta}_e + jcos\tilde{\theta}_e\right) - j\tilde{\omega}_e L_d \mathbf{i}_{\gamma\delta}, \tag{7.4}$$

where $\boldsymbol{u}_{\gamma\delta}$, $\boldsymbol{i}_{\gamma\delta}$, and $\boldsymbol{e}_{\gamma\delta}$ are the stator voltages, currents, and extended back-EMF in γ-δ axes. $\tilde{\theta}_e = \theta_e - \hat{\theta}_e$ and $\tilde{\omega}_e = \omega_e - \hat{\omega}_e$ are the position and speed errors. The extended back EMF model in γ-δ axes provides the position error instead of the rotor position, which is utilized to estimate the rotor position and speed.

It is noted that all mechanical dynamics are much slower than electrical ones, and the rotor speed can be considered to keep constant between two consecutive sampling instants. The dynamics of the extended back EMF in γ-δ axes could be obtained as

$$\dot{\mathbf{e}}_{\gamma\delta} = 0. \tag{7.5}$$

Since the changes of the rotor speed are small enough during a control period in most cases, the approximated assumption is suitable for position observer design.

According to (7.3) and (7.5), the extended back EMF-based state space model of PMSM can be readily constructed as follows:

$$\begin{aligned}\dot{\mathrm{x}} &= \mathrm{A}_c\mathrm{x} + \mathrm{B}_c\mathrm{u},\\ \mathrm{y} &= \mathrm{C}_c\mathrm{x},\end{aligned} \tag{7.6}$$

where

$$\dot{\mathrm{x}} = \left[\mathrm{i}_{\gamma\delta}\mathrm{e}_{\gamma\delta}\right]^{\mathrm{T}},\quad \mathrm{u} = \mathrm{u}_{\gamma\delta},\quad \mathrm{y} = \mathrm{i}_{\gamma\delta},$$

$$\mathrm{A}_c = \begin{bmatrix} -\dfrac{R}{L_d} - j\omega_e\dfrac{L_q}{L_d} & -\dfrac{1}{L_d} \\ 0 & 0 \end{bmatrix},\ \mathrm{B}_c = \left[\dfrac{1}{L_d} 0\right]^{\mathrm{T}},\ \mathrm{C}_c = [1,\ 0].$$

7.3.2 *Discrete-Time Domain Model*

In this section, the discretized effect due to the digital implementation is discussed. Then an exact ZOH equivalent of the continuous-time domain state-space model is formed.

Figure 7.1 illustrates the vector-controlled PMSM drive as a sampled data system, which is composed of the continuous-time domain PMSM model, the discrete-time controller, PWM, and sampler. The common in inverter-fed system is to sample the stator currents at the same instant that the voltage commands are updated, i.e., the sampling of the stator currents is synchronized with the PWM [13, 15].

Figure 7.2 shows the typical discrete timing sequence of the asymmetrical PWM with current sampling, control computation, and PWM update [15]. The sampling

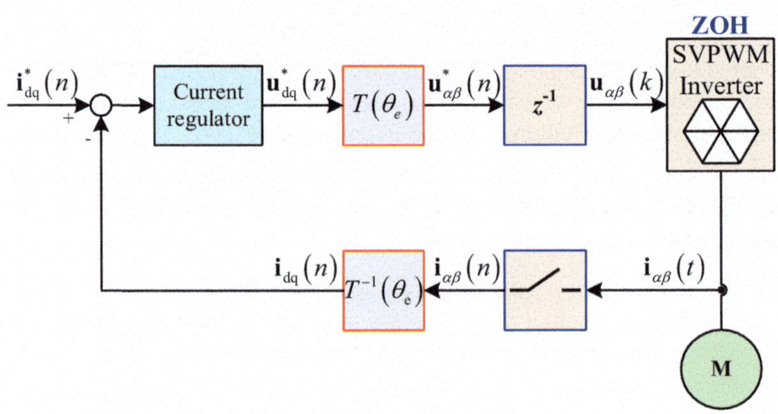

Fig. 7.1 Vector-controlled PMSM drive of a digital system

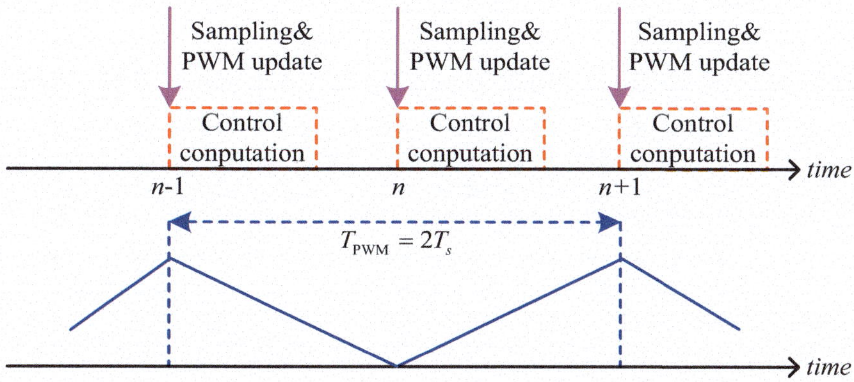

Fig. 7.2 Typical discrete timing sequence of asymmetrical PWM with current sampling, control computation, and PWM update

frequency is doubled compared to the PWM frequency. As can be seen, the controller output is delayed by one sampling period T_s due to the finite control computation time, i.e., $\mathrm{u}_{\alpha\beta}(n) = e^{-\omega_e T_s}\mathrm{u}^*_{\alpha\beta}(n-1)$ in the stationary frame, or transformed into the synchronous frame, $\mathrm{u}_{\alpha\beta}(n) = e^{-\omega_e T_s}\mathrm{u}^*_{\alpha\beta}(n-1)$. The computation delay compensation on the voltage angle can be easily implemented through advancing the voltage reference by one sampling period in the coordinate transformation as shown in Fig. 7.1, i.e., $\theta_{\text{adv}} = \omega_e T_s$.

Given that the sampling is synchronized with the PWM, the averaged quantities of the switching cycle are considered. Hence, the actual stator voltage $\mathrm{u}_{\alpha\beta}(t)$ in the stationary frame is piecewise constant between two consecutive sampling instants, which corresponds to the ZOH in the stationary frame. In the existing works, two different schemes to model the stator-voltage input have been utilized depending on the ZOH of the voltage in the synchronous frame or in the stationary frame. It is worth noting that the switching-cycle averaged quantities modeling the PWM-VSI as the ZOH neglects the HF switching harmonics created by the PWM of the power devices. Ignoring the switching harmonics is appropriate in the discrete-time design since the current ripple created by these harmonics can be eliminated from the sampled feedback signal [13, 14].

1. ZOH in Stationary Frame: As aforementioned, the output voltage of a PWM-VSI could be considered as hold constant at the commanded voltage vector location during the switching period T_s.

$$\mathrm{u}_{\alpha\beta}(t) = e^{j\theta(n\,T_s)}\mathrm{u}_{\text{dq}}(nT_s). \tag{7.7}$$

It is important to note that the voltage is latched in the stationary frame, where it physically is. Therefore, in the synchronous frame, the input voltage from the PWM-VSI can be presented as a vector (7.7) rotating in the reverse direction from the synchronous frame:

$$\mathrm{u_{dq}} = e^{-j\theta(t)}\mathrm{u}_{\alpha\beta} = e^{-j\theta(t-nT_s)}\mathrm{u_{dq}}(nT_s). \tag{7.8}$$

However, the discrete-time state-space model of (7.6) could be depicted by

$$\begin{aligned} \mathrm{x}(n+1) &= \mathrm{A}_d\mathrm{x}(n) + \mathrm{B}_d\mathrm{u}(n), \\ \mathrm{y}(n) &= \mathrm{C}_d\mathrm{x}, \end{aligned} \tag{7.9}$$

where A_d, B_d, and C_d and the system input and output matrices of the discrete-time state-space model of PMSM are presented as follows, and the closed-form expressions for the elements of the matrices are given as

$$\mathrm{A}_d = e^{\mathrm{A}_c T_s} = \begin{bmatrix} A_{d11} & A_{d12} \\ A_{d21} & A_{d22} \end{bmatrix}, \ \mathrm{B}_d = \left(\int_0^{T_s} e^{\mathrm{A}_c\tau} e^{-j\omega_e(T_s-\tau)} d\tau \right) \mathrm{B}_c, \ \mathrm{C}_d = \mathrm{C}_c. \tag{7.10}$$

The factor $e^{-j\omega_e(T_s-\tau)}$ inside the integral originates from the ZOH being modeled in stationary coordinate, i.e., the inverter output voltage, is time variant in synchronous coordinate during the switching period.

2. ZOH in Synchronous Frame: With the exception of one sampling period delay produced due to the control computation time, implementing a space-vector PWM in a discrete-time domain introduces further delay in the system. (7.11) shows the details:

$$\begin{aligned} \mathrm{u_{dq}}(n) &= \frac{1}{T_s} \int_{\tau=T_s}^{(n+1)T_s} e^{-j\omega_e(\tau-nT_s)} \cdot \mathrm{u}^*_{\mathrm{dq}}(nT_s) d\tau \\ &= \frac{2}{\omega_e T_s} \sin\left(\frac{\omega_e T_s}{2}\right) \cdot e^{\frac{-j\omega_e T_s}{2}} \cdot \mathrm{u}^*_{\mathrm{dq}}(nT_s). \end{aligned} \tag{7.11}$$

Equation (7.11) can be approximated as half a sampling period delay 0.5 T_s delay, resulting in a phase and magnitude errors which need to be compensated. Thus a total of 1.5 T_s delay is introduced in the discrete-time domain when the ZOH of the stator-voltage input is considered in the synchronous frame. The discrete-time domain model considering the ZOH in the stationary frame is preferred since it physically takes the ZOH delay properly into account [14].

7.4 Current Regulator Design

7.4.1 Introduction to Current Regulator Design

A common practice when designing digital controllers for continuous-time systems is to develop a continuous-time controller and then convert it to a discrete equivalent by using one of the approximations for the continuous s-transform to the discrete z-transform.

However, it is well known that the digital implementation of control systems introduces delays whose negative effects on dynamic performance decrease with the ratio of sampling to fundamental frequency. Highly oscillatory response, and even instability, may occur if the current regulator design does not properly incorporate the effects of the discreteness of the controller.

In this section, different design methodologies of current regulators are analyzed and compared. The aim of this section is to develop a current control scheme based on the discrete-time model of PMSM and complex vector notations.

In the analysis process, assume that the back-EMF can be fully compensated by the feedforward compensator at the output of the current regulator. The SPMSM is used as an example to demonstrate the discrete-time domain design of the current regulators. Assuming that $L = L_d = L_q$, we can get (7.12) according to (7.6):

$$G_{\mathrm{p}}(z) = \frac{\mathrm{i}_{\mathrm{dq}}(z)}{\mathrm{u}_{\mathrm{dq}}^{*}(z)} = \frac{1 - e^{-(R/L)T_s}}{R\left(ze^{j\omega_e T_s} - e^{-(R/L)T_s}\right)}. \tag{7.12}$$

In digital control system, the digital computation delay does exist. Taking one step digital delay into consideration,

$$G_{\mathrm{p\_d}}(z) = \frac{1 - e^{-(R/L)T_s}}{Rze^{j\omega_e T_s}\left(ze^{j\omega_e T_s} - e^{-(R/L)T_s}\right)}. \tag{7.13}$$

The discrete-time complex vector root locus of SPMSM considering digital delay is calculated and shown in Fig. 7.3 when the switching frequency is equal to 6 kHz. As can be seen from the figure, poles migrate and even approach the unit circle as the operating frequency increases from 0 to 600 Hz, which makes the system fragile.

7.4.2 Analysis and Comparison of Different Current Regulators

Compared with Euler transform, Tustin transform shows better performance for its smaller truncation error. The equivalent of the complex vector synchronous frame PI current regulator by using Tustin transform is shown as follows:

$$\begin{aligned} G_{\mathrm{cr}}(z) =& \frac{\left(K_{\mathrm{p}} + 0.5K_{\mathrm{i}}T_s\right)z + \left(0.5K_{\mathrm{i}}T_s - K_{\mathrm{p}}\right)}{z-1} \\ =& K_{\mathrm{p}} + \frac{K_{\mathrm{i}}T_s}{z-1} \cdot \frac{z+1}{2}. \end{aligned} \tag{7.14}$$

Compensate one-step delay caused by digital implementation, and the new transfer function of PI current regulator can be obtained:

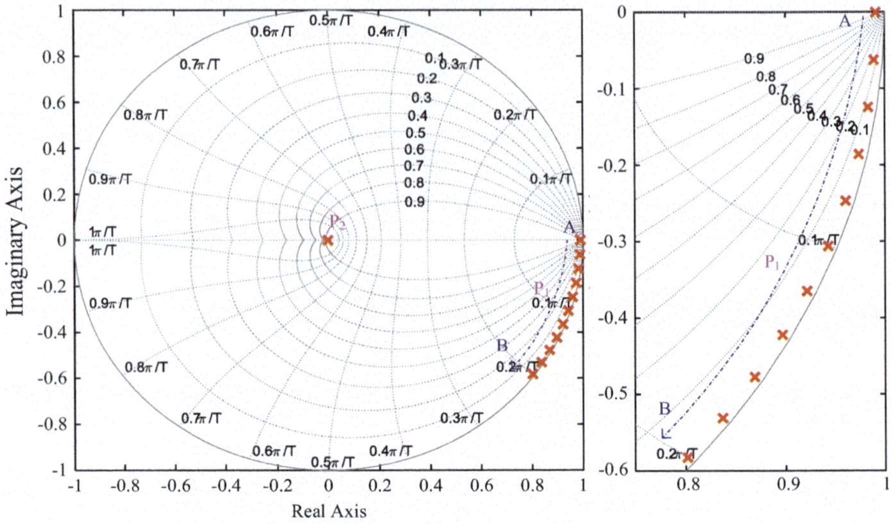

Fig. 7.3 *d-q* axis frame pole-zero migration considering digital delay

$$G_{\mathrm{cr}}(z) = \frac{(K_{\mathrm{p}} + 0.5K_{\mathrm{i}}T_s)z + (0.5K_{\mathrm{i}}T_s - K_{\mathrm{p}})}{z - 1} e^{j\omega_e T_s}, \tag{7.15}$$

where k_{p} and k_{i} are proportional gain and integral gain of the current regulator, respectively.

Thus, closed-loop transfer function of the current loop can be calculated by using

$$\frac{\mathrm{i}_{\mathrm{dq}}}{\mathrm{i}^*_{\mathrm{dq}}} = \frac{G_{\mathrm{cr}}(z) \cdot G_{\mathrm{p\_d}}(z)}{1 + G_{\mathrm{cr}}(z) \cdot G_{\mathrm{p\_d}}(z)}. \tag{7.16}$$

The gains of regulator are chosen to cancel out undesired poles of the motor, which makes the system to be stable. The most commonly adopted processing is to make sure that

$$\begin{cases} k_{\mathrm{p}} = \omega_{\mathrm{cr}} L \\ k_{\mathrm{i}} = \omega_{\mathrm{cr}} R, \end{cases} \tag{7.17}$$

where ω_{cr} is set as the bandwidth of current regulator.

Figure 7.4 shows the closed-loop pole-zero migration of current loop using discrete PI current regulator when the operating fundamental frequency increases.

As can be seen from the figure, the zeros of the regulator can cancel out the poles of the motor at standstill. However, it will not work with the increase of the fundamental frequency. It should be noted that a closed-loop pole migrating from point E to F approaches and eventually goes out of the unit circle, which makes the system without digital delay compensation unstable when $f_{\mathrm{ratio}} = 0.707$. After

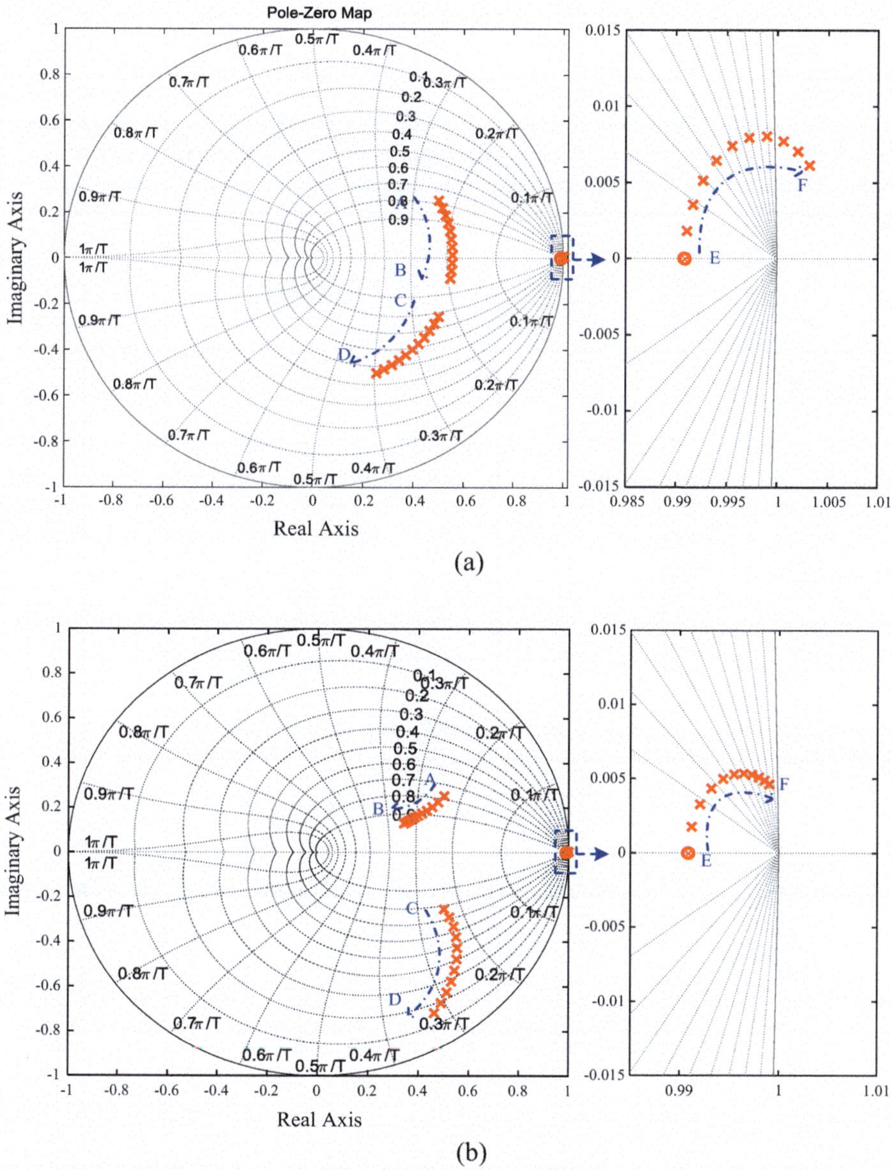

Fig. 7.4 Closed-loop pole-zero migration with operating frequency increase for the conventional PI control. (**a**) Without digital control delay compensation. (**b**) With digital control delay compensation

compensating one step delay, a closed-loop pole migrating from E to F approaches the unit circle, which causes the system oscillation.

7.4.3 Complex Vector PI Current Regulator

PI current regulator based on complex vector shows impressive performance on decoupling control for stator current in *d-q* axis frame. Figure 7.5a shows the block diagram of complex vector PI current regulator with digital delay compensation. Using Tustin transform, the transfer function of the current regulator can be calculated:

$$G_{\mathrm{cr}}(z) = \frac{\left(K_{\mathrm{p}} + 0.5K_{\mathrm{i}}T_s + j0.5w_e K_{\mathrm{p}} T_s\right)z}{z-1} e^{jw_e T_s} + \frac{0.5K_{\mathrm{i}}T_s + j0.5w_e K_{\mathrm{p}} T - K_{\mathrm{p}}}{z-1} e^{jw_e T_s}. \tag{7.18}$$

Another way of designing current regulator is also provided in this section. By designing the current regulator in the discrete-time domain using direct digital design methods, the truncation error caused by Tustin transform can be voided. Taking the computation delay and the pole-zero cancellation into consideration, the result can be obtained:

$$G_{\mathrm{cr}}(z) = \frac{k_{\mathrm{dq}}\left(ze^{j\omega_e T_s} - e^{-(R/L)T_s}\right)z}{z-1} e^{j\omega_e T_s}, \tag{7.19}$$

where k_{dq} is proportional gain of current loop in complex vector.

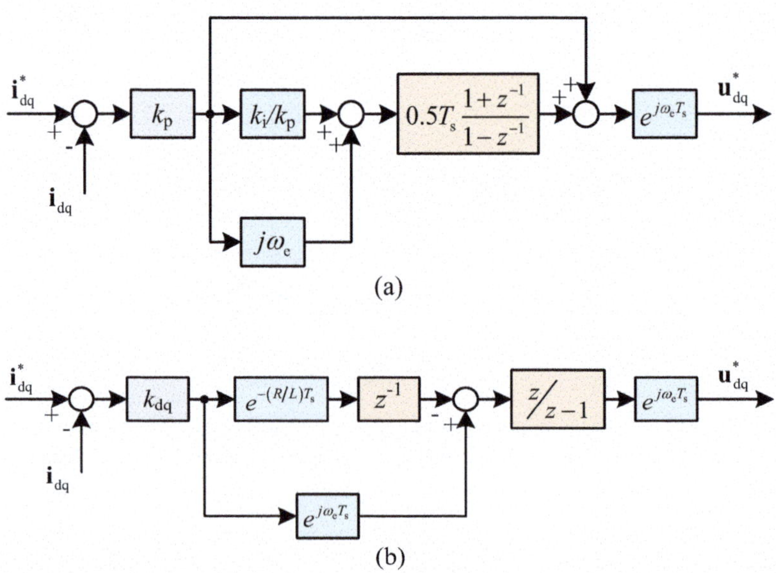

Fig. 7.5 Complex vector PI current regulator. (**a**) With Tustin transform. (**b**) Direct transform

Figure 7.6a shows the closed-loop pole-zero migration of current loop using complex vector PI current regulator including the two digital implementations above when the fundamental frequency increases from 0 to 600 Hz. Compared with the method of Tustin transform, the method using direct digital design shows better performance.

Figure 7.6b shows that this method can make sure the closed-loop pole-zero A-B canceled out completely regardless of the fundamental frequency changes. Besides, the decoupling control for current of *d-q* axis frame can be achieved.

The method provided above is designed for SPMSM drives, which means it is not suitable for IPMSM drives. Usually, it still works well and expected decoupling control is still valid when L_d is approximately equal to L_q in which case L is regarded as $(L_d + L_q)/2$. However, we cannot get consistent current response of the d and q-axes for the IPMSM drives with high saliency ratio, which means a new scheme is needed.

This section presents a modified complex vector PI current regulator, which is designed with direct digital design method to cancel out the undesired pole-zero. The control law for this controller is shown in Fig. 7.7.

The resistance voltage drop can be neglected when f_{ratio} is low, and PMSM can be approximated to an inductance coil whose mathematical model in discrete-time domain can be simplified as

$$G_{\text{p}}(z) = \frac{\text{i}_{\text{dq}}(z)}{\text{u}_{\text{dq}}^{*}(z)} = \frac{T_s}{L(ze^{j\omega_e T_s} - 1)}. \tag{7.20}$$

Consider one sample period delay caused by digital implementation, and then

$$G_{\text{p\_d}}(z) = \frac{T_s}{Le^{j\omega_e T_s}(ze^{j\omega_e T_s} - 1)}. \tag{7.21}$$

Taking computation delay and pole-zero cancellation into consideration, the complex vector PI current regulator can be designed with direct digital design methods whose transfer function can be expressed as

$$G_{\text{cr}}(z) = \frac{k_{\text{dq}}(ze^{j\omega_e T_s} - 1)}{T_s(z - 1)} e^{j\omega_e T_s}. \tag{7.22}$$

By adding $\boldsymbol{R}i_{\text{dq}}(n)$ at the output of the current regulator to compensate for the resistance voltage drop neglected before, u_{dq}^{*} is acquired. To get consistent current response in d and q-axes, it is a commonly adopted procedure to make sure that $k_{\text{d}}/L_d = k_{\text{q}}/L_q$.

Closed-loop pole-zero migration of the modified complex-vector current regulator considering digital delay compensation with different ω_{cr} is shown in Fig. 7.8.

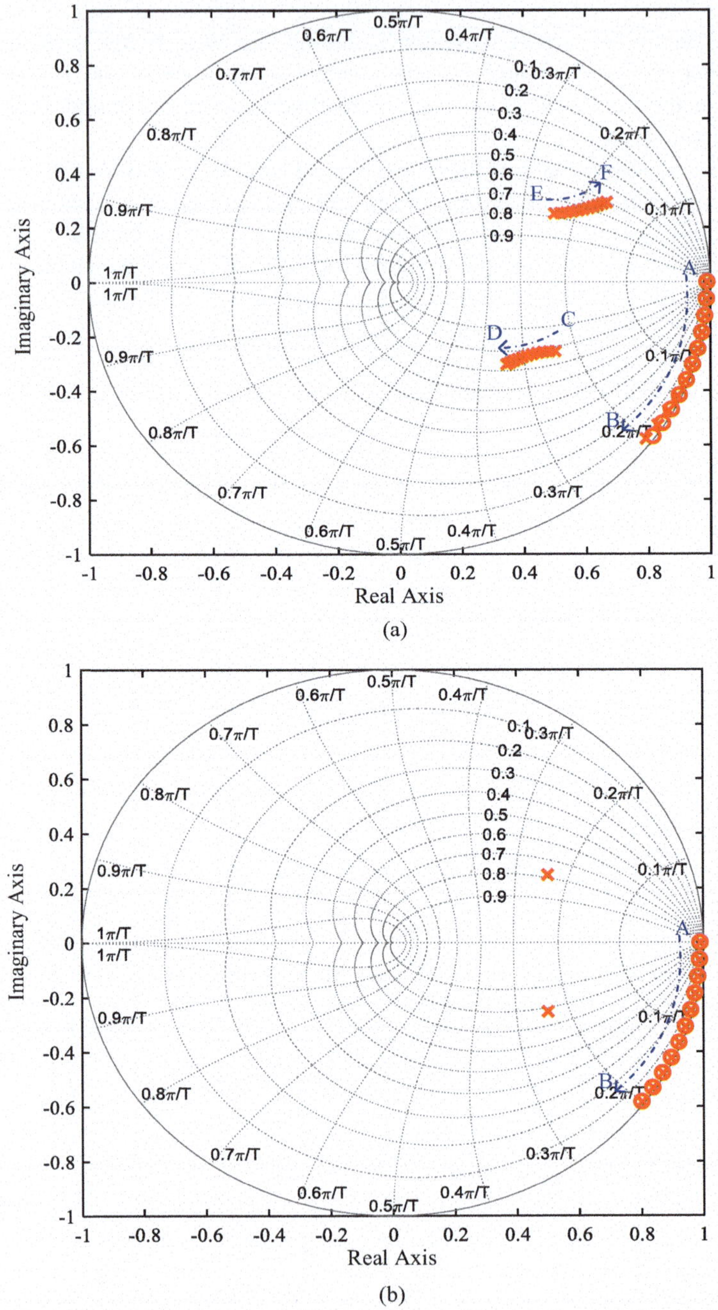

Fig. 7.6 Closed-loop pole-zero migration with a frequency increase for the complex-vector PI control. (**a**) With Tustin transform. (**b**) Direct transform

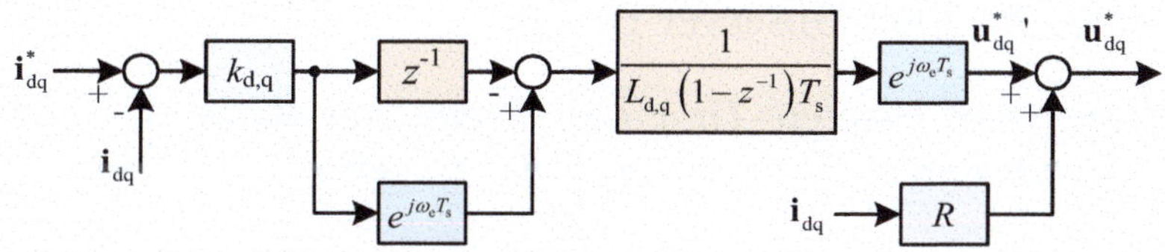

Fig. 7.7 Modified complex vector PI current regulator

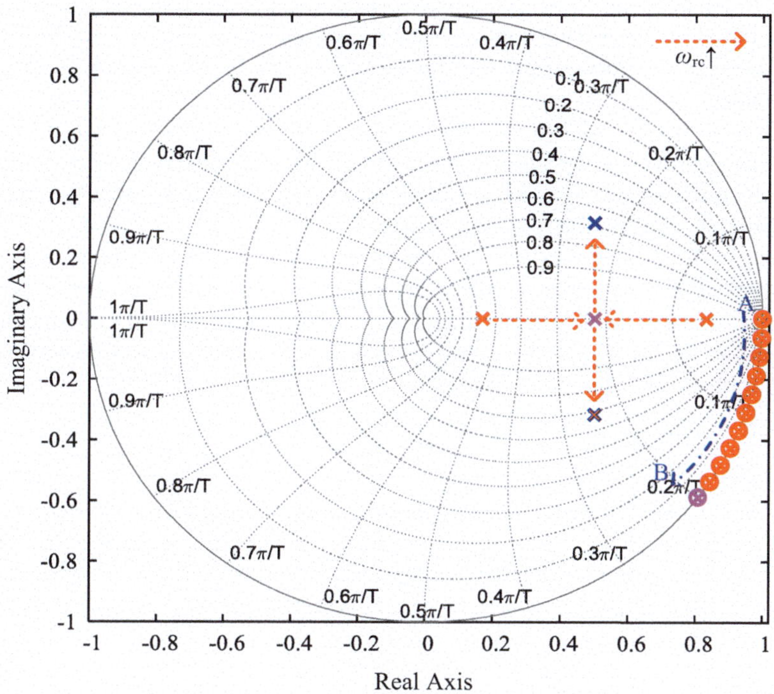

Fig. 7.8 Closed-loop pole-zero migration with a frequency increase for the modified complex-vector current regulator

The figure shows that the modified method can also make sure closed-loop pole-zero A-B is canceled out effectively when the fundamental frequency increases from 0 to 600 Hz and the stator current of d-axis decouples from that of q-axis, which shows the similar performance as the method provided for SPMSM.

7.5 Rotor Position Observer Design

7.5.1 Proposed Discrete-Time Low-Frequency-Ratio Synchronous-Frame Full-Order Observer

A discrete-time full-order observer in the d-q axis frame for position sensorless control of PMSM drives is proposed in this section to overcome some issues created by low-frequency ratio [13]. As shown in Fig. 7.9, the position sensorless control scheme based on the vector control principle which makes the synchronous frame oriented on the direction of the flux-linkage produced by the permanent magnet is implemented.

For conventional SMO in stationary frame, LPF is commonly used to filter HF components caused by the sign function. And if the state observer is constructed in α-β frame, phase delay exists when LPF is used to process AC signal, which makes it more difficult to design the position observer.

Since the EMF estimated in the synchronous frame is a DC signal, the phase delay caused by the intrinsic low-pass filtering characteristics of the position observer is evitable, which could facilitate the observer design and implementation.

As shown in Fig. 7.10, the proposed discrete-time domain synchronous-frame full-order observer can be constructed as

$$\hat{\mathrm{x}}(n+1) = \mathrm{A}_d\hat{\mathrm{x}}(n) + \mathrm{B}_d\mathrm{u}(n) + \mathrm{G}_d[\mathrm{y}(n) - \mathrm{C}_d\hat{\mathrm{x}}(n)], \tag{7.23}$$

where $\mathrm{G_d} = [g_1 g_2]^{\mathrm{T}}$ is the feedback gain matrix and ^ denotes the estimated value.

The eigenvalues (i.e., the closed-loop poles) of the proposed observer must be kept within the stability region through the optimally designed feedback gain matrix $\boldsymbol{G}_{\mathrm{d}}$. This can be obtained by assigning the observer poles as complex quantities

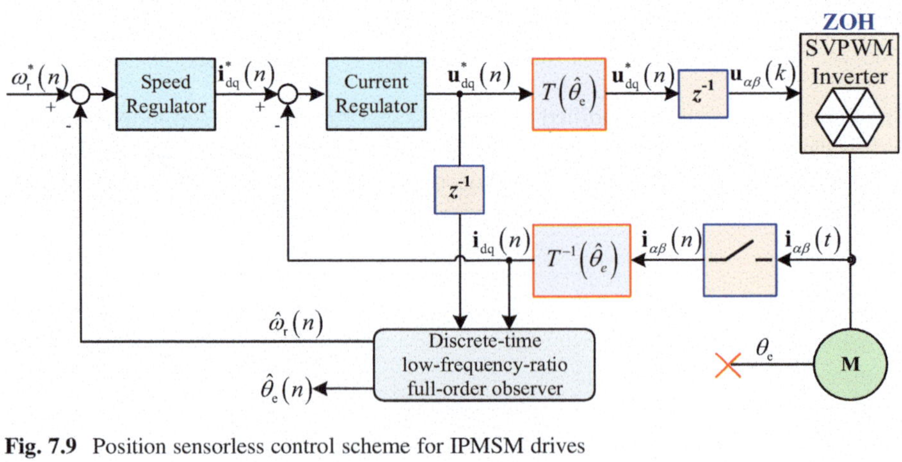

Fig. 7.9 Position sensorless control scheme for IPMSM drives

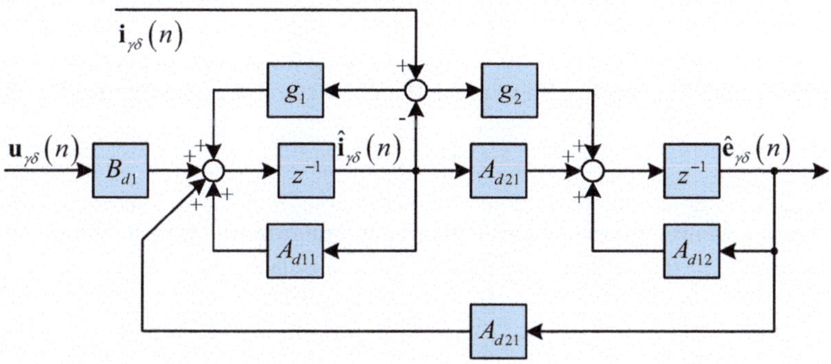

Fig. 7.10 Discrete-time synchronous-frame full-order observer

directly on the z-plane and then computing $g_1$and g_2 from the characteristic polynomial coefficients.

The characteristic polynomial of the observer is obtained as

$$\begin{aligned} P(z) &= det(z\mathrm{I} - \mathrm{A_d} + \mathrm{G_d C_d}) \\ &= z^2 - (A_E + \mathrm{A_{d22}} - g_1)z \\ &\quad + (\mathrm{A_{d22}}(\mathrm{A_{d11}} - g_1) - A_{\mathrm{d12}}A_{\mathrm{d21}} + \mathrm{A_{d12}}g_2), \end{aligned} \tag{7.24}$$

and the desired characteristic polynomial can be given as

$$P(z) = (z - p_1)(z - p_2), \tag{7.25}$$

where p_1 and p_2 are the desired closed-loop poles.

With (7.23) and (7.24), the feedback gain matrix $\boldsymbol{G}_\mathrm{d}$ can be obtained as

$$\begin{aligned} g_1 &= \mathrm{A_{d11}} + \mathrm{A_{d22}} - p_1 - p_2, \\ g_2 &= \mathrm{A_{d21}} + \frac{(\mathrm{A_{d22}} - p_1)(\mathrm{A_{d11}} - p_2)}{\mathrm{A_{d12}}}. \end{aligned} \tag{7.26}$$

As can be seen, the derivation of the feedback gain matrix $\boldsymbol{G}_\mathrm{d}$ has general validity. No assumption has been made yet on the discrete-time domain state-space model and on the desired closed-loop poles of the observer which means this method is suitable for different cases. It can be seen that the matrix exponential is required to implement the low-frequency-ratio full-order observer. And it can be achieved through a lookup table, which provides high discretization accuracy and low computational burden.

The desired closed-loop poles of the observer can be arbitrarily set within the limits of the accessible control effort and modeling precision. It is typically easier to specify the pole locations first in the continuous-time domain and then map them to the discrete-time domain via $z = e^{sT_s}$. When placing the poles in the continuous-time domain, it is often convenient to express the characteristic polynomial as

$$s^2 + 2\xi\omega_0 s + \omega_0^2, \tag{7.27}$$

where ξ is the damping ratio and ω_0 is the natural angular frequency of the observer, which is related to the desired bandwidth of the observer.

Thus, the feedback gain matrix $\boldsymbol{G}_\text{d}$ is fully determined by (7.26) and by the pole assignment dependent on ξ and ω_0. The damping ratio can be selected according to the pole location of the plant. From the system matrix shown in (7.6), the poles of the plant can be given as $s = -(R + j\omega_\text{q}L_\text{q})/L_\text{q}$. Therefore, if the damping ratio is selected as $\xi = R/\sqrt{R^2 + \omega_\text{e}^2 L_\text{q}^2}$, the poles of plant and observer have the same phase angle. But the drawback of this pole assignment approach is that when the operating fundamental frequency is increased, the poles would move closer to the unit circle, as shown in Fig. 7.11 (a). Therefore, to guarantee the stability of the observer even with low-frequency ratios, the damping ratio ξ is kept constant and set to $\sqrt{2}/2$ for

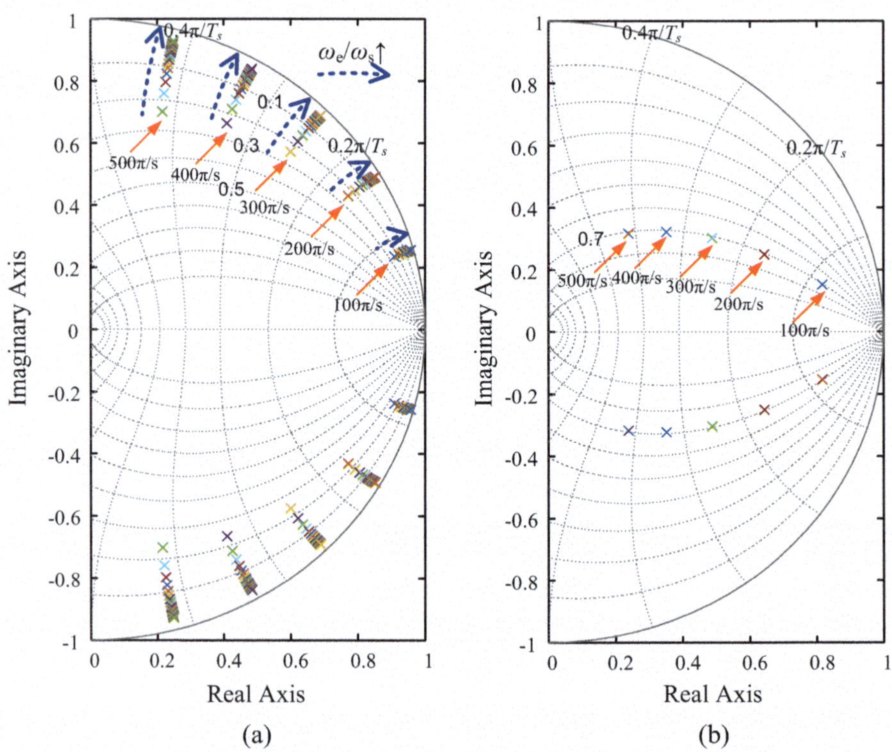

Fig. 7.11 Pole migration comparison of observer with operating fundamental frequency increase (f_ratio, 20 → 2.5). (**a**) Variable damping ratio. (**b**) Constant damping ratio

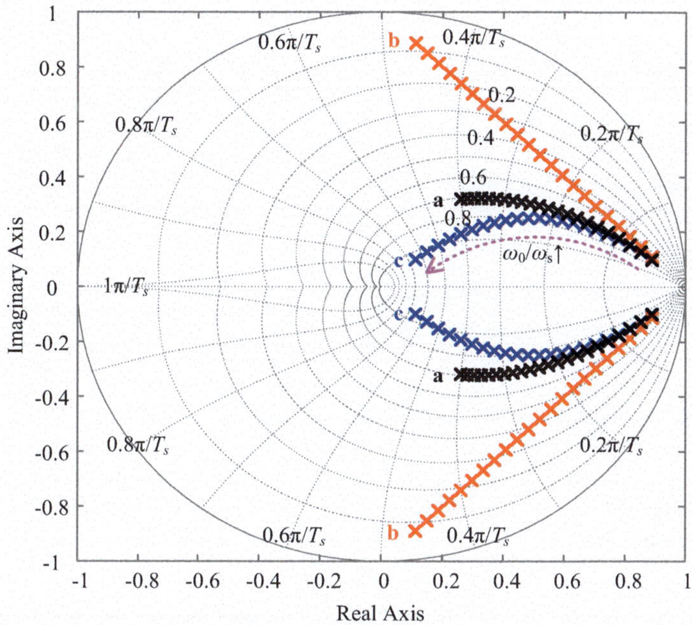

Fig. 7.12 Discrete-time pole migration comparison with ω_0 increase. (**a**: exact discretization; **b**: Euler approximation; **c**: Tustin approximation)

simplification. Due to the limit of Nquist frequency ($0.5f_{\text{sample}}$), the upper bound of the observer bandwidth is set two to five times lower than the PWM frequency.

Figure 7.12 shows the discrete-time pole migration comparison with increased ω_0. As can be seen, although the poles locate in the stable region, the system behavior of the discrete-time observer with Euler approximation is highly oscillatory when operating at high natural angular frequencies.

The system behavior of the discrete-time observer with Tustin approximation exhibits further improvement, whereas it suffers from the computational burden and performance degradation compared to the one with the exact ZOH equivalent model. In practical implementation, the natural angular frequency of the observer ω_0 is set to two-fifths of the PWM frequency.

As shown in Fig. 7.13, a tracking controller is implemented using a phase-locked loop with back EMF normalization to synchronize the estimated γ-δ axis frame to the actual d-q axis frame, and hence the rotor position and speed are estimated. This scheme can be achieved simply through regulating the estimated γ-axis back-EMF $\hat{e}_\gamma$ to zero. Besides, the feedback $\hat{e}_\gamma$ is multiplied by the sign of the back-EMF during the speed reversal to achieve four-quadrant operation. The back-EMF normalization can guarantee a constant linear dynamic response of the position tracking controller regardless of operating fundamental frequency.

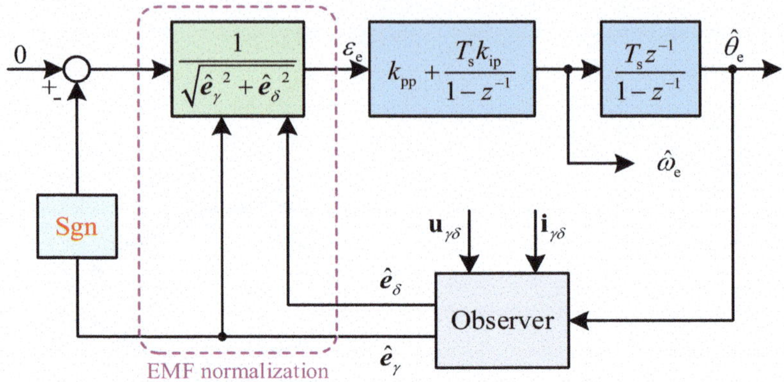

Fig. 7.13 Tracking controller with back EMF normalization

7.5.2 *Experimental Evaluation of the Proposed Position Observer*

To fully verify the validity and reliability of proposed method, experimental tests are implemented under various conditions.

Firstly, a 2.2 kW IPMSM test machine is experimentally evaluated to simulate high-power applications, where, notwithstanding there are many advances in power device technology, the PWM frequency is limited to values even lower than several hundreds of hertz, which keeps the switching losses of the power semiconductors at a tolerable level.

The rated speed of the tested IPMSM is 1500 r/min, which corresponds to 75 Hz. At the typical switching frequency 6 kHz, f_{ratio} is too large to investigate the proposed scheme. Therefore, the PWM frequency is reduced to 600 Hz, and asymmetric regularly sampled modulation is used. In this way the effect of low-frequency ratios could be emphasized.

Figure 7.14 shows the steady-state control performance using the proposed discrete-time synchronous-frame full-order observer with ω_0 set to two-fifths of the PWM frequency at 1500 r/min ($f_{\mathrm{ratio}} = 8$) under 50% rated load. From the experimental results, it can be seen that the estimated value of the γ- and δ-axis current (i_γ and i_δ) tracks the measured value well, as shown in Fig. 7.14b, and there is no visible difference between them. The γ- and δ-axis extended back-EMF estimation shown in Fig. 7.14c has good steady-state performance since the reasonably accurate estimates can be obtained. The rotor position estimation error is limited within 0.015π such that a stable low-frequency-ratio sensorless control can be maintained.

For comparison, as shown in Fig. 7.15, the steady-state control performance using the discrete-time observer with Euler approximation is also investigated at 1500 r/min ($f_{\mathrm{ratio}} = 8$) under 50% rated load.

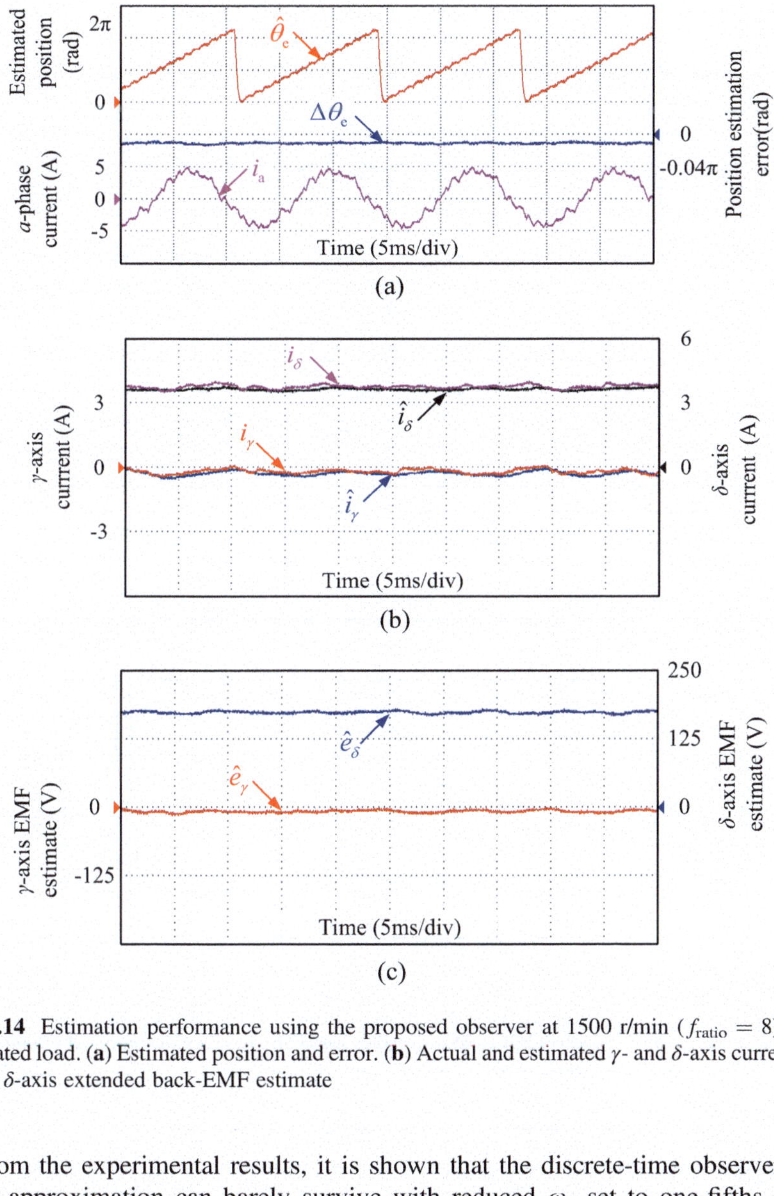

Fig. 7.14 Estimation performance using the proposed observer at 1500 r/min ($f_{\text{ratio}} = 8$) under 50% rated load. (**a**) Estimated position and error. (**b**) Actual and estimated γ- and δ-axis currents. (**c**) γ- and δ-axis extended back-EMF estimate

From the experimental results, it is shown that the discrete-time observer with Euler approximation can barely survive with reduced ω_0 set to one-fifths of the PWM frequency, which is not the limit of the discrete-time observer's stability but the limit of the experimental setup. Furthermore, it is worth noting that an increased ripple of position estimation error appears due to the first-order truncation error.

As shown in Fig. 7.16, the proposed discrete-time synchronous-frame full-order observer with a smaller f_{ratio} (4800 r/min, $f_{\text{ratio}} = 2.5$) at no load condition is further evaluated. The rotor position estimation error is limited within 0.02π such that a

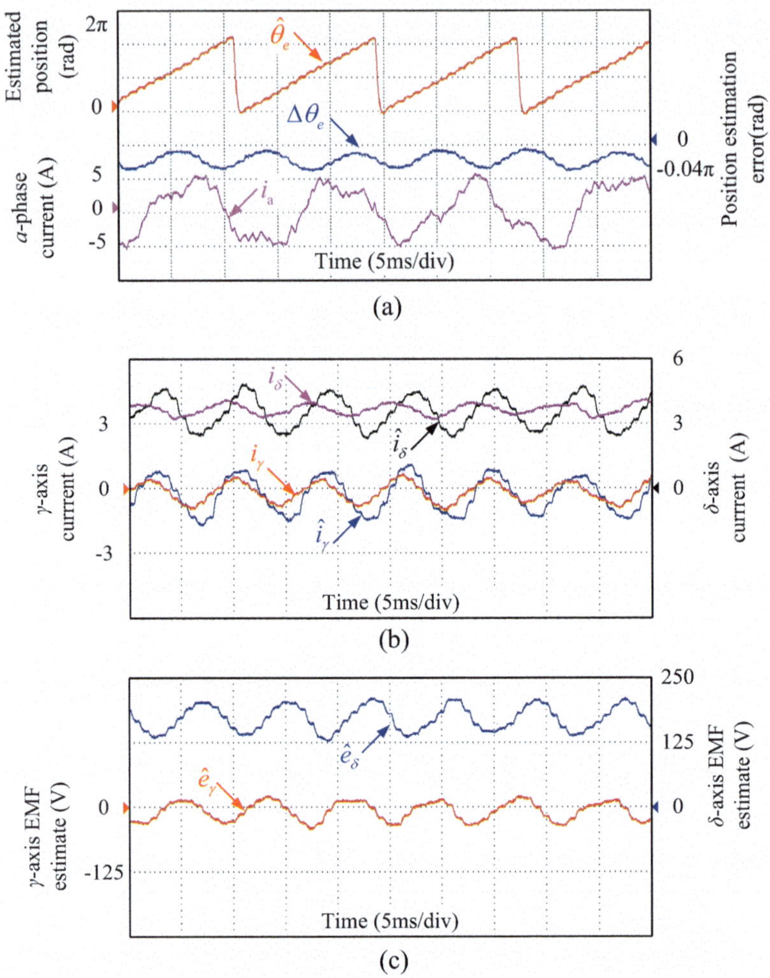

Fig. 7.15 Estimation performance using the proposed observer at 1500 r/min ($f_{\text{ratio}} = 8$) under 50% rated load. (**a**) Estimated position and estimation error. (**b**) Actual and estimated γ- and δ-axis currents. (**c**) γ- and δ-axis extended back-EMF estimate

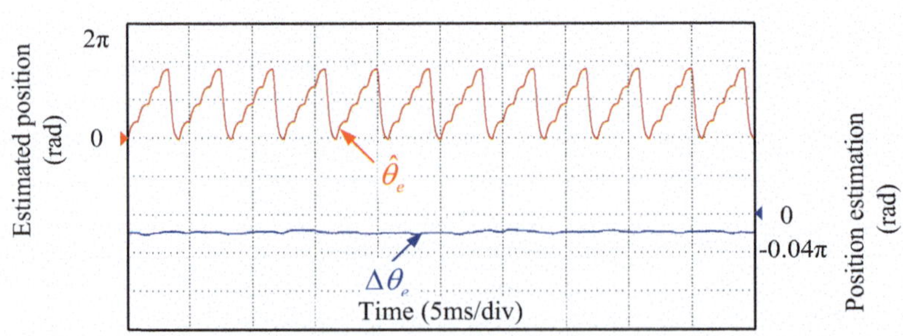

Fig. 7.16 Further evaluation of the proposed observer with a smaller frequency ratio ($f_{\text{ratio}} = 2.5$)

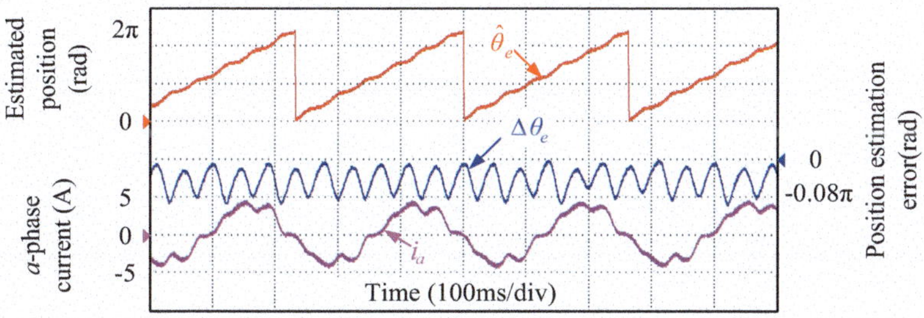

Fig. 7.17 Experimental tests on the estimated position and error for low-speed operation

stable low-frequency-ratio sensorless control can be maintained. It is worth noting that the observer with Euler approximation cannot survive with such a smaller f_{ratio}.

As shown in Fig. 7.17, the experimental tests on the estimated position and error at 75 r/min (5% rated speed, $f_{\text{ratio}} = 160$) under 50% rated load are carried out to investigate the low-speed performance of the proposed discrete-time low-frequency-ratio synchronous-frame full-order observer.

It can be found from the experimental results that the estimated rotor position is distorted with an undesired sixth harmonic fluctuation up to 0.1π due to the model uncertainties and the inverter nonlinearities, which deteriorates low-speed performance.

As shown in Fig. 7.18, the experimental tests on the effect of parameter variations at 75 r/min (5% rated speed) under 50% rated load are carried out to further investigate the proposed discrete-time low-frequency-ratio synchronous-frame full-order observer. Within the tests, a 30% step increase has been applied to the stator resistance R, and a 30% step decrease has been applied to the d-q axis inductances L_d and L_q. It can be seen from the experimental results that the proposed observer provides good tolerance against parameter variations. The estimated rotor position, however, is distorted with an undesired sixth harmonic fluctuation due to the inverter nonlinearities. Besides, a DC-offset is induced with the decrease in L_q, and the harmonic fluctuation amplitude is increased with the increase in R. Future works are expected to deal with both model uncertainties and inverter nonlinearities by working on several improvement strategies.

As shown in Fig. 7.19, the proposed discrete-time low-frequency-ratio synchronous-frame full-order observer has been investigated for the four-quadrant sensorless operation at 1500 r/min ($f_{\text{ratio}} = 8$) under 50% rated load, covering a wide speed range. As can be seen, the sensorless drive system is always stable, and the position estimation error is limited within 0.03π under an acceptable range.

To further investigate the proposed discrete-time synchronous-frame full-order observer versus the observer with Euler approximation, the position estimation error comparison over different frequency ratios has been explored as shown in Fig. 7.20. As can be seen from the experimental results, the peak-to-peak value of position estimation error obtained through the proposed observer can be reduced effectively over different frequency ratios, which contributes to providing better position

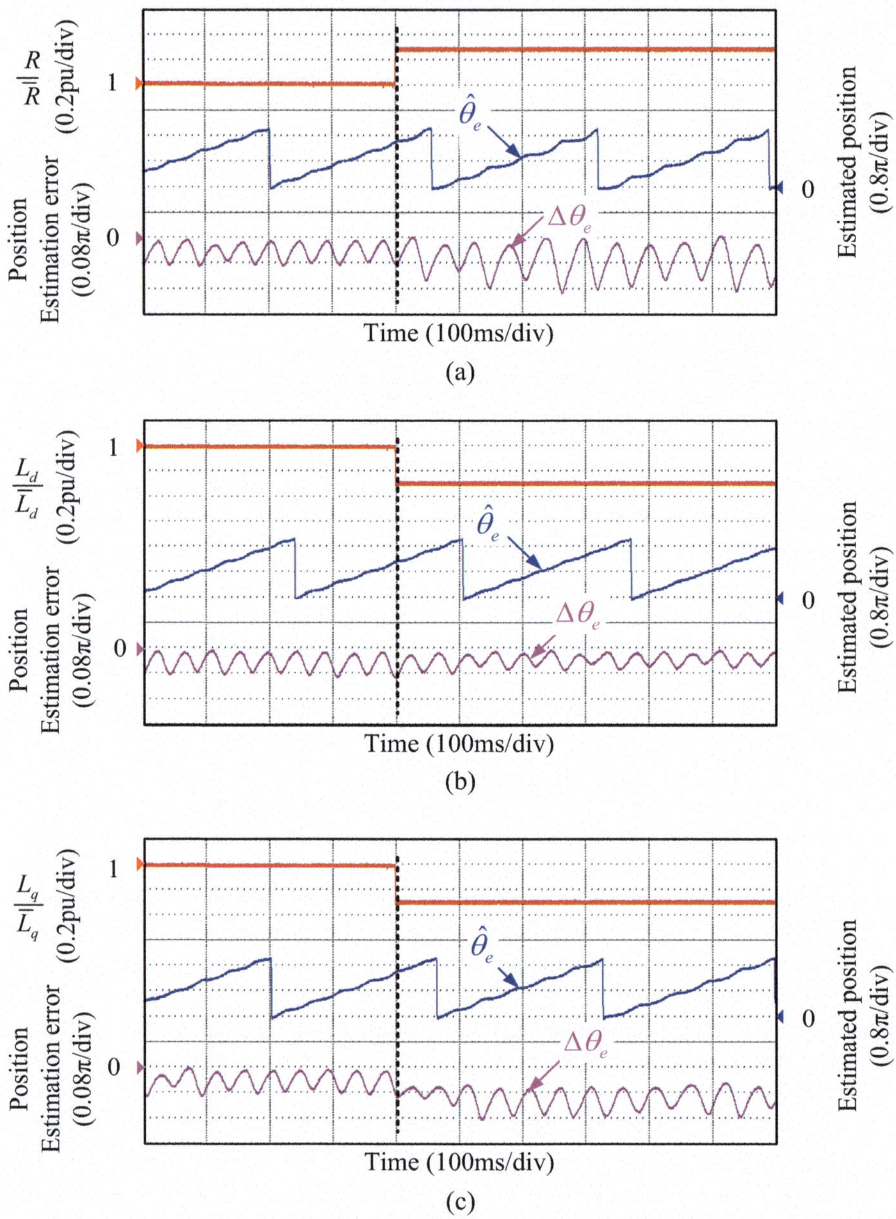

Fig. 7.18 Experimental tests on the effect of parameter variations for low-speed operation. (**a**) A 30% step decrease applied to L_q. (**b**) A 30% step decrease applied to L_d. (**c**) A 30% step increase applied to R

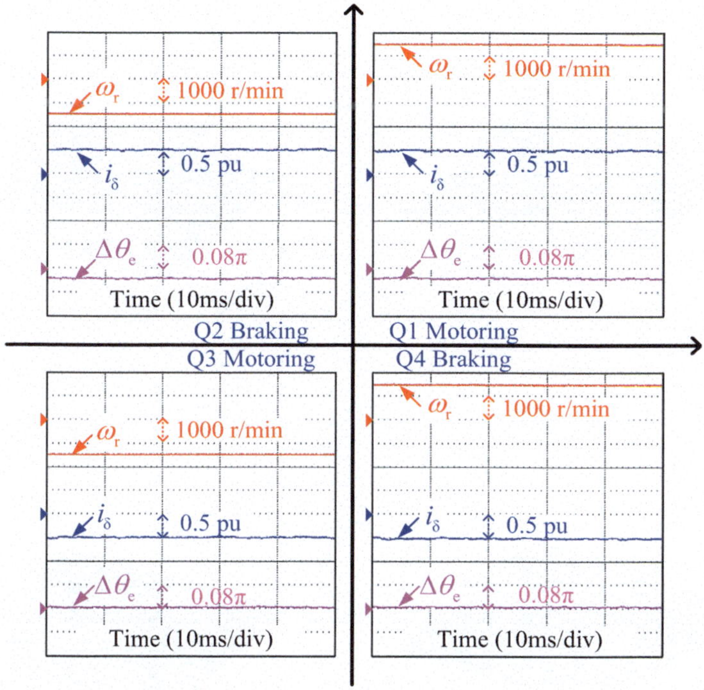

Fig. 7.19 Performance of the sensorless drive under four-quadrant operations

estimation accuracy. It is worth noting that the discrete-time observer with Euler approximation cannot survive as ω_0 increases to one-thirds of the PWM frequency, and it becomes unstable with a large load disturbance when compared with the proposed observer.

Figure 7.21 evaluates the step load disturbance response from 20% to 80% rated load at 1500 r/min ($f_{\text{ratio}} = 8$). From the experimental results, it can be noticed that the proposed discrete-time synchronous-frame full-order observer behaves well and the position estimation error can be limited within 0.05π during transients. In contrast, the observer with Euler approximation becomes unstable with a large load disturbance.

Figure 7.22 shows the response with speed variations with constant feedback gains of observer. From the experimental results, it can be seen that the proposed discrete-time synchronous-frame full-order observer behaves well and the position estimation error can be limited under an acceptable range during the speed variations from 1500 r/min ($f_{\text{ratio}} = 8$) to 4800 r/min ($f_{\text{ratio}} = 2.5$). When the speed rises continuously to 5160 r/min ($f_{\text{ratio}} \approx 2.32$), the system goes unstable because the

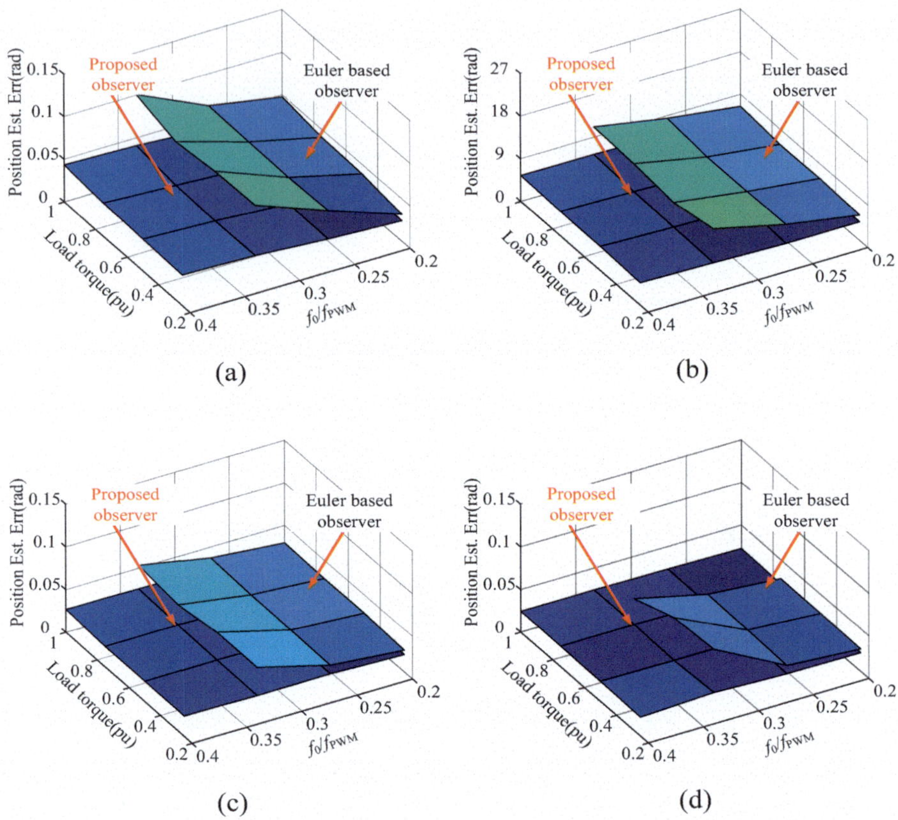

Fig. 7.20 Position estimation error experimental comparison results over different frequency ratios. (**a**) $f_{\text{ratio}} = 15$ (800 r/min) (**b**) $f_{\text{ratio}} = 12$ (1000 r/min) (**c**) $f_{\text{ratio}} = 10$ (1200 r/min) (**d**) $f_{\text{ratio}} = 8$ (1500 r/min)

harmonics noise due to the PWM voltage harmonics and the modeling error of PMSM is difficult to be eliminated. And meanwhile, the response with speed variations in open-loop estimation operation using the discrete-time observer with Euler approximation is also comparatively investigated. As can be seen, the observer becomes unstable as the speed increases up to 1700 r/min, corresponding to $f_{\text{ratio}} \approx 7$.

Next part evaluates high-speed EMF estimation on a high-speed IPMSM and a high-speed SPMSM, respectively, using proposed discrete-time synchronous-frame full-order observer.

A 4-pole high speed IPMSM with 18-kr/min rated dehydration speed used in a washing machine is tested first, as shown in Fig. 7.23. Its parameters of it are listed in Table 7.1.

The PWM frequency is set as 5 kHz for the switching control. Figure 7.24 shows experimental results at the speed of 18,000 r/min ($f_{\text{ratio}} \approx 4.167$) and 15,000 r/min ($f_{\text{ratio}} \approx 5$).

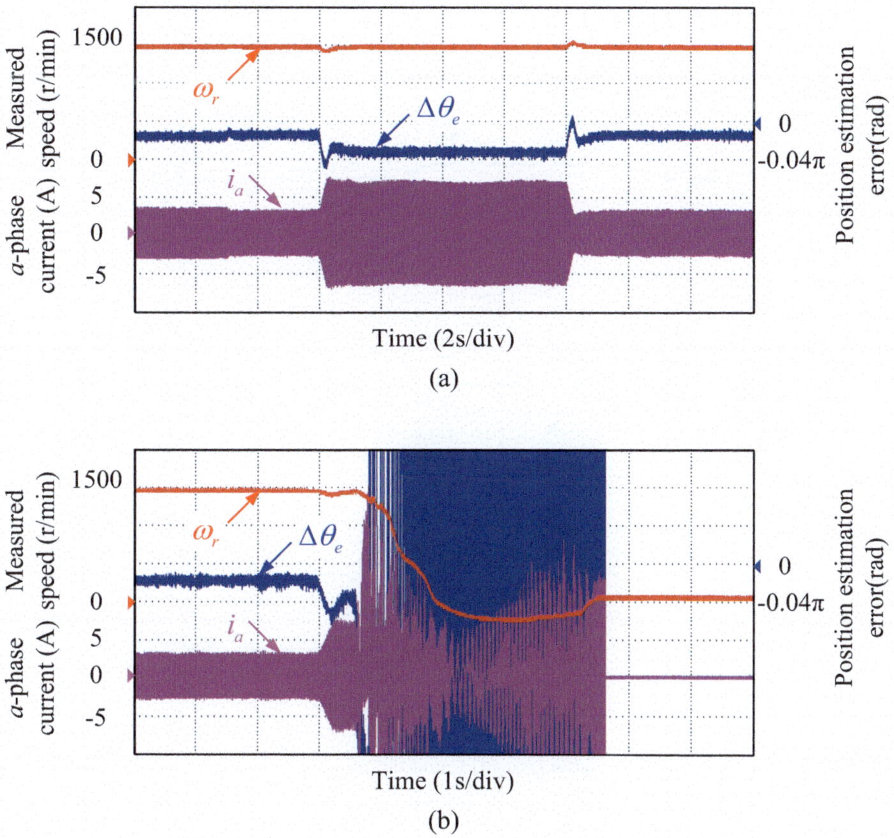

Fig. 7.21 Step load disturbance response results. (**a**) Using the proposed observer. (**b**) Using the Euler based observer

As seen in Fig. 7.24, the proposed discrete-time synchronous-frame full-order observer shows satisfactory position and speed estimation performance in different working conditions, which means the proposed method is suitable for different IPMSM.

Next, a 4-pole high-speed INVT SPMSM with 9000 r/min rated speed and 18,000 r/min maximum speed is tested, as shown in Fig. 7.25. Its parameters are listed in Table 7.2.

The PWM frequency is set as 8 kHz. Then the PWM frequency is reduced to 2.5 kHz ($f_{\text{ratio}} \approx 4.167$). Figure 7.26 shows experimental results at 18,000 r/min maximum speed ($f_{\text{ratio}} \approx 4.167$). Obviously, the proposed discrete-time synchronous-frame full-order observer works well in estimating position and speed. The maximum error in estimating position is merely 5.4° at the condition of $f_{\text{PWM}} = 4$ kHz, which means the proposed method is also suitable for SPMSM.

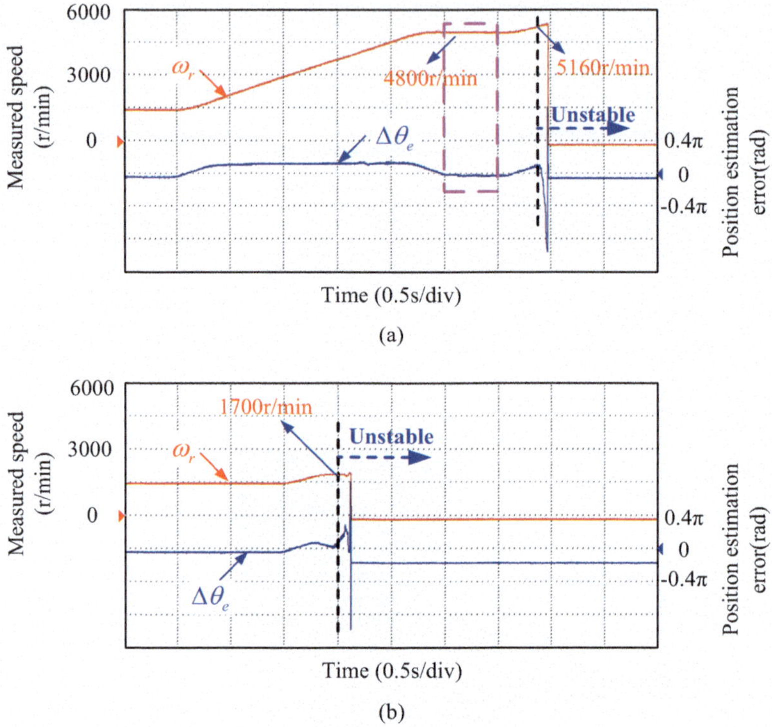

Fig. 7.22 Response with speed variations. (**a**) Using the proposed observer. (**b**) Using the Euler-based observer

Fig. 7.23 Experimental test setup of washing machine

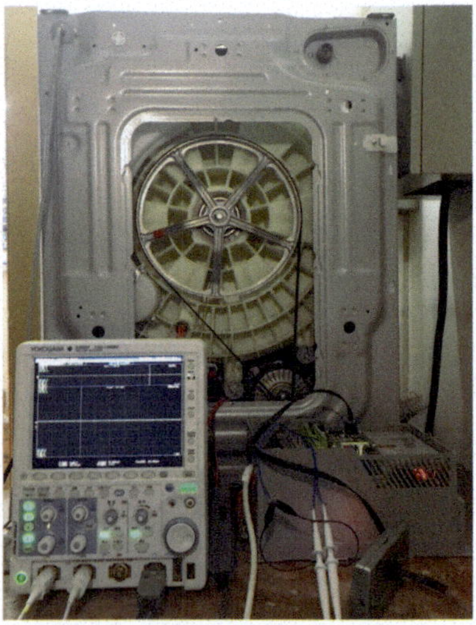

Table 7.1 IPMSM parameters of the washing machine

Parameter	Value	Parameter	Value
Rated spin-dry power (kW)	0.5	Pole-pairs	4
Rated spin-dry speed (r/min)	18,000	Stator resistance (Ω)	3.24
Rated spin-dry frequency (kHz)	1.2	*d*-axis inductance (mH)	18
Rated spin-dry current (A)	2.3	*q*-axis inductance (mH)	23

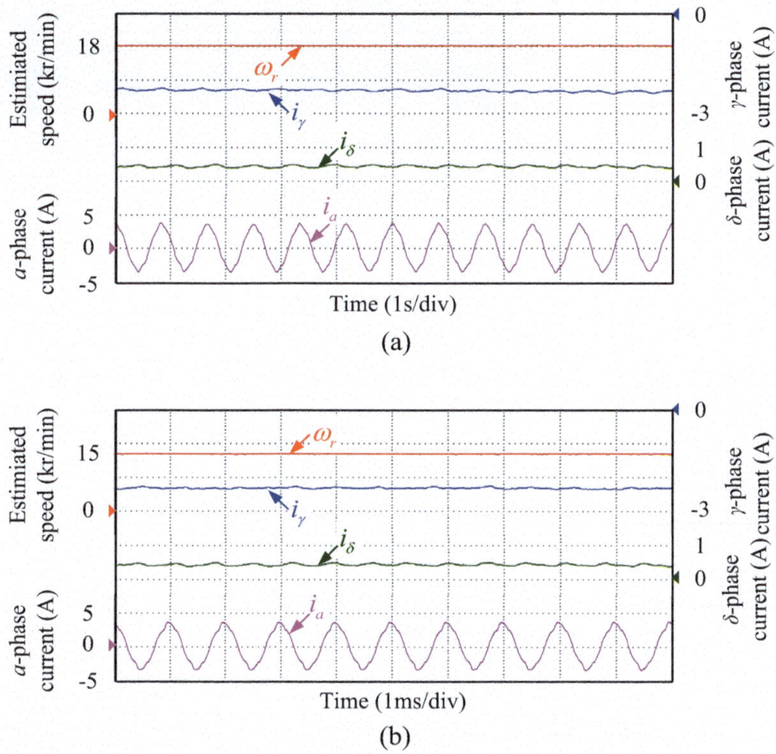

Fig. 7.24 Experimental tests with Haier washing machine. (**a**) Experimental results at 18000 r/min ($f_{\text{ratio}} \approx 4.167$). (**b**) Experimental results at 15,000 r/min ($f_{\text{ratio}} = 5$)

Figure 7.27 shows the experimental results with speed variations using the proposed discrete-time synchronous-frame full-order observer. From the experimental results, it can be seen that the acceleration time can be limited under 0.5 s and the deceleration time can be limited under 2 s during the speed that varies from 0 to 18,000 r/min ($f_{\text{PWM}} = 4$ kHz, $f_{\text{ratio}} = 6.67$) indicating that the proposed method works well for SPMSM drive system.

Fig. 7.25 Experimental test setup of high-speed SPMSM

Table 7.2 SPMSM parameters

Parameter	Value	Parameter	Value
Rated power (kW)	6	Rated speed (r/min)	9000
Pole-pairs	2	Maximum speed (r/min)	18,000
Stator resistance (Ω)	0.17	Rated frequency (Hz)	300
Stator inductance (mH)	0.75	Maximum frequency (Hz)	600

7.6 Summary

In this chapter, the complex vector PI current regulator, which is designed with direct digital design methods to cancel out the undesired pole-zero, is firstly proposed. Then the complex-vector-based discrete-time synchronous-frame full-order observer is proposed for position sensorless PMSM drives with low-PWM-to-operating fundamental frequency ratio. Taking the stationary frame voltage latch and the control computation delay into consideration, the complex-vector-based discrete-time domain ZOH equivalent model of IPMSM is developed explicitly. On this basis, the full-order observer is designed and implemented using direct pole assignment in the discrete-time domain. Experimental results using the proposed scheme on different PMSM platforms are carried out to evaluate the performance and demonstrate how the desired properties have been achieved.

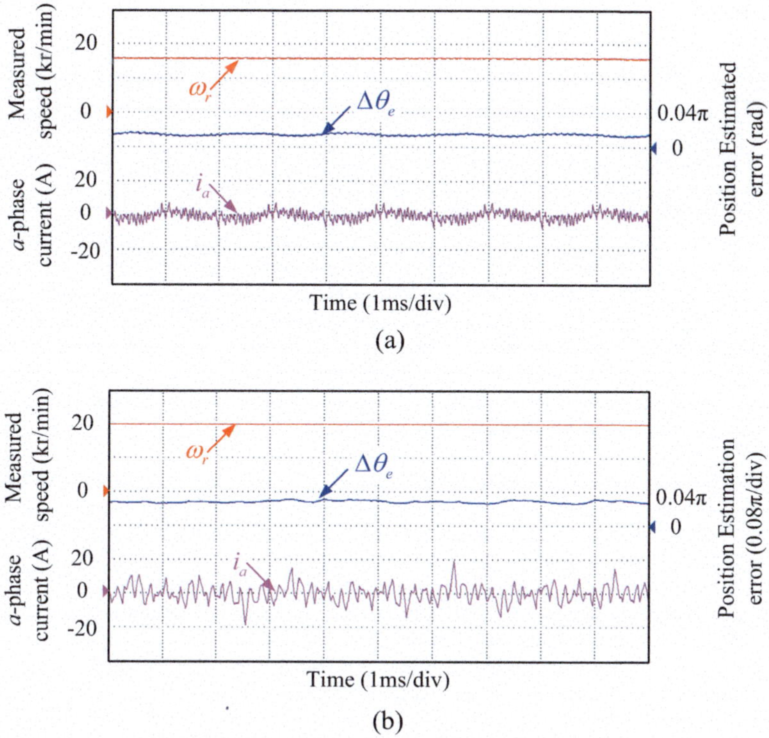

Fig. 7.26 Experimental tests at the highest fundamental operating speed 18,000 r/min. (**a**) Experimental results at $f_{\mathrm{PWM}} = 8$ kHz ($f_{\mathrm{ratio}} \approx 13.3$). (**b**) Experimental results at $f_{\mathrm{PWM}} == 4$ kHz ($f_{\mathrm{ratio}} \approx 6.67$)

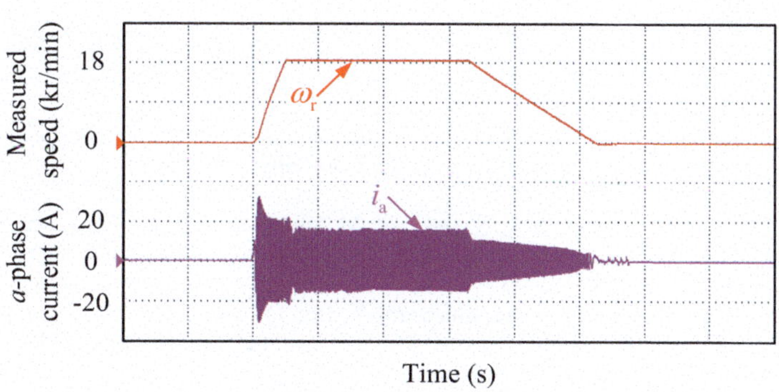

Fig. 7.27 Experimental result of operating speed variations

References

1. D. Gerada, A. Mebarki, N. L. Brown, C. Gerada, A. Cavagnino, A. Boglietti, High-speed electrical machines: technologies trends and developments. IEEE Trans. Ind. Electron. **61**(6), 2946–2959 (June 2014)
2. J.-S. Yim, S.-K. Sul, B.-H. Bae, N. R. Patel, and S. Hiti, Modified current control schemes for high-performance permanent-magnet AC drives with low sampling to operating frequency ratio. IEEE Trans. Ind. Appl. **45**(2), 763–771 (March/April 2009)
3. M. Abrate, G. Griva, F. Profumo, A. Tenconi, High speed sensorless fuzzy-like Luenberger observer. Proc. IEEE PESC. **1**, 477–481 (September 1999)
4. G. Griva, F. Profumo, L. Rosell, R. Bojoi, Optimization of fuzzy-like Luenberger observer for high speed sensorless induction motor drives using genetic algorithms. Proc. IAS. **2**, 1268–1274 (October 2000)
5. Y. Zhao, W. Qiao, L. Wu, Improved rotor position and speed estimators for sensorless control of interior permanent-magnet synchronous machines. IEEE J. Emerg. Sel. Topics Power Electron. **2**(3), 627–639 (September 2014)
6. B.-H. Bae, S.-K. Sul, A compensation method for time delay of full-digital synchronous frame current regulator of PWM AC drives. IEEE Trans. Ind. Appl. **39**(3), 802–810 (May 2003)
7. J. Shen, S. Schröder, H. Stagge, R. W. De Doncker, Precise modeling and analysis of DQ-frame current controller for high power converters with low pulse ratio. Proc. IEEE Energy Convers. Congr. Expo., 61–68 (September 2012)
8. J. Holtz, J. Quan, J. Pontt, J. Rodriguez, P. Newman, H. Miranda, Design of fast and robust current regulators for high-power drives based on complex state variables. IEEE Trans. Ind. Appl. **40**(5), 1388–1397 (September 2004)
9. W. Xu, R. D. Lorenz, Low-sampling-frequency stator flux linkage observer for interior permanent-magnet synchronous machines. IEEE Trans. Ind. Appl. **51**(5), 3932–3942 (September/October 2015)
10. Y. Wang, S. Tobayashi, R. D. Lorenz, A low-switching-frequency flux observer and torque model of deadbeat–direct torque and flux control on induction machine drives. IEEE Trans. Ind. Appl. **51**(3), 2255–2267 (May/June 2015)
11. Y. Zhao, W. Qiao, L. Wu, Sensorless control for IPMSMs based on a multilayer discrete-time sliding-mode observer. Proc. IEEE Energy Convers. Congr. Expo, 1788–1795 (September 2012)
12. M. Hinkkanen, H. A. A. Awan, Z. Qu, T. Tuovinen, F. Briz, Current control for synchronous motor drives: direct discrete-time pole-placement design. IEEE Trans. Ind. Appl. **52**(2), 1530–1541 (March/April 2016)
13. G. Zhang, G. Wang, D. Xu, Y. Yu, Discrete-time low-frequency-ratio synchronous-frame full-order observer for position sensorless IPMSM drives. IEEE J. Emerging Select. Topics Power Electron. **5**(2), 870–879 (June 2017)
14. H. A. A. Awan, T. Tuovinen, S. E. Saarakkala, M. Hinkkanen, Discrete-time observer design for sensorless synchronous motor drives. IEEE Trans. Ind. Appl. **52**(5), 3968–3979 (September/October 2016)
15. K. Hongrae, M. W. Degner, J. M. Guerrero, F. Briz, R. D. Lorenz, Discrete-time current regulator design for AC machine drives. IEEE Trans. Ind. Appl. **46**(4), 1425–1435 (July/August 2010)

Chapter 8
Motor Parameter Self-Commissioning at Standstill

The self-commissioning method for PMSMs at standstill is presented in this chapter, which requires no additional hardware but only a VSI. Considering detailed identification error analysis of inverter nonlinearity influence at different rotor positions, the inductance parameter identification method is introduced in detail. After estimating the initial rotor position, the inductances are identified by injecting high frequency signal in the estimated *d*- and *q*-axes, respectively. The proposed inductance identification belongs to the frequency-domain approach. Additionally, a supplementary DC current is injected into the *d*-axis to attenuate the influence of the inverter nonlinearities and keep the rotor at the initial position.

The identified inductance results are compensated for a nonlinear model considering the parasitic capacitance effect and the influence of cross coupling. Different cases of the voltage error distribution with different DC offsets and HF amplitudes are classified, which can describe the negative effects of inverter nonlinearities effectively. In order to diminish the influence of the inverter nonlinearities during the resistance identification, a linear regression-based resistance identification methodology is proposed by injecting a linearly increasing current in *d*-axis. Due to the induction machine winding resistance identification methods by injecting current in any two phases or three phases cause the alignment torque, the space current vector injected in *d*-axis is proposed to address this issue.

8.1 Initial Rotor Position Detection

8.1.1 Rotor Magnetic Pole Position Detection with HF Pulse Voltage Signal Injection Method

Accurate detection of the initial rotor position at standstill is the key to the normal starting operation of the sensorless control for PMSM drives, which can avoid

G. Wang et al., *Position Sensorless Control Techniques for Permanent Magnet Synchronous Machine Drives*, https://doi.org/10.1007/978-981-15-0050-3_8

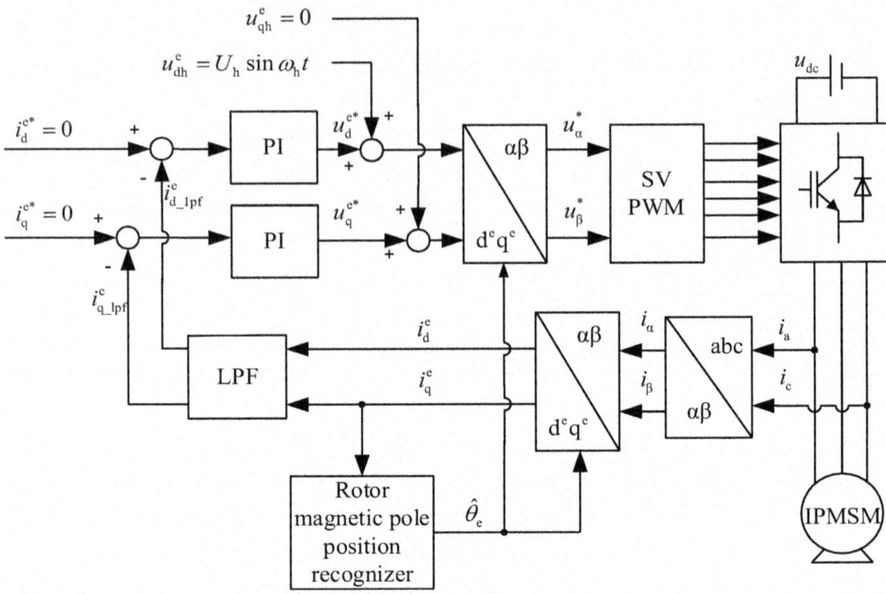

Fig. 8.1 Block diagram of rotor position information extraction

unnecessary reversal and start failure. Besides, it can also ensure to provide sufficient starting torque. According to the saliency of IPMSMs, the rotor position information can be obtained by adopting the composite signal injection method in the stationary state. Firstly, the rotor magnetic pole position information can be detected by the pulse HF voltage signal injection method, and then the rotor magnetic pole polarity information can be extracted by the pulse voltage signal injection method based on the stator core saturation nonlinear characteristics [1].

Before the motor startup, the pulse HF voltage signal injection method is used to detect the rotor magnetic pole position information. The HF sinusoidal voltage signal is injected into the estimated d^e-axis, and then the HF response current in the d^m-q^m frame can be used to obtain rotor magnetic pole position information. Figure 8.1 shows the block diagram of the rotor position detected by the pulse HF voltage signal injection method in the d^e-q^e frame.

As shown in Fig. 8.1, the HF voltage signal injected into the d^e-q^e axis frame can be expressed as

$$\begin{bmatrix} u^{e}_{dh} \\ u^{e}_{qh} \end{bmatrix} = \begin{bmatrix} U_h \sin \omega_h t \\ 0 \end{bmatrix}, \tag{8.1}$$

where U_h and ω_h are the amplitude and the frequency of the injected HF voltage signal, respectively. Then

$$\begin{bmatrix} u_{\mathrm{dh}}^{\mathrm{m}} \\ u_{\mathrm{qh}}^{\mathrm{m}} \end{bmatrix} = \frac{\sqrt{2}}{2} U_{\mathrm{h}} \sin \omega_{\mathrm{h}} t \begin{bmatrix} 1 \\ 1 \end{bmatrix}. \tag{8.2}$$

The HF response current signal will contain corresponding rotor position information, and the HF response current in the d^{m}-q^{m} frame can be expressed as

$$p\begin{bmatrix} i_{\mathrm{dh}}^{\mathrm{m}} \\ i_{\mathrm{qh}}^{\mathrm{m}} \end{bmatrix} = \frac{\sqrt{2}}{2} \frac{U_{\mathrm{h}} \sin \omega_{\mathrm{h}} t}{(\Sigma L^2 - \Delta L^2)} \begin{bmatrix} \Sigma L - \Delta L(\cos 2\tilde{\theta}_{\mathrm{e}} + \sin 2\tilde{\theta}_{\mathrm{e}}) \\ \Sigma L - \Delta L(\cos 2\tilde{\theta}_{\mathrm{e}} - \sin 2\tilde{\theta}_{\mathrm{e}}) \end{bmatrix}. \tag{8.3}$$

By simplifying (8.3), (8.4) can be obtained as

$$p\left(i_{\mathrm{qh}}^{\mathrm{m}} - i_{\mathrm{dh}}^{\mathrm{m}}\right) = \frac{U_{\mathrm{h}} \sin \omega_{\mathrm{h}} t}{\sqrt{2}(\Sigma L^2 - \Delta L^2)} \cdot 2\Delta L \sin\left(2\tilde{\theta}_{\mathrm{e}}\right). \tag{8.4}$$

Assuming the position error is close to zero, (8.4) can be approximated as

$$p\left(i_{\mathrm{qh}}^{\mathrm{m}} - i_{\mathrm{dh}}^{\mathrm{m}}\right) \approx \frac{2\sqrt{2}\Delta L U_{\mathrm{h}} \sin \omega_{\mathrm{h}} t}{\Sigma L^2 - \Delta L^2} \tilde{\theta}_{\mathrm{e}}. \tag{8.5}$$

Therefore, according to (8.5), the rotor magnetic pole position error signal ε_{h} can be obtained by extracting the amplitude of the HF exciting current in the d^{m}-q^{m} frame, as shown in (8.6).

$$\varepsilon_{\mathrm{h}} = \frac{\omega_{\mathrm{h}}\left(\Delta L^2 - \Sigma L^2\right)}{2\sqrt{2}\Delta L U_{\mathrm{h}}}\left(I_{\mathrm{qh}}^{\mathrm{m}} - I_{\mathrm{dh}}^{\mathrm{m}}\right) = k_{\mathrm{h}}\left(I_{\mathrm{qh}}^{\mathrm{m}} - I_{\mathrm{dh}}^{\mathrm{m}}\right), \tag{8.6}$$

where $k_{\mathrm{h}} = \frac{\omega_{\mathrm{h}}\left(\Delta L^2 - \Sigma L^2\right)}{2\sqrt{2}\Delta L U_{\mathrm{h}}}$.

Figure 8.2 shows the signal processing for extracting the rotor magnetic pole position information with the HF pulse signal injection method.

As shown in Fig. 8.2, after the bandpass filter filtering process, the HF response current in the d^{m}-q^{m} axis frame are multiplied by 2 cos $\omega_{\mathrm{h}} t$ separately. The LPF is used to filter out the HF noise to obtain the amplitude of the HF response current in the d^{m}-q^{m} axis frame, $I_{\mathrm{dh}}^{\mathrm{m}}$ and $I_{\mathrm{qh}}^{\mathrm{m}}$.

Then the rotor magnetic pole position error signal ε_{h} is extracted according to (8.6), and the rotor magnetic pole position information can be tracked by a software phase-locked loop. In the implementation of the estimation scheme, according to the amplitude of the HF response current $I_{\mathrm{h}}^{\mathrm{m}} = \sqrt{I_{\mathrm{dh}}^{\mathrm{m}\,2} + I_{\mathrm{qh}}^{\mathrm{m}\,2}}$ in the d^{m}-q^{m} axis frame, the rotor magnetic pole position error signal ε_{h} is normalized to reduce the effect of motor parameters and improve the accuracy of initial position estimation.

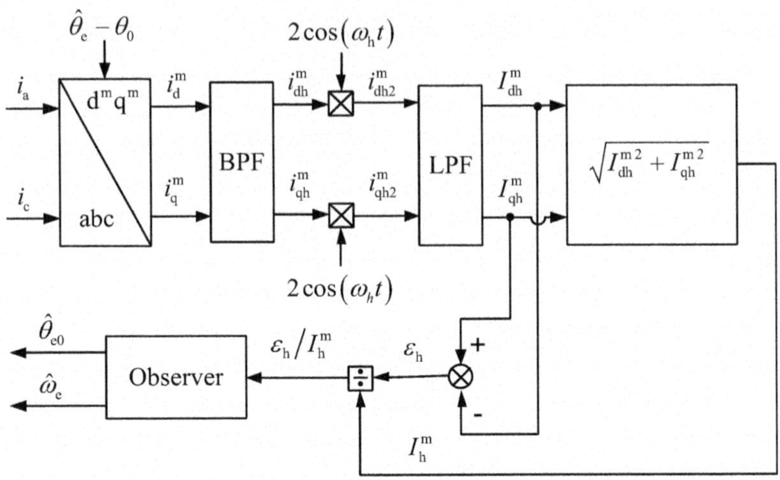

Fig. 8.2 Signal processing for extracting the rotor magnetic pole position information

Fig. 8.3 Schematic diagram of stator core saturation nonlinearity

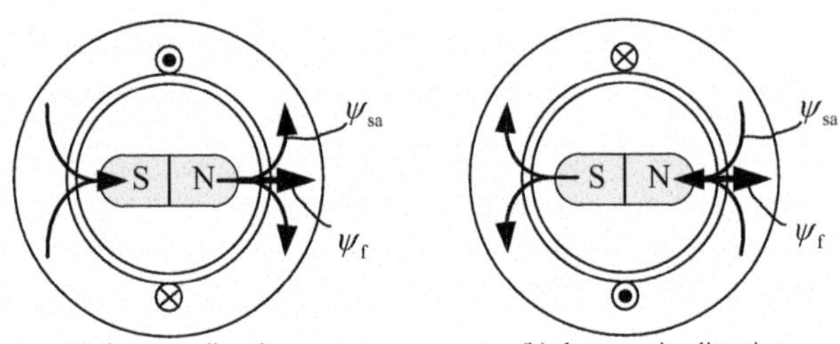

8.1.2 *Rotor Magnetic Pole Polarity Information Detection with Pulse Voltage Signal Injection Method*

The air gap magnetic field of PMSMs is the synthetic magnetic field formed by the stator armature reaction magnetic field and the permanent magnet excitation magnetic field in the air gap [2], namely,

$$\psi_{sg} = \psi_f + \psi_{sa}, \tag{8.7}$$

where ψ_{sg} is the air gap magnetic field and ψ_{sa} is the stator armature reaction magnetic field.

The stator core has saturation nonlinearity, as shown in Fig. 8.3. If the armature reaction magnetic field ψ_{sa} and the permanent magnet excitation magnetic field ψ_f are in the same direction, the armature reaction magnetic field ψ_{sa} will act as a magnetization. The stator core saturation will enhance, and the permeance and the stator inductance will decrease. Conversely, if ψ_{sa} and ψ_f are opposite in direction, the armature reaction magnetic field ψ_{sa} will act as a demagnetization. In this case,

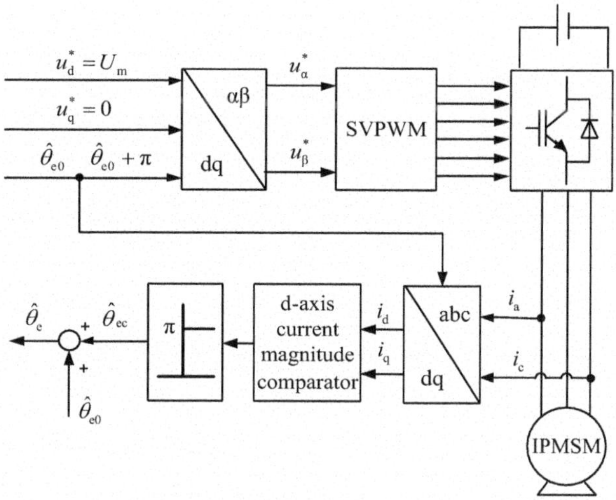

Fig. 8.4 Block diagram of pulse signal injection method for detecting rotor pole polarity

the stator core saturation will decrease, and the permeance and the stator inductance will increase [3, 4].

After obtaining the rotor magnetic pole position information $\hat{\theta}_{e0}$ by the HF pulse voltage signal injection method, based on the abovementioned stator core saturation nonlinear principle, the pulse voltage signal injection method is used to detect the rotor magnetic pole polarity. The block diagram is shown in Fig. 8.4. Two pulse voltage vectors are injected into the direction $\hat{\theta}_{e0}$ and the reverse direction $\hat{\theta}_{e0} + \pi$ separately. Then the rotor pole polarity can be identified by comparing the amplitude of the i_d. If the d-axis current of the direction $\hat{\theta}_{e0}$ is larger than the direction $\hat{\theta}_{e0} + \pi$, the direction $\hat{\theta}_{e0}$ is the N pole, and the compensation value is $\hat{\theta}_{ec} = 0$. Otherwise, the direction $\hat{\theta}_{e0} + \pi$ is the N pole, and the compensation value is $\hat{\theta}_{ec} = \pi$.

Besides, the second harmonic method is also a conventional rotor magnetic pole polarity detection technique [5]. In the case of IPMSM, the north magnetic pole position can be estimated by the hysteresis due to rotor flux saturation and its physical response on the iron material. If the HF signal is injected at the d-axis, the injected signal is distorted due to the hysteresis and then the distorted waveform has the even-order harmonics. Since the second-order harmonic is especially effective, this can be used to detect the d-axis. This method has the fast dynamics compared to the conventional method since the HF signal is used.

8.1.3 Experimental Results of Initial Rotor Position Estimation

With the above signal injection method, a 2.2 kW IPMSM ($L_d = 45$ mH, $L_q = 60$ mH) is used to verify the effectiveness. The amplitude of the pulse HF voltage signal is 57 V, and the injection frequency is 1 kHz. The amplitude of the injected pulse

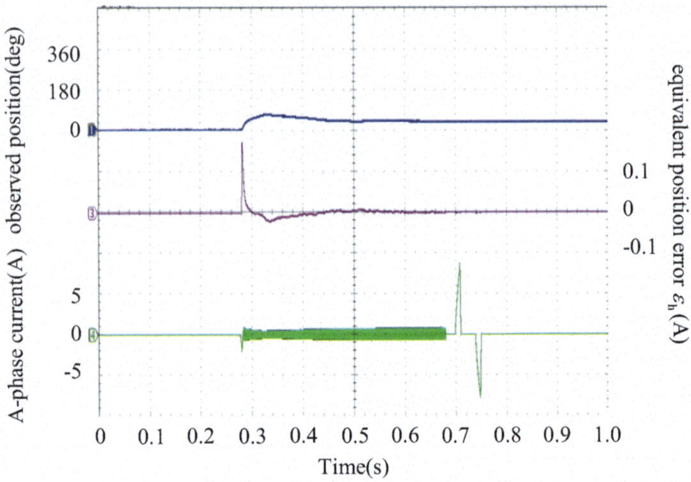

Fig. 8.5 Waveforms of rotor initial position estimation ($\theta_e = 35.7^{\circ}$)

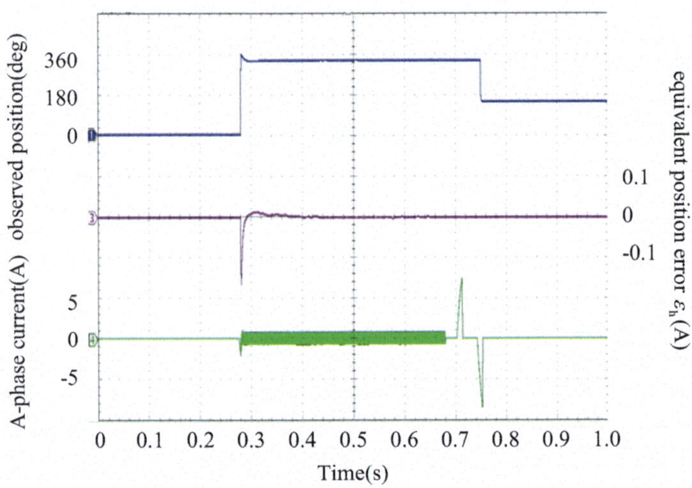

Fig. 8.6 Waveforms of rotor initial position estimation ($\theta_e = 153.2^{\circ}$)

voltage vector is 190 V, and its pulse width time is 900 μs. The experimental results are shown in Figs. 8.5, 8.6, 8.7, and 8.8. The experimental waveforms are the rotor observation position, the position error, and the *a-phase* current from top to bottom.

Figure 8.5 shows the waveforms when the actual rotor position is 35.7°. The estimated magnetic pole position is $\hat{\theta}_{e0} = 31.9^{\circ}$, and the directions of the injected pulse voltage vector are $\hat{\theta}_{e0} = 31.9^{\circ}$ and 211.9°. Since the amplitude of the stator current excited by the first voltage pulse is larger than the second one, the final rotor position estimation value is 31.9°, and the estimation error is 3.8°.

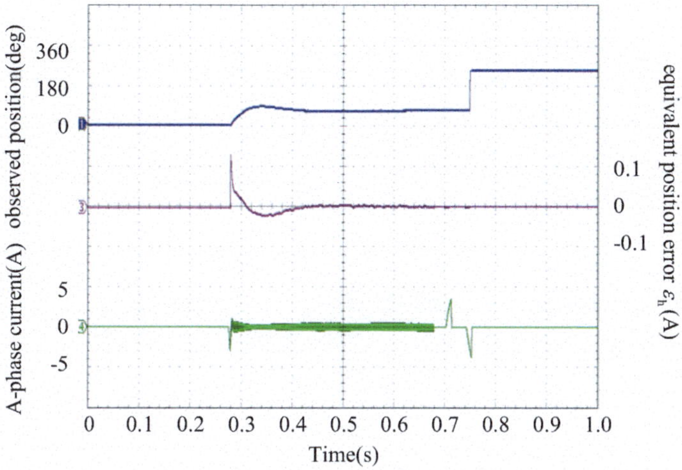

Fig. 8.7 Waveforms of rotor initial position estimation ($\theta_e = 237.5^{\circ}$)

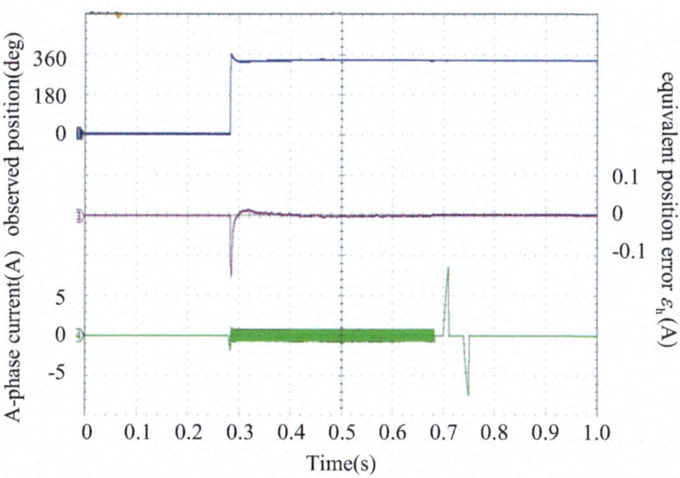

Fig. 8.8 Waveforms of rotor initial position estimation ($\theta_e = 329.3^{\circ}$)

Figure 8.6 shows the waveforms when the actual rotor position is 153.2°. The estimated magnetic pole position is $\hat{\theta}_{e0} = 328.7^{\circ}$, and the direction of the injected pulse voltage vector are 328.7° and 148.7°. Since the amplitude of the current excited by the first voltage pulse is smaller than the second one, the final rotor position estimation value is 148.7°, and the estimation error is 4.5°.

Figures 8.7 and 8.8 show the waveforms when the actual rotor initial positions are 237.5° and 329.3°, respectively. Through the above compensation strategy, the rotor initial position estimation values are 243.1° and 330.5°, and the position estimation errors are 5.6° and 1.2°, respectively.

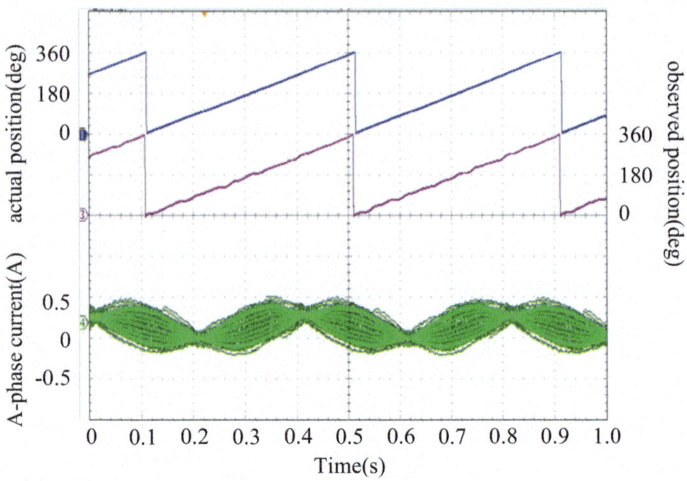

Fig. 8.9 Waveforms of HF response current and the rotor position

Figure 8.9 shows the experimental waveforms of the HF excited current and the rotor position when the motor operates at low speed. From top to bottom, the actual rotor position, the observed position, and the *a*-phase current are shown. It can be seen that the HF current is modulated by the rotor position and the rotor position information can be obtained by the HF response current sampling.

Figure 8.10 shows the relationship between the actual rotor position and the corresponding position error. The absolute value of the maximum rotor position estimation error is 9.1°, and the absolute value of the average estimation error is 4.1°. The algorithm has higher accuracy and can meet the requirement of large starting torque output capability at motor startup.

8.2 Stator Resistance Offline Identification

8.2.1 Proposed Scheme of Stator Resistance Identification

Injecting two different DC currents into the stator windings is the conventional way to counteract the influence of the inverter nonlinearities. Then according to the output current difference, the resistance identification value can be calculated from dividing the reference voltage difference. All the above assumes that the equivalent voltage error of inverter nonlinearities is constant. The block diagram of the proposed resistance identification is shown in Fig. 8.11. The main contribution is taking advantage of linear regression to enhance the robustness [6]. Besides, it is also against inverter nonlinearities and undesired perturbations during the identification. A DC current with the gradually increasing amplitude is injected into the *d*-axis to identify resistance at standstill. And the current command in discrete form is selected as

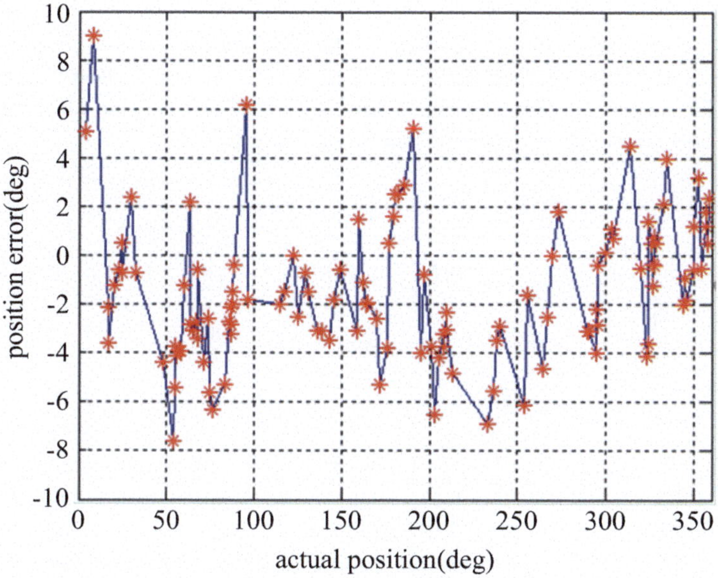

Fig. 8.10 Experimental testing result of the initial position estimation error

$$i_{sd}^{*}(k+1) = i_{sd}^{*}(k) + \Delta i \tag{8.8}$$

where Δi denotes the incremental value of the current reference, $\Delta i = I_{max}/n$, n is the number of the sampling points, and I_{max} is the maximum injected current.

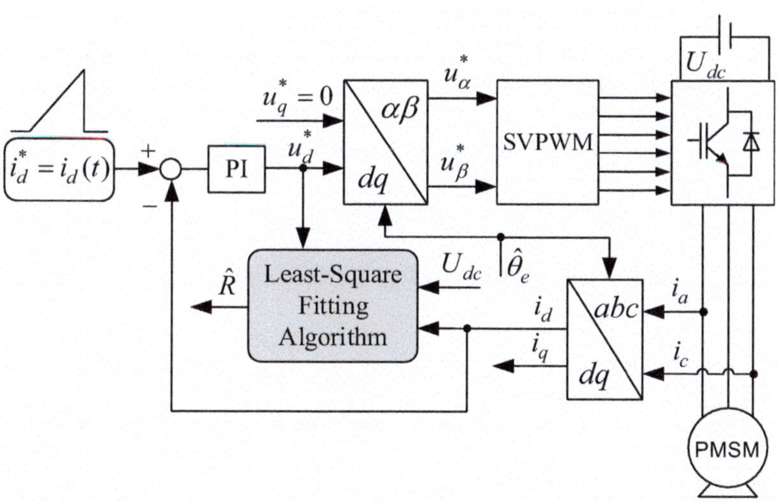

Fig. 8.11 Diagram of the stator resistance identification

The voltage error caused by the inverter nonlinearities will change obviously in the small current range due to the effect of the parasitic capacitance [7]. On account of the VSI nonlinearities, the starting sampling current point should be high enough to avoid this influence. The regression line is $u=f(i)=Ri+\Delta u$, and after the regression, the corresponding coefficients can be obtained as

$$\begin{cases} \Delta u = \dfrac{\left(\sum_{j=1}^{n} u_j\right)\left(\sum_{j=1}^{n} i_j^2\right) - \left(\sum_{j=1}^{n} i_j\right)\left(\sum_{j=1}^{n} i_j u_j\right)}{n\left(\sum_{j=1}^{n} i_j^2\right) - \left(\sum_{j=1}^{n} i_j\right)^2} \\ \hat{R} = \dfrac{n\left(\sum_{j=1}^{n} i_j u_j\right) - \left(\sum_{j=1}^{n} i_j\right)\left(\sum_{j=1}^{n} u_j\right)}{n\left(\sum_{j=1}^{n} i_j^2\right) - \left(\sum_{j=1}^{n} i_j\right)^2}, \end{cases} \tag{8.9}$$

where Δu is the equivalent voltage error of the VSI and i_j and u_j are the injected current and the reconstructed output voltage at the sampling point, respectively.

The robustness of the R identification can be improved by using the linear regression for the gradient calculation according to the multipoint information. In this way, the gradient of the identification line is the stator resistance and the error induced from inverter nonlinearities could be resisted effectively.

8.2.2 *Error Analysis of the Stator Resistance Identification*

There is only DC signal injected, so the phase voltage error caused by the parasitic capacitance can be referred in [8]. The relation between the voltage error vector angle γ and the amplitude $\Delta U_{err\text{-}syn}$ in vector form can be obtained by modeling the synthetic vector, where γ is defined as the phase lead of the voltage error vector to the a-axis. The voltage error vector is defined as

$$\begin{aligned} u_{err-syn} &= \frac{2}{3}\left(\Delta U_{err-a} + \Delta U_{err-b} \cdot e^{j120^\circ} + \Delta U_{err-c} \cdot e^{j240^\circ}\right) \\ &= \Delta U_{err-syn} \cdot e^{j\gamma}, \end{aligned} \tag{8.10}$$

where $\Delta U_{err\text{-}syn}$ and γ satisfy (8.11) and (8.12), respectively.

$$\begin{cases} \Delta U_{err-syn} = \dfrac{2}{3}\left[\left(\Delta U_{err-a} - \dfrac{1}{2}\Delta U_{err-b} - \dfrac{1}{2}\Delta U_{err-c}\right)^2 + \left(\dfrac{\sqrt{3}}{2}\Delta U_{err-b} - \dfrac{\sqrt{3}}{2}\Delta U_{err-c}\right)^2\right]^{0.5}, \\ \gamma_1 = \tan^{-1}\dfrac{\sqrt{3}\Delta U_{err-b} - \sqrt{3}\Delta U_{err-c}}{2\Delta U_{err-a} - \Delta U_{err-b} - \Delta U_{err-c}} \end{cases} \tag{8.11}$$

$$\gamma = \begin{cases} \gamma_1, & \text{if } \left(\sqrt{3}\Delta U_{err-b} - \sqrt{3}\Delta U_{err-c}\right) > 0 \\ \gamma_1 + \pi, & \text{if } \left(\sqrt{3}\Delta U_{err-b} - \sqrt{3}\Delta U_{err-c}\right) < 0 \end{cases}. \tag{8.12}$$

Based on (8.11) and (8.12), the voltage error vector angle γ and amplitude $\Delta U_{err\text{-}syn}$ are shown in Fig. 8.12a, b, respectively. According to the simplified error model, the phase voltage error $U_{err\text{-}x}$ can be described by a sign function. Then the voltage error vector is one of the six constant space vectors. Using the accurate voltage error vector, $U_{err\text{-}x}$ can be described by a sigmoid function. And the accurate voltage error vector is no more than one of the six constant space vectors. As shown in Fig. 8.12, the dashed lines describe the angle and amplitude of the voltage error vector at different rotor positions.

The corresponding components of the voltage error vector decomposed into the *d*-*q* synchronous rotating coordinate are shown in Fig. 8.13. They describe the *d*-*q* axis voltage errors at different current values and rotor positions. The amplitude of *d*-axis error tends to be constant if the injected *d*-axis DC current is high enough. The same tendency occurs in the *q*-axis error component. The characteristic shown in Fig. 8.13 properly explains why the resistance calculation should start from an enough high current point at a certain rotor position.

8.2.3 Experimental Results of Signal Processing

The proposed stator resistance identification algorithm was validated on a DSP-based 22 kW IPMSM drive system, as shown in Fig. 8.14. The rated parameters of the IPMSM are listed as follows: 380 V, 37.2 A, 50 Hz, 210 Nm, and 1000 r/min. The nominal stator resistance measured by Micro-Ohm Resistance Meter (RM3544) is 0.135 Ω. The intelligent power module PM100RSE120 whose typical switching time are $t_{on} = 1.0\mu s$ and $t_{off} = 2.5\mu s$ is used. The typical rated voltage drop of IGBT is 2.4 V and the typical rated voltage drop of diode is 2.5 V. The digital signal processor TMS320F2808 is adopted to execute the whole parameter identification algorithm. The PWM switching frequency of the inverter is 10 kHz, and the dead time is set to 3.2 μs. The frequency of the injected HF voltage signal is 500 Hz. The current reference increases linearly at 0.0665 pu/s during the resistance identification, and the number of sampling points is 26. An absolute encoder (ECN1113) with 13bit resolution is used to obtain the actual initial position that is solely used for comparison and not for parameter identification.

The initial rotor position estimation adopts the scheme in [9] by injecting HF carrier signal and two short voltage pulses. After obtaining the initial rotor position, the parameter identification can be performed and the rotor can be kept at standstill.

In order to ensure the accuracy of the identification, the initial rotor position estimation should be precise enough. Figure 8.15 shows the accuracy of the estimated initial rotor position. The electrical angle error is less than 5°, which means the

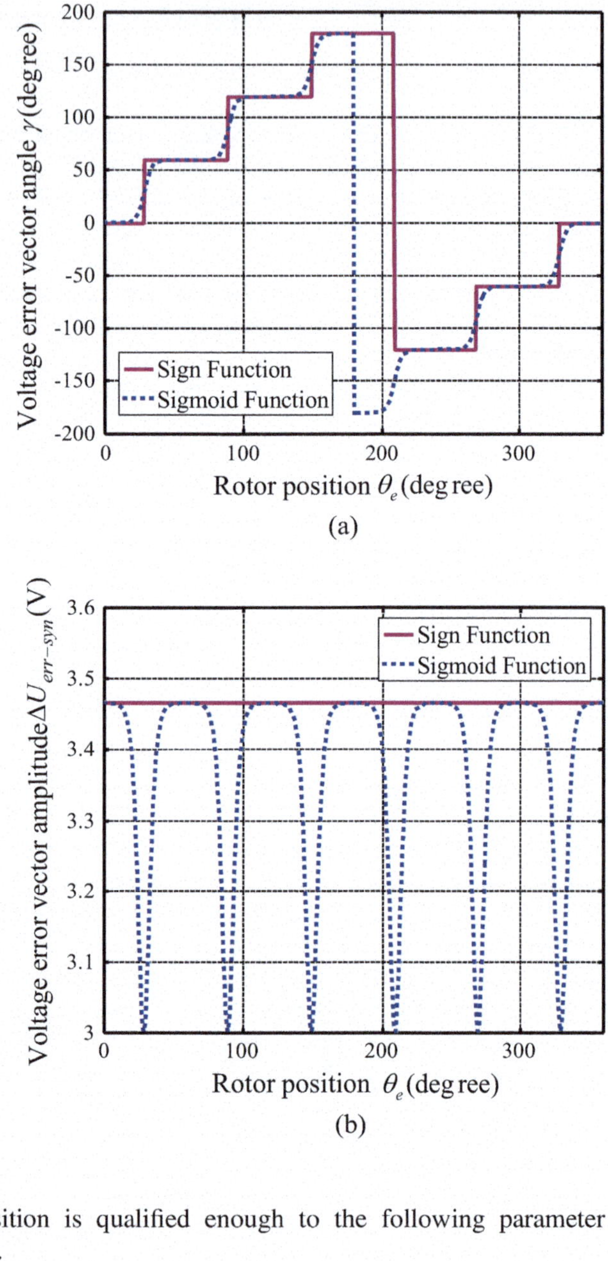

Fig. 8.12 Variation of voltage error vector versus rotor position. (**a**) The voltage error vector angle γ, (**b**) the voltage error vector amplitude $\Delta U_{err\text{-}syn}$

estimated initial rotor position is qualified enough to the following parameter identification methodology.

Figure 8.16 shows the experimental waveforms of the stator resistance identification by adopting the proposed linear regression. From top to bottom, the d-axis voltage of the current regulator output, the identified resistance, and the a-phase current are given. Figure 8.17 shows the identified stator resistance at different initial rotor positions. Figure 8.16a shows the identified resistance is 0.131 Ω at $\theta_e = 108°$,

Fig. 8.13 Variation of voltage error vector decomposed into *d*-*q* axes versus rotor position and injected DC current. (**a**) The amplitude of voltage error vector decomposed in *d*-axis, (**b**) the amplitude of voltage error vector decomposed in *q*-axis

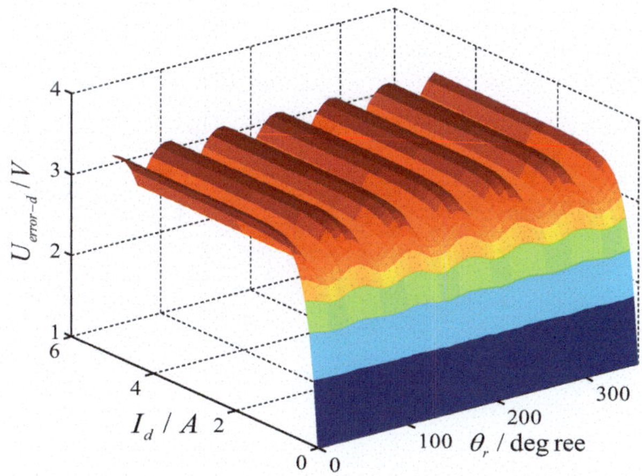

(a)

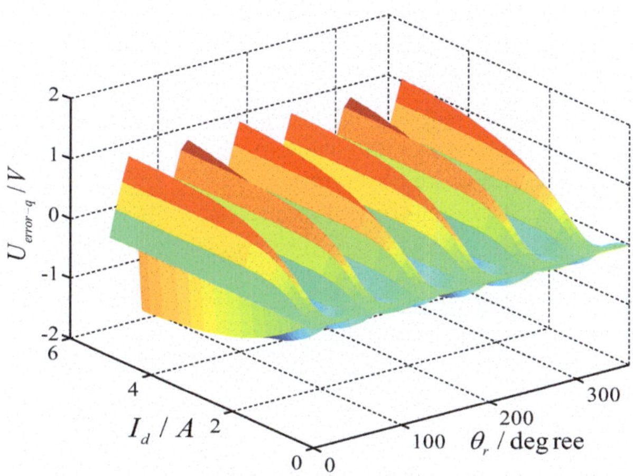

(b)

Fig. 8.14 Test platform of 22 kW PMSM

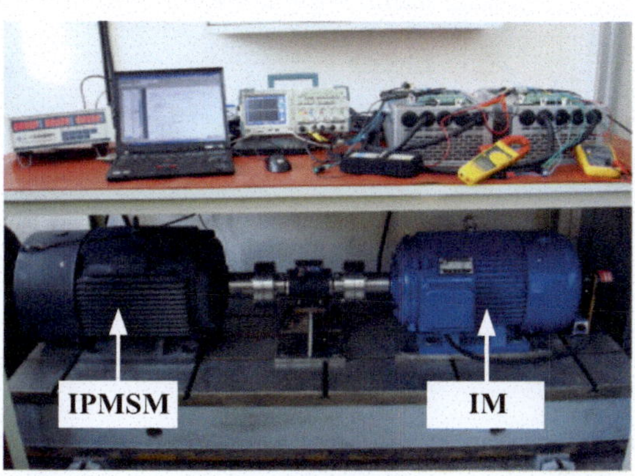

Fig. 8.15 Estimation error of the initial rotor position in electrical degree

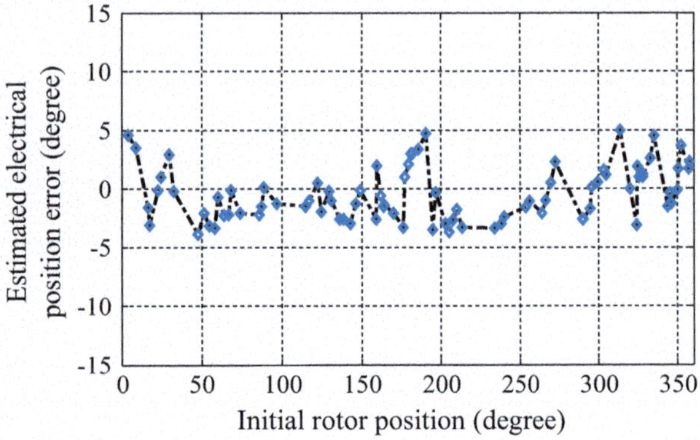

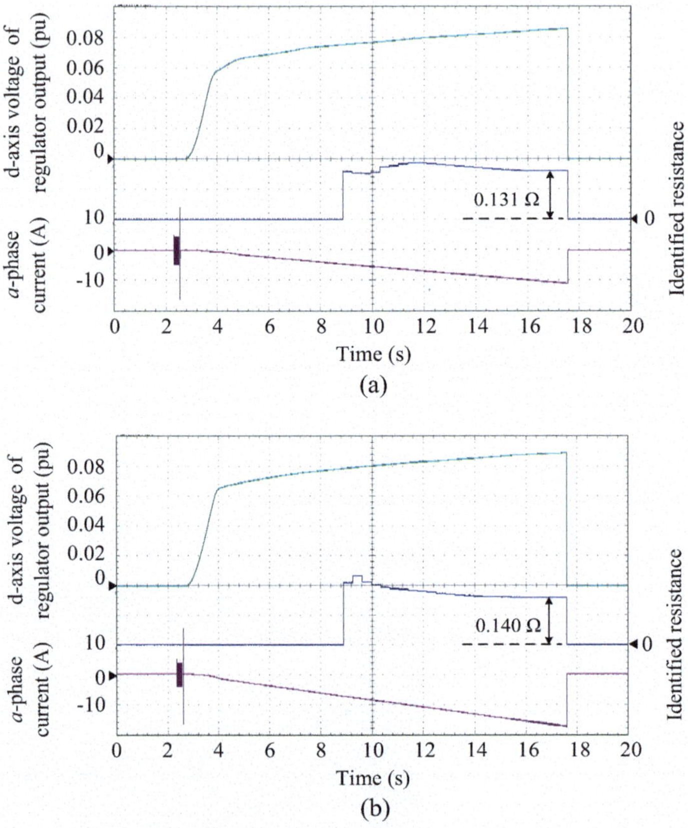

Fig 8.16 Experimental waveforms of the stator resistance identification. (**a**) $\theta_e = 108°$, (**b**) $\theta_e = 60°$

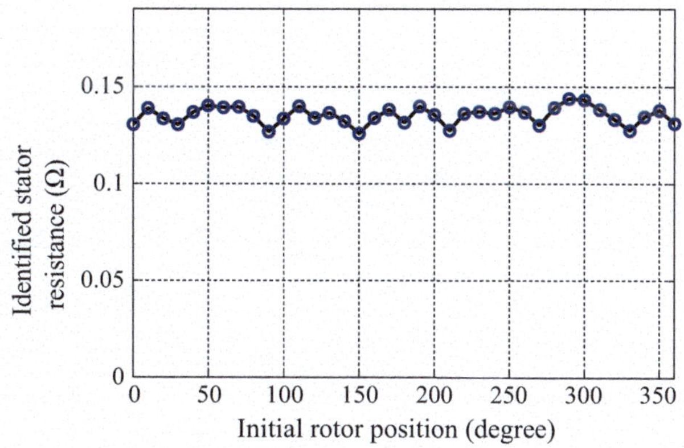

Fig. 8.17 Identified stator resistance at different rotor positions

and Fig. 8.16b shows the identified resistance is 0.140 Ω at $\theta_e = 60°$.The average deviation of the proposed method is 3.35%. The phase voltage error is 17.2 V. The voltage drop of the nominal resistance 0.135 Ω is 5.022 V at rated current 37.2 A. Although the resistance voltage drop is much smaller than the phase voltage error, the latter can still be eliminated by using linear regression no matter how large it is. In practical applications, the saturation value of voltage will vary slightly with the current considering the nonideal properties of the device.

8.3 Stator Inductance Self-Identification

8.3.1 Proposed Scheme of d-q Axis Inductance Identification

The VSI is used for the self-commissioning at standstill. For a position sensorless PMSM drive, the initial rotor position should be estimated before parameter identification. In this scheme, the initial position estimation method proposed in [9] is adopted. HF carrier signal injection is employed to identify the magnetic pole position, and two short voltage pulses are injected to determine the rotor polarity. Ultimately, the rotor position θ_e can be acquired at standstill before the parameter identification.

The proposed offline inductance identification scheme [6] is shown in Fig. 8.18. The *d*- or *q*-axis inductance identification is selected through terminal 1 or 2 correspondingly. For the *d*-axis inductance L_d identification, a supplementary DC current is injected into the estimated *d*-axis using the current closed-loop control. And an HF sinusoidal voltage signal is superposed upon the output of *d*-axis current regulator. The DC current can attenuate the influence of inverter nonlinearities.

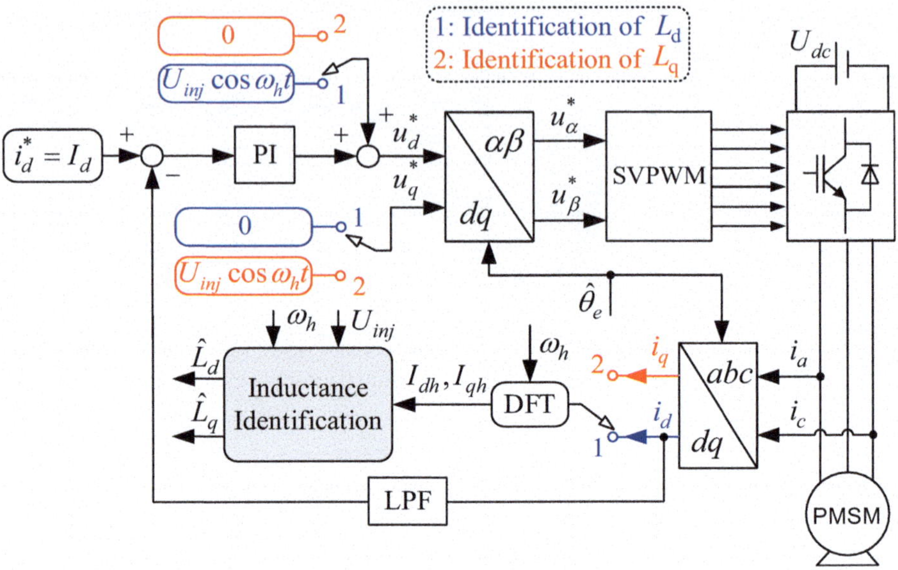

Fig. 8.18 The proposed scheme of d-q axis inductance identification

The *d*-axis voltage equation in the synchronous rotating reference frame can be represented as

$$u_d = Ri_d + p\varphi_d - \omega_e \varphi_q, \tag{8.13}$$

where R is the stator resistance, u_d and i_d are the *d*-axis voltage and current, φ_d and φ_q are the *d*-*q* axis flux linkages, p is the differential operator, ω_e is the electrical angular velocity, and it is equal to zero because the rotor is kept at standstill. Then (8.13) can be expressed as

$$u_d = \frac{d\varphi_d}{dt} + Ri_d = \frac{d\left(L_d i_d + \varphi_f\right)}{dt} + Ri_d = \frac{d(L_d i_d)}{dt} + Ri_d. \tag{8.14}$$

The *d*-axis current response should be extracted according to (8.14), and its amplitude can be calculated by adopting Discrete Fourier Transform (DFT). The amplitude of excited HF current can be obtained from

$$\begin{cases} R_1 = \dfrac{2}{N}\left[x(0) + \sum\limits_{i=1}^{N-1} x[i]\cos\dfrac{2\pi i}{N}\right] \\ I_1 = \dfrac{2}{N}\left[-\sum\limits_{i=1}^{N-1} x[i]\sin\dfrac{2\pi i}{N}\right] \\ I_{dh} = \sqrt{{R_1}^2 + {I_1}^2}, \end{cases} \tag{8.15}$$

where R_1 and I_1 are the real and imaginary components of the HF current, respectively.

Therefore, d-axis inductance can be estimated as

$$\hat{L}_d = U_{inj}/(I_{dh}\omega_h), \tag{8.16}$$

where the symbol "^" means the estimated value and U_{inj} and ω_h are the amplitude and frequency of the injected HF signal, respectively. This estimated value is not the final identified inductance because of the stator resistance and the current distortion during the injection process.

During L_q identification, it might cause a biggish pulsating torque by injecting an HF current in q-axis. But the self-lock can be guaranteed by injecting a DC current in d-axis to keep the rotor at standstill, and then an HF sinusoidal voltage signal is superposed upon the output of q-axis current regulator to acquire the inductance parameter. The signal process of L_q identification is similar to that of L_d, which are shown in (8.15) and (8.16).

8.3.2 Error Analysis of the d-Axis Inductance Identification

Concerning the phase current, voltage error characteristic of the inverter nonlinearities at fundamental-frequency operation can be generally formulated through the saturation function [10]. For HF signal injection, this description can not completely indicate the nonlinearity characteristic. A more accurate model was proposed in [11] where the HF equivalent resistance was introduced. And this model employed small signal linearization to simplify the analysis of nonlinearities. The influence is depicted more precisely when the phase current is close to zero. The voltage error $\Delta U_{err\text{-}x}$ in arbitrary phase can be expressed as

$$\Delta u_{err-x} \approx f(i_{xf}) + f'(i_{xf}) \cdot i_{xh} \approx sign(i_{xf}) \cdot \Delta U + R_{xh} \cdot i_{xh}, \tag{8.17}$$

where i_{xf} denotes the fundamental current of arbitrary phase x, ΔU denotes the saturation value of the voltage error in region II of Fig. 8.19, and R_{xh} denotes the HF equivalent resistance that is a nonlinear variable related to i_{xf}.

As in [8], the voltage error of the fundamental component can be expressed as

$$\Delta u_{err-x} = f(i_{xf}) = 2\Delta U\left(\frac{1}{1+e^{-ki_{xf}}} - \frac{1}{2}\right). \tag{8.18}$$

It offers more information of the parasitic capacitance affecting inductance identification in region I of Fig. 8.19. R_{xh} depends on the characteristic of IGBT, and it can be approximated as

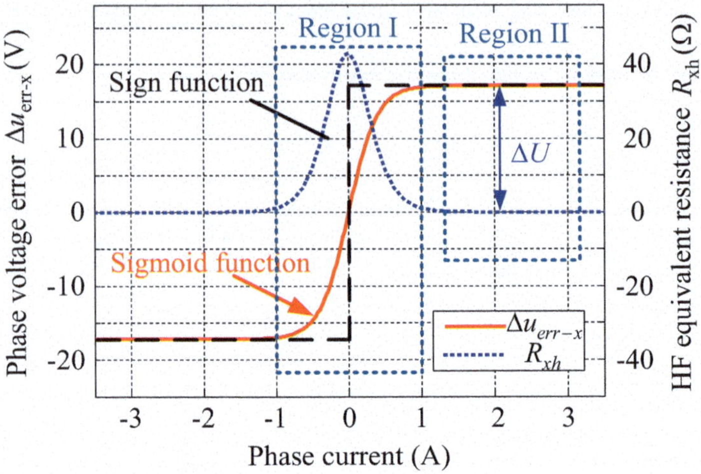

Fig 8.19 The phase voltage error and the HF equivalent resistance

$$R_{xh} = f'(i_{xf}) = \frac{2\Delta U \cdot k \cdot e^{-ki_{xf}}}{(1 + e^{-ki_{xf}})^2}, \tag{8.19}$$

where k is a model parameter only determined by the dead time and the device characteristic. Under the condition of $U_{dc} = 537$V and the dead time 3.2 μs, $k = 0.6$ and $\Delta U = 17.2$ V can be obtained. Then $\Delta u_{err\text{-}x}$ and R_{xh} are described in Fig 8.19. The way to obtain the coefficients k and ΔU is the same as the thought in [5, 10].

The HF voltage signal superposed on the d-axis winding can be expressed as

$$u_d^* = U_{inj} \cos(\omega_h t). \tag{8.20}$$

And the current response of d-axis will be

$$i_h = I_h \sin(\omega_h t + \theta_s), \tag{8.21}$$

where θ_s is the phase shift caused by the inverter nonlinearities and the winding resistance which will be addressed later.

The voltage error generated from R_{xh} can be calculated by i_h, and the d-axis voltage error can be derived through rotating coordinate transformation. The analysis of L_d estimation error due to inverter nonlinearities is shown as follows.

The DC current offset superposed in the d-axis can improve the identification accuracy by pulling the stator current out of the ZCC zone. The whole current response at the d-q axes can be given as

$$\begin{cases} i_d = i_h + I_d = I_h \sin(\omega_h t + \varphi) + I_d \\ i_q = 0 \end{cases}. \tag{8.22}$$

Since the DC component keeps constant at the same rotor position, the corresponding voltage error is also constant. Thus, the fundamental voltage error has no impact on the result. Only the influence of the voltage error induced from i_h needs to be considered. The *d*-*q* axis HF voltage errors can be represented as

$$\begin{cases} \Delta U_{err-dh} = \frac{2}{3} i_h \left[R_{ah} \cos^2(\theta_e) + R_{bh} \cos^2\left(\theta_e - \frac{2}{3}\pi\right) + R_{ch} \cos^2\left(\theta_e + \frac{2}{3}\pi\right) \right] \\ \Delta U_{err-qh} = \frac{1}{3} i_h \left[R_{ah} \sin(2\theta_e) + R_{bh} \sin 2\left(\theta_e - \frac{2}{3}\pi\right) + R_{ch} \sin 2\left(\theta_e + \frac{2}{3}\pi\right) \right] \end{cases}. \tag{8.23}$$

Therefore, the HF equivalent resistances in *d*-*q* axes are

$$\begin{cases} R_{dh} = \frac{2}{3} \left[R_{ah} \cos^2(\theta_e) + R_{bh} \cos^2\left(\theta_e - \frac{2}{3}\pi\right) + R_{ch} \cos^2\left(\theta_e + \frac{2}{3}\pi\right) \right] \\ R_{qh} = \frac{1}{3} \left[R_{ah} \sin(2\theta_e) + R_{bh} \sin 2\left(\theta_e - \frac{2}{3}\pi\right) + R_{ch} \sin 2\left(\theta_e + \frac{2}{3}\pi\right) \right] \end{cases}. \tag{8.24}$$

Only the *d*-axis voltage error needs to be concerned for L_d identification; the voltage error in *q*-axis would not affect the identification if the cross-coupling effect is neglected. By substituting (8.19) into (8.24), R_{dh} is portrayed in polar coordinate with an electrical cycle as shown in Fig 8.20. According to Fig 8.20a, R_{dh} keeps constant if there is no additional DC current injected into the *d*-axis. From Fig. 8.20b, R_{dh} decreases gradually with the increase of the injected *d*-axis DC current.

Considering the nonlinearities of inverter, the actual output voltage on the *d*-axis winding amounts to

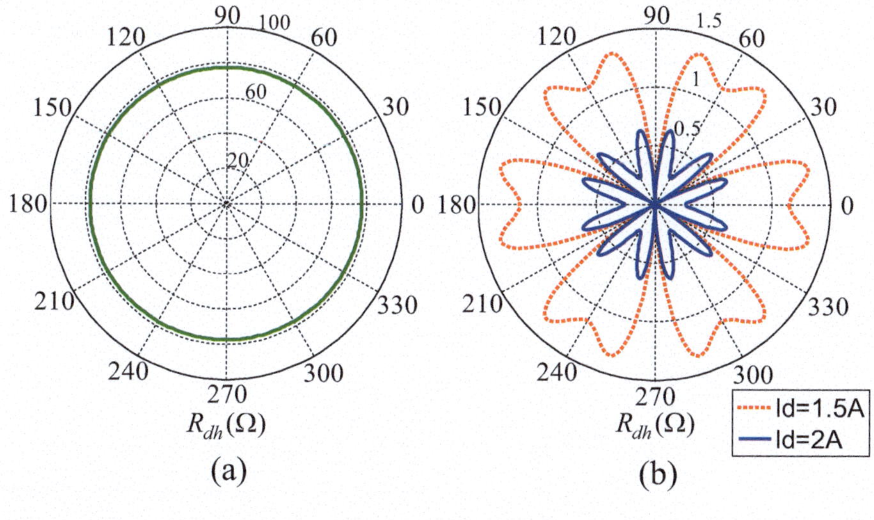

Fig. 8.20 HF equivalent resistance R_{dh} in polar coordinate with an electrical cycle. (**a**) $I_d = 0$, (**b**) $I_d = 1.5$ A and 2 A

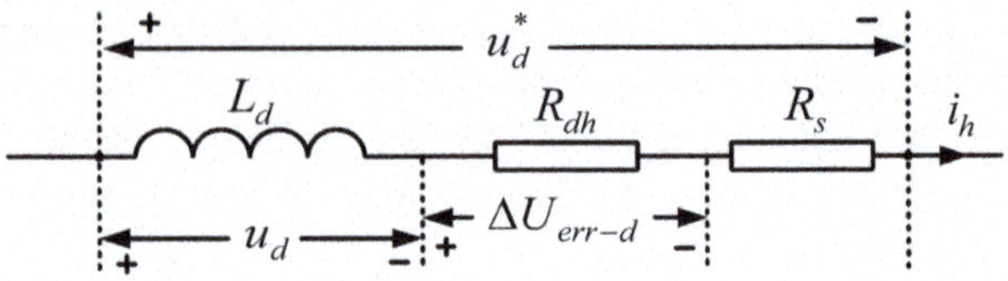

Fig. 8.21 HF equivalent circuit of L_d identification

$$u_d = u_d^* - \Delta U_{err-d}. \tag{8.25}$$

Therefore, the HF equivalent circuit of L_d identification can be described as shown in Fig. 8.21.

Only u_d^* and i_h can be known when the identification algorithm is plunged into the commercial general-purpose drives. Considering the influence of R_{dh} as shown in Fig 8.21, the obtained L_d can be given as

$$\hat{L}_d = \frac{\sqrt{\left(\omega_h^2 L_d^2 + R_{dh}^2 + R_s^2 + 2R_{dh}R_s\right)}}{\omega_h}. \tag{8.26}$$

Then the estimation error of L_d can be obtained:

$$\Delta L_d = L_d - \hat{L}_d = L_d - \frac{\sqrt{\left(\omega_h^2 L_d^2 + R_{dh}^2 + R_s^2 + 2R_{dh}R_s\right)}}{\omega_h}. \tag{8.27}$$

The phase-shifted angle θ_s in (8.22) is caused by the existence of R and R_{dh} due to the inverter nonlinearities. Assuming that the inverter nonlinearities do not exist, there will be no equivalent HF resistance in equivalent circuit. As shown in Fig. 8.22, if the DC component of an arbitrary phase current turns to zero, the HF component i_{xh} also becomes zero, which means the voltage drop on R_{xh} is zero. And the HF equivalent resistances of the other two phases are also nearly zero when their DC components are large enough to make them far away from the ZCC zone. Therefore, the HF current would not be distorted. In this case, no estimation error is induced for L_d identification.

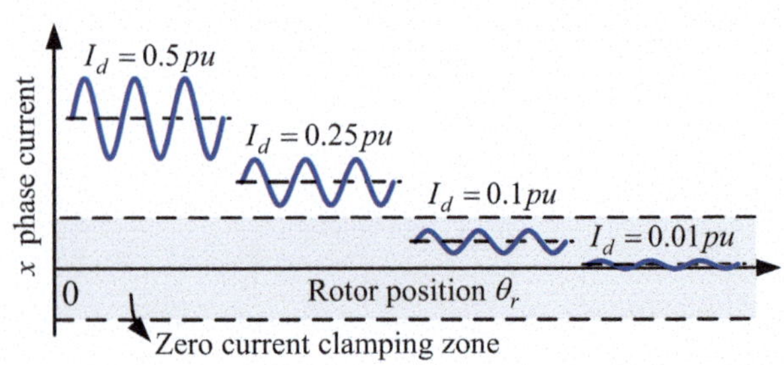

Fig. 8.22 HF current response of arbitrary phase during d-axis inductance identification

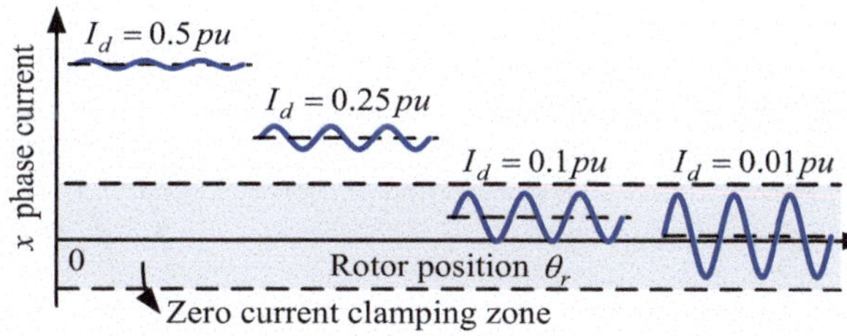

Fig. 8.23 HF current response of arbitrary phase during q-axis inductance identification

8.3.3 Error Analysis of the q-Axis Inductance Identification

The same as the way to estimate the d-axis inductance, the identified q-axis inductance can be represented as

$$\hat{L}_q = U_{inj}/\left(I_{qh}\omega_h\right). \tag{8.28}$$

The injected d-axis DC component can partly let the HF current stay far away from ZCC zone. As shown in Fig. 8.23, the amplitude of the HF current will be modulated by the rotor position, which is in an opposite way compared to Fig. 8.22.

The amplitude of HF current i_{xh} reaches the maximum if the DC component of that phase turns to zero. And i_{xh} diminishes with the DC component increase. Although i_{xh} will not be disturbed by the nonlinearities if the corresponding DC component reaches the maximum, its proportion of the q-axis HF current reduces nearly to zero and makes no contribution to L_q estimation. The ratios of HF currents in the other two phases, parts of the d-axis HF current response staying in ZCC zone, are higher. In this case, the HF currents are distorted, which deteriorate the identification results.

As the estimation error analysis of L_d, the HF equivalent resistances can be expressed as

$$\begin{cases} R_{dh} = -\frac{1}{3}\left[R_{ah}\sin\left(2\theta_e\right) + R_{bh}\sin 2\left(\theta_e - \frac{2}{3}\pi\right) + R_{ch}\sin 2\left(\theta_e + \frac{2}{3}\pi\right)\right] \\ R_{qh} = \frac{2}{3}\left[R_{ah}\sin^2(\theta_e) + R_{bh}\sin^2\left(\theta_e - \frac{2}{3}\pi\right) + R_{ch}\sin^2\left(\theta_e + \frac{2}{3}\pi\right)\right] \end{cases}. \tag{8.29}$$

Neglecting the cross-coupling effect, the voltage error of d-axis would not affect the L_q identification. And R_{qh} portrayed in polar coordinate can be described as shown in Fig. 8.24, where $k = 0.6$, $\Delta U = 17.2$ V, and $I_d = 0$, 0.5 A and 2 A, respectively. When I_d is zero, R_{qh} stay constant in different positions, which will cause larger estimation error. Besides, a sixth pulsation exists in an electrical circle when I_d is not zero. Theoretically, if I_d is large enough, the zero error points will appear at $\theta_e = k\pi/3$, where $k \in \{0,1,2,3,4,5\}$.

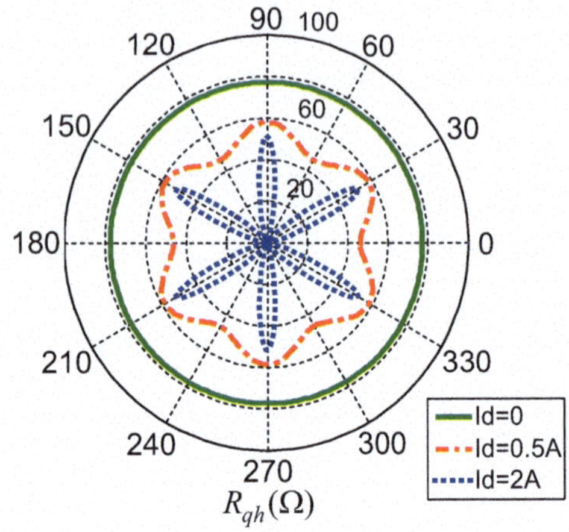

Fig. 8.24 HF equivalent resistance R_{qh} in polar coordinate with an electrical cycle

Considering the inverter nonlinearities, the actual voltage superposed on q-axis winding is

$$u_q = u_q^* - \Delta U_{err-q}. \tag{8.30}$$

So, the HF equivalent circuit of L_q identification can be described similarly as shown in Fig. 8.21. Meanwhile the identification error can be expressed as

$$\Delta L_q = L_q - \hat{L}_q = L_q - \frac{\sqrt{\left(\omega_h^2 L_q^2 + R_{qh}^2 + R_s^2 + 2R_{qh}R_s\right)}}{\omega_h}. \tag{8.31}$$

8.3.4 Analysis of the Cross-Coupling Effects

According to the proposed identification method, for d-axis inductance identification, the HF voltage equation can be expressed as

$$\begin{cases} u_{dh} = L_d \dfrac{di_{dh}}{dt} + L_{dq}\dfrac{di_{qh}}{dt} + i_{dh}R_s \\ 0 = L_q \dfrac{di_{qh}}{dt} + L_{qd}\dfrac{di_{dh}}{dt} + i_{qh}R_s, \end{cases} \tag{8.32}$$

where L_{dq} and L_{qd} are the cross-coupling inductances.

Based on (8.32), the identified d-axis inductance can be represented as

$$\hat{L}_d \approx L_d - L_{dq}^2/L_q, \tag{8.33}$$

where L_{dq}=L_{qd} is assumed and the influence of cross-coupling inductance is concerned. In the same way, the identified q-axis inductance can be represented as

$$\hat{L}_q \approx L_q - L_{dq}^2/L_d. \tag{8.34}$$

It indicates that the cross-coupling effects influence the identified results. The extent of cross-coupling effects depends on the electrical machine design and operating current condition.

8.3.5 Design of PI Coefficients for d-Axis Current Regulator

As shown in Fig. 8.18, there is a current closed-loop for injecting the DC current component in d-axis. Usually, in order to design the PI coefficients optimally, the machine parameters are demanded. However, the estimated machine parameter can be obtained after the parameter identification execution. In the proposed scheme, the rough parameters are derived according to the basic information from the rated values on the machine nameplate, including the rated power P_N, the rated current I_N, and the rated voltage U_N. Then the practicable PI coefficients for d-axis current regulator can be obtained through the design.

The rated current I_N and the rated power P_N are known when an electric machine is given. Then the resistance can be roughly calculated according to the loss equation:

$$P_N \frac{1-\eta}{\eta} \gamma = 3 I_N^2 R, \tag{8.35}$$

where η denotes the efficiency and γ denotes the copper loss percentage of the total loss at rated operation point. When the efficiency of machine is not known exactly, it can be roughly estimated according to the rated power of the test machine. The general scope of γ is 1/2 to 2/3, which has been investigated in many papers and motor design books, such as [12, 13]. For the 22 kW PMSM used in experiment, the efficiency is chosen as 95% and the rated current is 37.2 A. The percentage coefficient γ is regarded as 0.5, and the resistance can be roughly estimated as 0.139 Ω which is close to the real value.

The rough inductance can be estimated according to the voltage equations in the *d-q* axes. Assume L_m=L_d=L_q so a rough value which is only used for regulator parameter design in the parameter identification process can be estimated. The quadratic sum of the *d-q* axis voltage equations is

$$U_d^2 + U_q^2 = \left(R_s I_d + X_q I_q\right)^2 + \left(E_0 - I_d X_d + I_q R\right)^2, \tag{8.36}$$

where X_d and X_q are *d-q* axis inductive reactances. Due to I_d=0, $U_d{}^2$+$U_q{}^2$=$U_N{}^2$ and R have been obtained before, the back-EMF can be estimated as

$$E_0 = P_N/(3I_N), \tag{8.37}$$

where P_N is 22 kW and the estimated back-EMF is 197.13 V. Since L_m=L_d=L_q, X_d=X_q=X_m, the final equation for rough inductance calculation is

$$U_N^2 = \left(X_m I_q\right)^2 + \left(E_0 + I_q R_s\right)^2, \tag{8.38}$$

where I_q is 37.2 A and U_N=220 V for the tested 22 kW IPMSM. So the derived inductance is 7.40 mH.

According to the obtained rough resistance and inductance, the PI coefficients of the *d*-axis current regulator can be designed for the offline parameter identification. The equivalent diagram of *d*-axis current loop is shown in Fig. 8.25. The PWM period and the closed-loop transfer function of d-axis current loop can be expressed as

$$\mathbf{G}(s) = \frac{i_d}{i_d^*} = \frac{K_p}{T_s L_d s^2 + L_d s + K_p}, \tag{8.39}$$

where K_i=RK_p/L_d, and the term $T_s L_d s^2$ is so small in this situation that the closed-loop transfer function can be regarded as first-order system.

$$\mathbf{G}(s) = \frac{i_d}{i_d^*} = \frac{K_p}{L_d s + K_p} = \frac{K_p/L_d}{s + K_p/L_d}. \tag{8.40}$$

The term K_p/L_d in (8.40) indicates the bandwidth of the current loop. When there is a desired bandwidth of $\omega_{cb}{}^*$, then the K_p can be obtained.

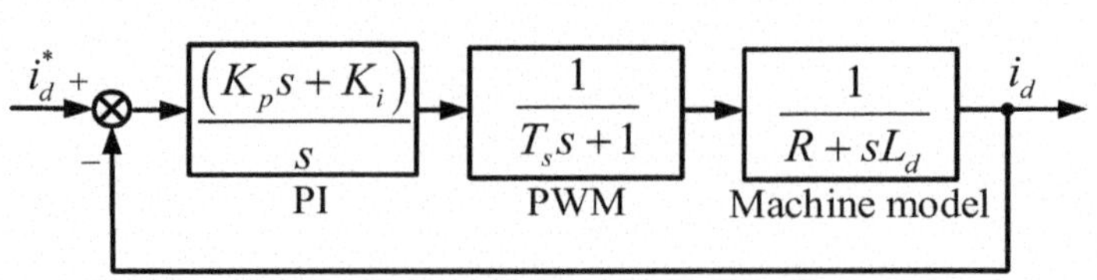

Fig. 8.25 Equivalent diagram of the *d*-axis current loop

$$\omega_{cb}^{*} = K_p / L_d \Rightarrow K_p = \omega_{cb}^{*} L_d, \tag{8.41}$$

So

$$K_i = \frac{R_s}{L_d} \cdot K_p = \frac{R_s}{L_d} \cdot \omega_{cb}^{*} L_d = R_s \omega_{cb}^{*}. \tag{8.42}$$

In the identification strategy, ${\omega_{cb}}^{*}$ is selected as 100∗2π rad/s because high control performance for the transient process is not required. Therefore, the PI coefficients can be obtained as $K_p = 4.9$ and $K_i = 83.33$. The current response of a step input is shown in Fig. 8.26. According to the estimated rough machine parameters, effectiveness of the design for the *d*-axis current regulator can be verified.

From top to bottom of Fig. 8.26, the *d*-axis current command, the *d*-axis current response, and the *a*-phase current response are given, respectively. The *d*-axis current response is acceptable that it can be applied in this application. The derived resistance and inductance are rough and the initial PI coefficients are merely for the parameter identification. Once the more accurate parameters are identified, the more refined PI coefficients could be calculated for the high-performance vector control system.

8.3.6 Experimental Results of Inductance Identification

The proposed self-commissioning algorithm was validated on a DSP-based 22 kW IPMSM drive system, as shown in Fig. 8.14. The rated parameters of the IPMSM are listed as follows: 380 V, 37.2 A, 50 Hz, 210 Nm, and 1000 r/min.

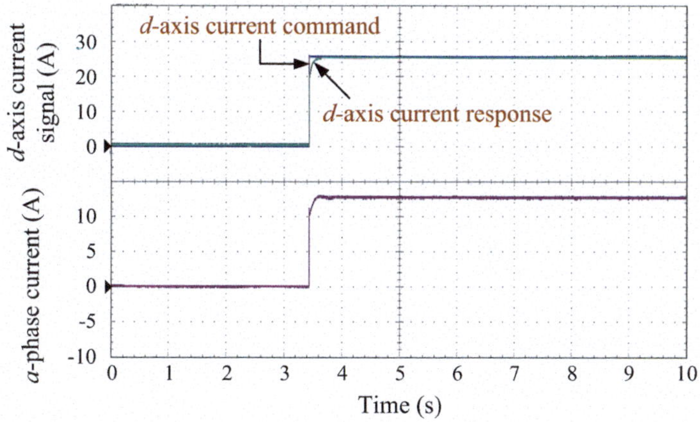

Fig. 8.26 Experimental results of step response of *d*-axis current loop for parameter identification

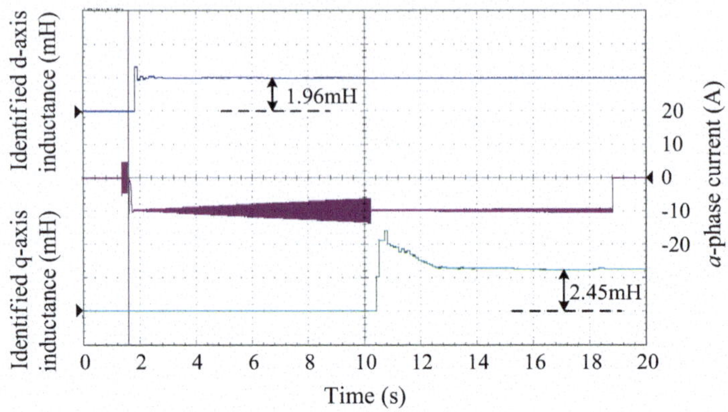

Fig. 8.27 The waveforms of the inductance identification results

Figure 8.27 shows the *d-q* axis inductance identification results at the initial position of 108°. From top to bottom, the estimated L_d, the a-phase current, and the estimated L_q are given. The supplementary *d*-axis DC current is 0.25 pu. The HF injected voltage increases gradually to avoid producing overcurrent and ensure the intensity of signals. The *d*- and *q*-axis inductances converge to stable values with the increase of the injected current.

Figure 8.28 shows the identified *d-q* axis inductances at different initial angles (0-360°) with different injected *d*-axis DC currents. The injected *d*-axis DC component is set to 0, 0.3 pu and 0.6 pu, respectively, and the injected HF voltage is 0.1 pu.

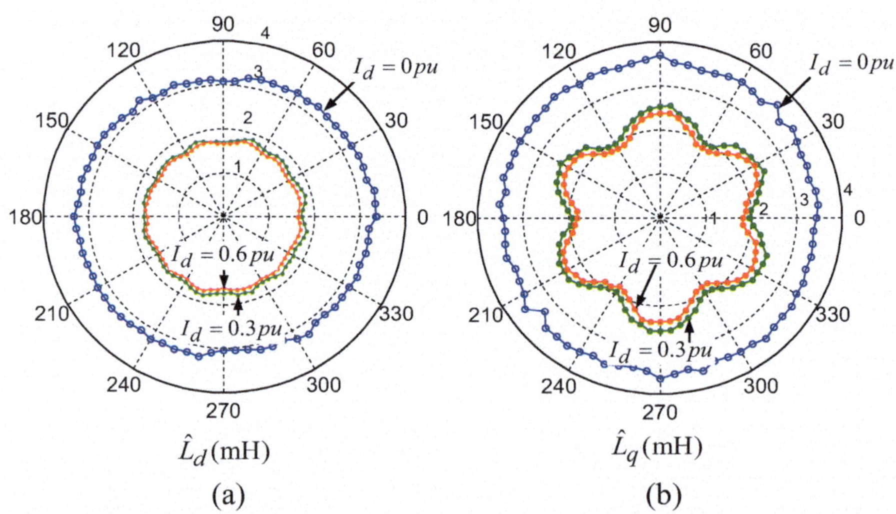

Fig. 8.28 Estimated *d-q* axis inductances in an electrical cycle. (**a**) The estimated *d*-axis inductance and (**b**) estimated q-axis inductance

From the result of Fig. 8.28a, the HF equivalent resistance becomes constant when I_d equals zero, which means the error in polar coordinate is constant during a circle. When I_d is not zero, there are 12 pulsations in an electrical circle, which is consistent with Fig. 8.20b. The estimation error approaches zero at $\theta_e=(2k+1)\pi/6$, $k \in \{0,1,2,3,4,5\}$. Figure 8.28b describes the identification results of q-axis inductance with different injected d-axis DC currents. When the injected DC current keeps constant, the maximum estimation error will be introduced at $\theta_e=k\pi/6$, $k \in \{0,1,2,3,4,5\}$. The experimental results verify the influence of the HF equivalent resistance on the L_d and L_q estimation errors.

Figure 8.29 shows the estimated d-q axis inductances before and after the model-based compensation. The compensated d-axis inductance is almost the same as the uncompensated value in Fig. 8.29a because the influence of nonlinearities is relatively small. For q-axis inductance identification results, the effect of compensation is distinct. However, there are still small fluctuations in the compensated identification results. But the identification accuracy is improved by the model-based compensation. It should be noted that the parameters of the nonlinearity model need to be accurate. The coefficients used in the compensation model are $I_d = 0.3$ pu, $k = 0.6$, $\Delta U = 17.2$ V, and $R = 0.135\Omega$.

The cross-coupling effects between the two axes are analyzed by injecting constant current in one axis and HF signal in the other. Figure 8.30 shows the identified q-axis inductance when the fundamental current component in d-axis varies. A constant current injected in d-axis determining its effect on q-axis inductance does not produce torque. However, it is not the same for q-axis; when constant current in q-axis is required to identify cross-saturation effect in d-axis, the machine tends to rotate due to the alignment torque. Since the estimated q-axis inductance almost keeps constant when the injected current is constant in d-axis varies. The

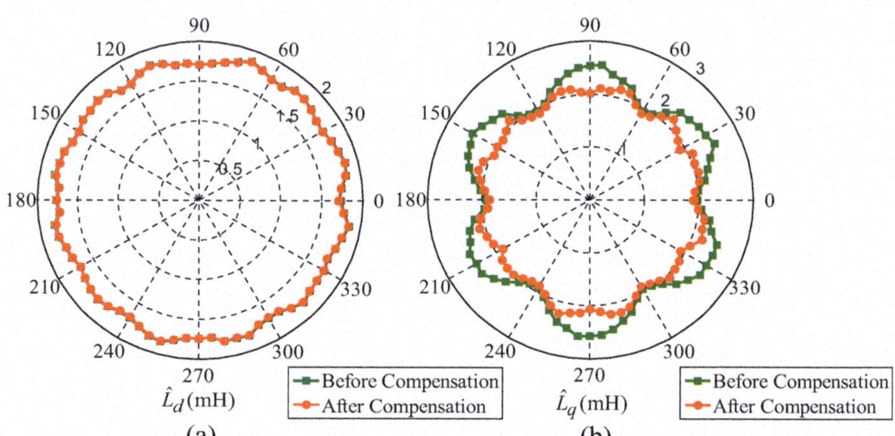

Fig. 8.29 Estimated d-q axis inductances in an electrical cycle before and after compensation. (**a**) The estimated d-axis inductance before and after compensation ($I_d = 0.3$ pu), (**b**) the estimated q-axis inductance before and after compensation ($I_d = 0.3$ pu)

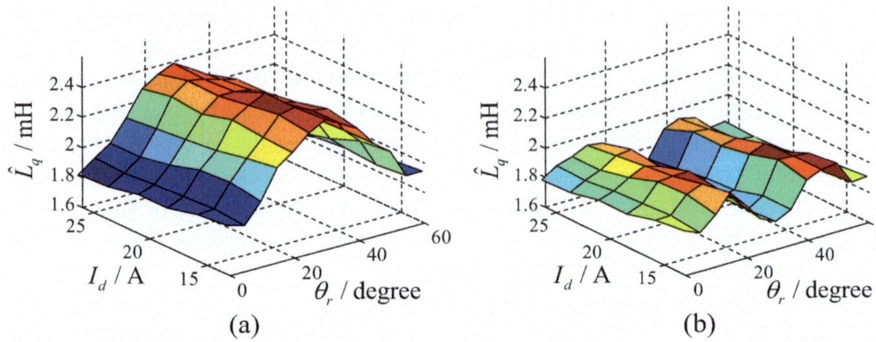

Fig. 8.30 Experimental results of the influence of cross-coupling effects. (**a**) Before compensation, (**b**) after compensation

experimental results indicate that the cross-coupling effects does not affect the inductance identification at the same rotor position on the tested PMSM. The cross-coupling effects impacting on the identification depend on different electric machine designs.

The experimental results considering different injected HF currents are shown in Fig. 8.31.

It can be seen that the identified inductances, especially for the *q*-axis inductance, tend to be affected less by the inverter nonlinearities since the increase of signal-to-noise ratio. However, when the *q*-axis inductance is identified particularly, the limit of the injected HF current exists avoiding the rotor vibrating. The estimated *d*-axis inductance almost stays the same after the model compensation since the VSI nonlinearity effect imposes on *d*-axis slightly. Overcompensation might occur because the small-signal model cannot be guaranteed at large HF current injection. The minimum variation percentage of *d*-axis inductance decreases from 3.08% to 2.90% and the maximum variation percentage of *q*-axis inductance decreases from 32.31% to 12.77% after compensation.

Figure 8.32 shows the measured result of *a*-phase inductance using the inductance bridge method. The measured phase inductance can be converted into *d*-*q* axes and the measured L_d=1.5(L_a-L_b) = 1.703 mH and L_q=1.5(L_a+L_b) = 2.025 mH can be a reference to verify the proposed method.

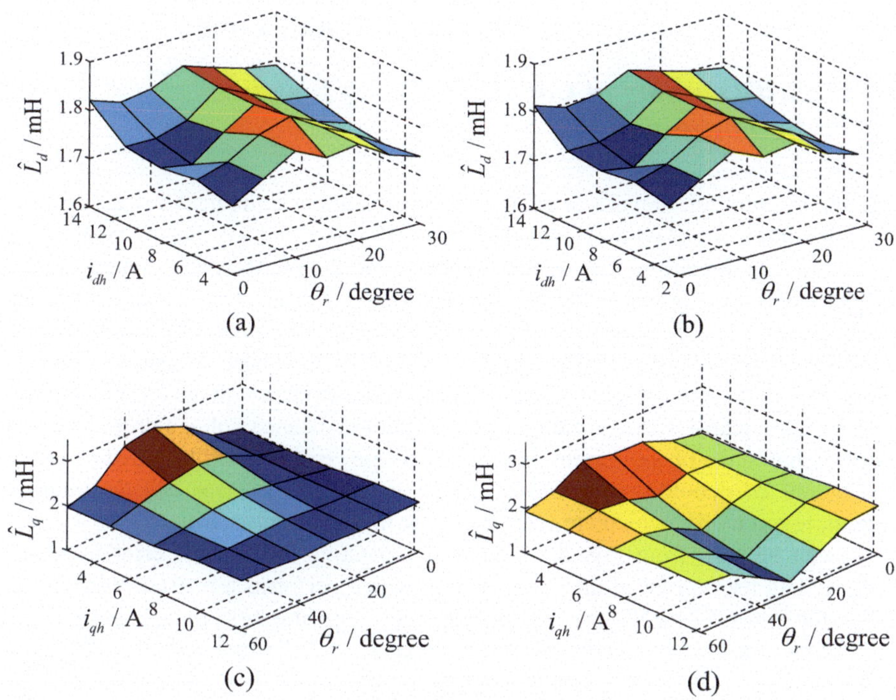

Fig. 8.31 Experimental results of estimated inductance considering different injected HF currents. (**a**) Estimated L_d before compensation, (**b**) estimated L_d after compensation, (**c**) estimated L_q before compensation, and (**d**) estimated L_q after compensation

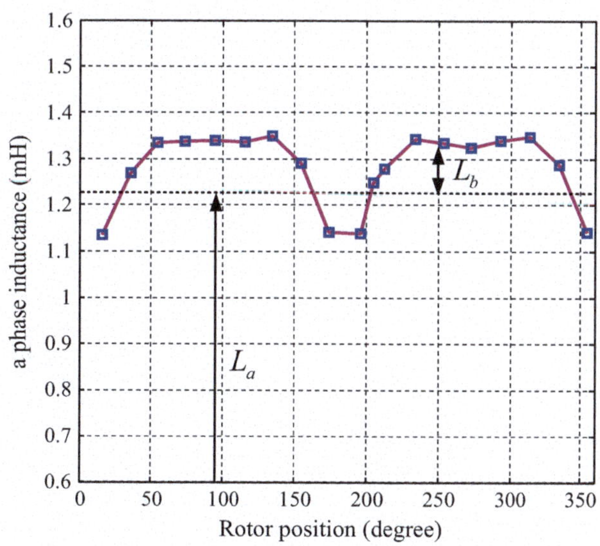

Fig. 8.32 Measure result of a-phase inductance using the inductance bridge method

8.4 Inverter Nonlinearity Effect and Compensation

8.4.1 Self-Learning of Characteristics of Inverter Nonlinearities

Usually, a saturation function is adopted to formulate the equivalent voltage error of inverter nonlinearities without considering the HF injection [10]. Due to the effect of the inverter nonlinearities, the relationship between the phase voltage reference and the phase current is shown in Fig. 8.33. Curve 1 indicates the phase voltage reference which increases nonlinearly with the linear increasing of phase current considering the influence of the inverter nonlinearities. Normally, the phase voltage reference is composed of the stator resistance voltage drop and the phase voltage error caused by inverter nonlinearities. After subtracting the stator resistance voltage drop, which is proportional to the phase current, the phase voltage error can be illustrated by curve 2. ΔU and ΔI denote the saturation values of the voltage error and the induced current, respectively, which are relevant to the inverter nonlinearities.

As can be seen from Fig. 8.33, curve 2 indicating the voltage error increases linearly with the current in the linear region, which performs a resistance behavior. When the current exceeds the linear region into the saturation region, the voltage error keeps constant. In this period, the gradient of curve 1 which indicates the phase voltage reference decreases, and it contains the sum of the stator resistance and the device on resistance. Additionally, the gradient K_2 in the linear region includes the gradient K_1 in the saturation region. Thus, the relationships can be expressed as follows:

$$\begin{cases} K_1 = R + R_{on} \\ K_2 = K_1 + K \end{cases}. \tag{8.43}$$

In Fig. 8.33, the intercept ΔU of line 1 can be derived by using linear regression method; then

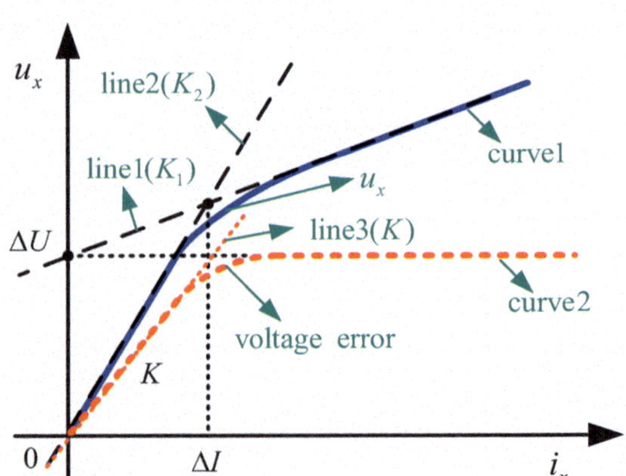

Fig. 8.33 Relationship between the phase voltage and the phase current

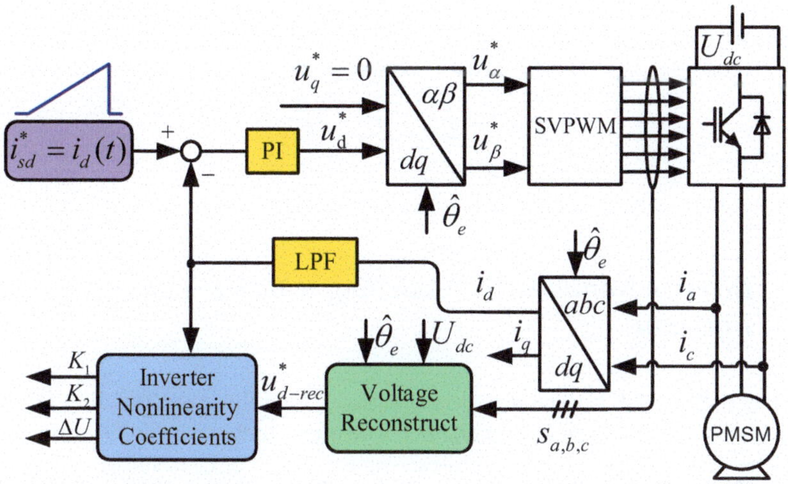

Fig. 8.34 Self-learning method for acquiring the inverter nonlinearity characteristics at standstill

$$\begin{cases} u_x = K_2 i_x, & \text{in linear region} \\ u_x = K_1 i_x + \Delta U, & \text{in saturation region} \end{cases}. \tag{8.44}$$

Therefore, $K = K_2 - K_1$, and the inverter nonlinearity parameters are finally obtained. In order to analyze the harmonic contents in the induced current, K and ΔU should be obtained automatically before the inductance identification by using a self-learning solution. In addition, the way to get K and ΔU still requires keeping the rotor at standstill.

The self-learning method for acquiring K and ΔU is shown in Fig. 8.34. Firstly, a linearly increasing current is injected into the d-axis by closed-loop control. Then the reconstructed phase voltage obtained from the PWM signals $S_{a,b,c}$ and the DC bus voltage U_{dc} contains the nonlinearity information. The gradient and the intercept of phase voltage reference used to estimate K and ΔU can be calculated by using the linear regression algorithm.

The formulas of the linear regression algorithm used to calculate ΔU and K_1 are shown as follows:

$$\begin{cases} \Delta U = \dfrac{\left(\sum_{j=1}^{n} u_j\right)\left(\sum_{j=1}^{n} i_j^2\right) - \left(\sum_{j=1}^{n} i_j\right)\left(\sum_{j=1}^{n} i_j u_j\right)}{n\left(\sum_{j=1}^{n} i_j^2\right) - \left(\sum_{j=1}^{n} i_j\right)^2} \\ K_1 = \dfrac{n\left(\sum_{j=1}^{n} i_j u_j\right) - \left(\sum_{j=1}^{n} i_j\right)\left(\sum_{j=1}^{n} u_j\right)}{n\left(\sum_{j=1}^{n} i_j^2\right) - \left(\sum_{j=1}^{n} i_j\right)^2} \end{cases}. \tag{8.45}$$

The calculation of the coefficient K_2 is similar to K_1, and the saturation value of phase current also can be obtained:

$$\Delta I = \frac{\Delta U}{K_2 - K_1} = \frac{\Delta U}{K}. \tag{8.46}$$

8.4.2 Inverter Nonlinearity Modelling with HF Voltage Injection

Assuming that the HF phase current is $I_1\sin(\omega_h t)$, then the corresponding HF phase voltage can be described as $U_m\cos(\omega_h t\text{-}\theta_s)$. The phase inductance is $L_s(\theta_e)$ depending on θ_e. The nonlinearity model of arbitrary phase x (a, b or c) is shown in Fig. 8.35.

The nonlinearity model of Fig. 8.35 can be described by the following equation:

$$L_x \frac{di_x}{dt} + i_x R_x = u_x - f(i_x), \tag{8.47}$$

where $f(i_x)$ is the voltage error caused by the inverter nonlinearities. It is difficult to obtain the general solution of (8.47) due to the existence of the nonlinear function $f(i_x)$. So a practical method is adopted to analyze the HF current harmonics. There are four cases considering the different relationships between the voltage error and the HF current.

8.4.3 Partial HF Current Located in Linear Region: Case 1

The voltage error induced by the harmonic components of the HF current is neglected since its value is very small, and only the fundamental component of the

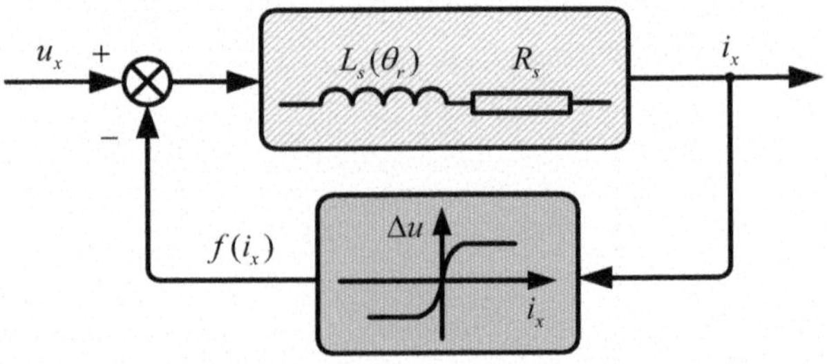

Fig. 8.35 Nonlinearity model of arbitrary phase x

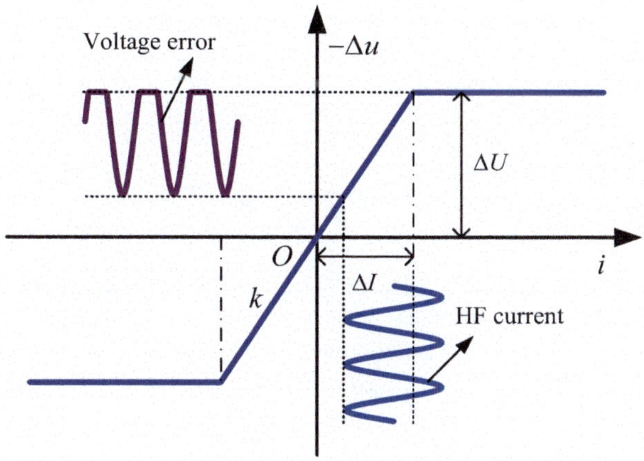

Fig. 8.36 Relationship between the voltage error and the HF current (case 1)

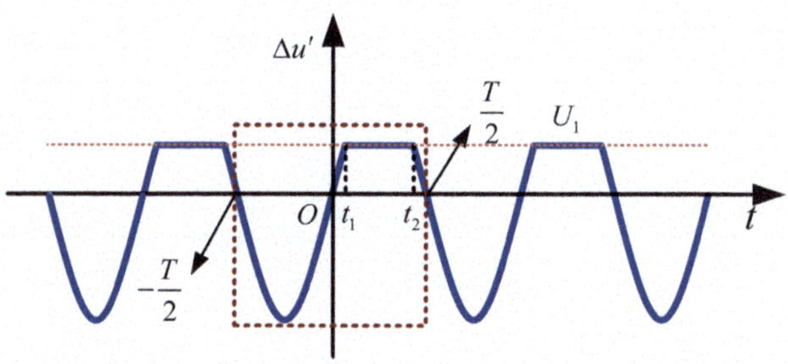

Fig. 8.37 HF voltage error component in case 1

HF current is used for the voltage error analysis. In order to simplify the harmonic analysis, the sigmoid function is approximately replaced by the saturation function.

The first case, which is the most complicated one, is that partial HF current locates in the linear region. The relationship between the voltage error and the HF current is shown in Fig. 8.36.

As can be seen from Fig. 8.37, the voltage error is a clipped sine wave which can be expressed as

$$\begin{aligned}\Delta u =& [U_0 \sin(\omega_h t) - U_1] \cdot u[U_0 \sin(\omega_h t) - U_1] \\ & - U_0 \sin(\omega_h t) + \Delta U_1,\end{aligned} \tag{8.48}$$

where ΔU_1 is the DC voltage, $U_0\sin(\omega_h t)$ is the HF sinusoidal voltage, and $u(\cdot)$ is the unit step function. Only the first two terms in (8.48) induce the HF harmonic current. The first term is the HF fundamental component, so the second term defined as $\Delta u'$ is solely analyzed into a Fourier series to describe the HF harmonics.

The relationship of the Fourier series in Fig. 8.37 can be expressed as

$$\begin{aligned} F(\Delta u'') &= \frac{1}{T}\int_{-T/2}^{T/2} (U_0 \sin(\omega_h t) - U_1) \cdot u(U_0 \sin(\omega_h t) - U_1) e^{-jn\omega_h t} dt \\ &= \frac{1}{T}\int_{t_1}^{t_2} U_0 \sin(\omega_h t) \cdot e^{-jn\omega_h t} dt + \frac{1}{T}\int_{t_1}^{t_2} - U_1 \cdot e^{-jn\omega_h t} dt, \end{aligned} \tag{8.49}$$

where the term $\frac{1}{T}\int_{t_1}^{t_2} U_0 \sin(\omega_h t) \cdot e^{-jn\omega_h t} dt$ is deduced as follows:

$$\begin{aligned} &\frac{1}{T}\int_{t_1}^{t_2} U_0 \sin(\omega_h t) \cdot e^{-jn\omega_h t} dt \\ &= \frac{1}{T} \cdot \frac{1}{-jn\omega_h}\int_{t_1}^{t_2} U_0 \sin(\omega_h t) d e^{-jn\omega_h t} \\ &= \frac{-1}{jn\omega_h T}\left[U_1\left(e^{-jn\omega_h t_2} - e^{-jn\omega_h t_1}\right)\right] \\ &\quad + \frac{U_0 \cos(\omega_h t_1)}{jn}\left(e^{-jn\omega_h t_2} + e^{-jn\omega_h t_1}\right) \\ &\quad + \frac{\omega_h}{jn}\int_{t_1}^{t_2} U_0 \sin(\omega_h t) d e^{-jn\omega_h t}]. \end{aligned} \tag{8.50}$$

Define

$$\begin{cases} \frac{1}{T}\int_{t_1}^{t_2} U_0 \sin(\omega_h t) \cdot e^{-jn\omega_h t} dt = Y \\ e^{-jn\omega_h t_2} - e^{-jn\omega_h t_1} = A \\ e^{-jn\omega_h t_2} + e^{-jn\omega_h t_1} = B. \end{cases} \tag{8.51}$$

Then (8.50) can be simplified into

$$Y = \frac{-U_1}{Tjn\omega_h} A - \frac{U_0 \cos(\omega_h t_1)}{Tn^2 \omega_h} B + \frac{1}{n^2} Y. \tag{8.52}$$

Thus Y can be obtained:

$$\begin{cases} Y = \dfrac{A[U_1/(Tnj\omega_h)] + B\left[U_0 \cos(\omega_h t_1)/\left(Tn^2 \omega_h\right)\right]}{1/n^2 - 1} & (n \neq 1) \\ Y = \dfrac{U_1 \cos(\omega_h t_1)}{2\pi j} & (n = 1). \end{cases} \tag{8.53}$$

Especially for A and B, there are two situations:

1. When n is even, there is

$$\begin{cases} A = 2j\sin(n\omega_h t_1) = -2j\sin(n\omega_h t_2) \\ B = 2\cos(n\omega_h t_1) = 2\cos(n\omega_h t_2). \end{cases} \tag{8.54}$$

2. When n is odd, there is

$$\begin{cases} A = -2\cos(n\omega_h t_1) = 2\cos(n\omega_h t_2) \\ B = -2j\sin(n\omega_h t_1) = -2j\sin(n\omega_h t_2). \end{cases} \tag{8.55}$$

As a result,

$$\frac{1}{T}\int_{t_1}^{t_2} -U_1 \cdot e^{-jn\omega_h t}dt = \frac{AU_1}{2\pi jn}. \tag{8.56}$$

The conclusion demonstrates that when $n \neq 1$, there is $F(\Delta u'') = Y + \frac{AU_1}{2\pi jn}$. Then it can be expressed as follows:

$$F(\Delta u'') = \begin{cases} \dfrac{AU_1 + jnBU_0\cos(\omega_h t_1)}{2j\pi(n - n^3)} & (n \neq 1) \\ -\dfrac{U_1\cos(\omega_h t_1)}{2\pi j} & (n = 1). \end{cases} \tag{8.57}$$

8.4.4 Partial HF Current Located in Linear Region: Case 2

When the amplitude of HF current is high enough compared with the DC component, both the maximum and minimum values of the induced current exceed the positive and negative saturation values, respectively. The voltage error is shown in Fig. 8.38.

As can be seen from Fig. 8.38, the voltage error is a clipped sine wave which can be expressed as

$$\Delta u = \begin{cases} U_0\sin(\omega_h t) - [U_0\sin(\omega_h t) - U_2]\cdot u[U_2 - U_0\sin(\omega_h t)]t \in [-T/2, 0] \\ U_0\sin(\omega_h t) - [U_0\sin(\omega_h t) - U_1]\cdot u[U_0\sin(\omega_h t) - U_1]t \in [0, T/2] \end{cases}. \tag{8.58}$$

According to the same analytical method introduced in case 1, case 2 just analyzes the second term of voltage error which causes the distortion of induced current in (8.58). Thus, the clipped sine wave removing DC component is shown in

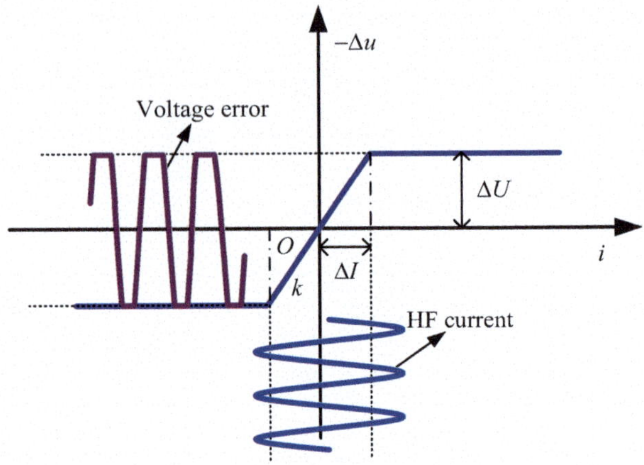

Fig. 8.38 Relationship between the voltage error and the HF current (case 2)

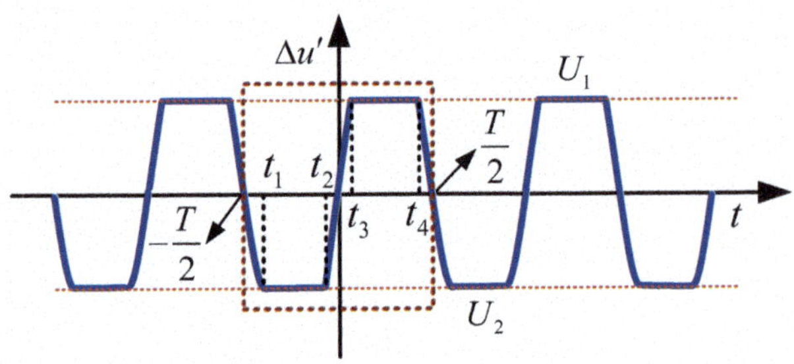

Fig. 8.39 HF voltage error component in case 2

Fig. 8.39, where t_1, t_2, t_3, and t_4 are the clipped time and U_1 and U_2 are the saturation values.

The resolution of the Fourier series of Fig. 8.39 is

$$
\begin{aligned}
F(\Delta u'') = & \frac{1}{T}\left\{\int_{-T/2}^{0}[U_0\sin(\omega_h t) - U_2]\cdot u[-U_0\sin(\omega_h t) + U_2]e^{-jn\omega_h t}dt\right. \\
& \left.+\int_{0}^{T/2}[U_0\sin(\omega_h t) - U_1]\cdot u[U_0\sin(\omega_h t) - U_1]\cdot e^{-jn\omega_h t}dt\right\} \\
= & \frac{1}{T}\int_{t_1}^{t_2}U_0\sin(\omega_h t)\cdot e^{-jn\omega_h t}dt - \frac{1}{T}\int_{t_1}^{t_2}U_2\cdot e^{-jn\omega_h t}dt \\
& +\frac{1}{T}\int_{t_3}^{t_4}U_0\sin(\omega_h t)\cdot e^{-jn\omega_h t}dt - \frac{1}{T}\int_{t_3}^{t_4}U_1\cdot e^{-jn\omega_h t}dt.
\end{aligned}
\tag{8.59}
$$

According to the same way deduced as case 1, the result is:

1. When $n \neq 1$, (8.59) can be expressed as

$$F(\Delta u'') = \frac{AU_1 + jnBU_0\cos(\omega_h t_1) + CU_2 + jnDU_0\cos(\omega_h t_3)}{2j\pi(n - n^3)}, \tag{8.60}$$

where A, B, C, and D can be expressed as

$$\begin{cases} A = 2j\sin(n\omega_h t_1) \\ B = 2\cos(n\omega_h t_1) \\ C = 2j\sin(n\omega_h t_3) \\ D = 2\cos(n\omega_h t_3) \end{cases} n \text{ is even}, \begin{cases} A = -2\cos(n\omega_h t_1) \\ B = -2j\sin(n\omega_h t_1) \\ C = -2\cos(n\omega_h t_3) \\ D = -2j\sin(n\omega_h t_3) \end{cases} n \text{ is odd}. \tag{8.61}$$

2. When n=1, (8.59) can be expressed as

$$F(\Delta u'') = -\frac{U_1\cos(\omega_h t_1) + U_2\cos(\omega_h t_3)}{2\pi j}. \tag{8.62}$$

8.4.5 HF Current Located in Linear Region: Case 3

Similarly to Fig. 8.39 when the whole HF current is located in the linear region, the induced voltage error has the same phase with the HF current as shown in Fig. 8.40. In this case, the voltage error only contains the sinusoidal component without other harmonics, which can be expressed as Δu=-$U_0\sin(\omega_h t)$+ΔU_1. The equivalent HF resistance reaches the maximum value that is equal to the gradient of the linear region.

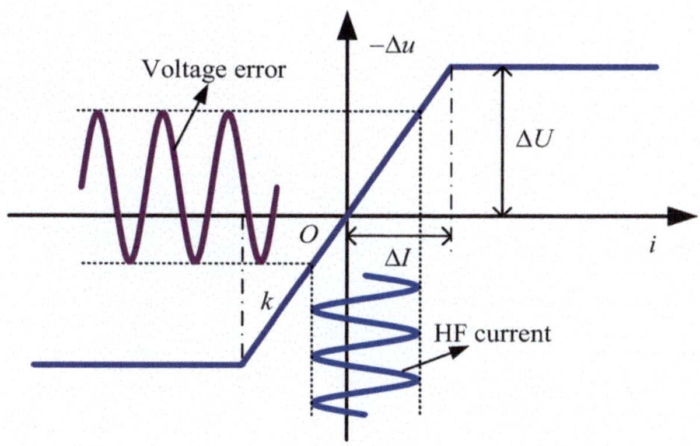

Fig. 8.40 Relationship between the voltage error and the HF current (case 3)

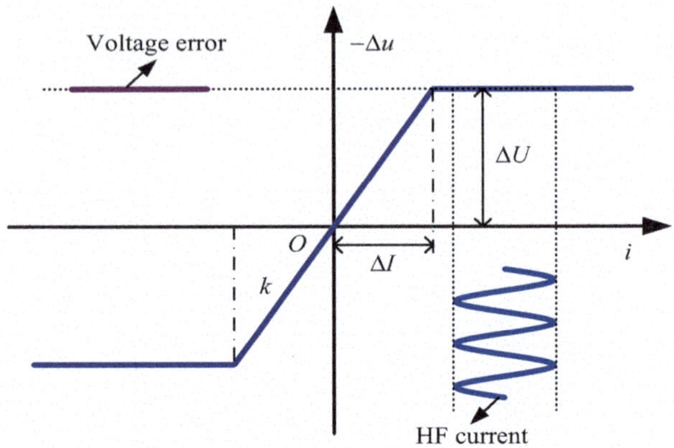

Fig. 8.41 Relationship between the voltage error and the HF current (case 4)

8.4.6 Whole HF Current Located in Saturation Region: Case 4

When the DC current component is high enough, the HF current will be far away from the linear region and the voltage error is constant as shown in Fig. 8.41. The equivalent HF resistance reaches the minimum value that is equal to zero. Thus the voltage error does not affect the identification results in this case.

According to the inverter nonlinearity parameters which are obtained by the self-learning method, a two-dimensional region can be established based on the HF amplitude and the DC offset of the induced current. Figure 8.42 shows the regions of the four cases in the first quadrant, where the current saturation value ΔI is 0.1 pu

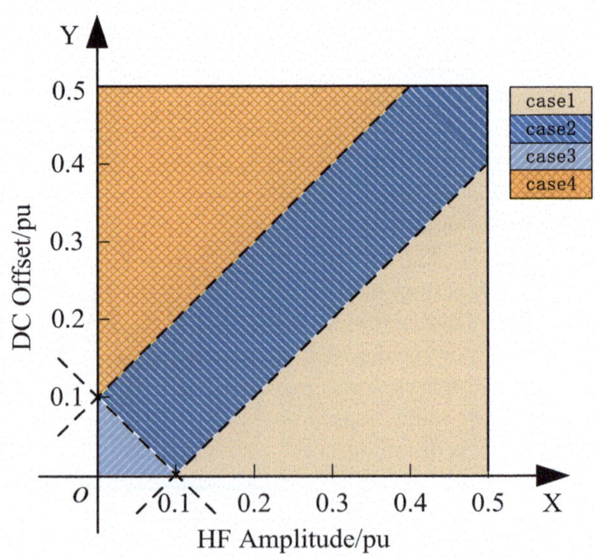

Fig. 8.42 Different effects of the inverter nonlinearities described by a two-dimensional region with four cases

(current base value is selected as 7.9 A), and the voltage error saturation value ΔU is 0.044 pu (voltage base value is selected as 311 V).

The HF amplitude and the DC offset of induced current are defined as P and Q, respectively. If P and Q meet $-0.1<-P+Q<0.1$ and $-0.1<P+Q<0.1$, it belongs to case 3; if P and Q meet $-P+Q>0.1$, it belongs to case 4. In these two cases, the voltage error only contains the sinusoidal component without other harmonics. While P and Q meet the conditions of $-0.1<-P+Q<0.1$ and $P+Q>0.1$, it belongs to case 1; thus $U_0=0.4P$ and $U_1=0.04–0.4Q$. While P and Q meet the conditions of $-P+Q<-0.1$ and $P+Q>0.1$, it belongs to case 2; thus $U_0=0.4P$, $U_1=0.04-0.4Q$, and $U_2=-0.04–0.4Q$.

After analyzing the distribution of the four cases in a two-dimensional region, the second and third harmonics are selected to analyze because the amplitudes of high-order harmonics are so small that it can be neglected. According to the harmonic analysis above, the amplitudes of the second and third harmonics which are represented by A_2 and A_3 can be obtained as

$$A_2 = \begin{cases} \frac{2}{3\pi} U_0 \cdot (\cos \omega_h t_1)^3, & \text{case 1} \\ \frac{2}{3\pi} \left[U_0 \cdot (\cos \omega_h t_1)^3 - U_0 \cdot (\cos \omega_h t_3)^3 \right], & \text{case 2} \\ 0, & \text{case 3} \\ 0, & \text{case 4} \end{cases}, \tag{8.63}$$

$$A_3 = \begin{cases} \frac{2}{3\pi} U_1 \cdot (\cos \omega_h t_1)^3, & \text{case 1} \\ \frac{2}{3\pi} \left[U_1 \cdot (\cos \omega_h t_1)^3 - U_2 \cdot (\cos \omega_h t_3)^3 \right], & \text{case 2} \\ 0, & \text{case 3} \\ 0, & \text{case 4} \end{cases}. \tag{8.64}$$

8.4.7 *Inverter Nonlinearity Effects on Offline Inductance Identification*

According to the analysis above, the inverter nonlinearities cause voltage error during the offline inductance identification process. The actual voltage drop of *d-q* inductances and stator resistance is $U_{inj}\sin(\omega_h t)-\Delta u$. So the induced current can be expressed as follows:

$$I_{d(q)\text{real}} = \frac{u_{inj} \sin(\omega_h t) - \Delta u}{\omega_h L_{d(q)}}. \tag{8.65}$$

Besides, the estimated values of *d-q* inductances are larger than the theoretical values due to the effects of inverter nonlinearities. The relationship between the estimated and theoretical values can be obtained as

$$\hat{L}_{d(q)} = \frac{U_{inj}}{\left|U_{inj}\sin(\omega_h t) - \Delta u\right|} L_{d(q)}, \tag{8.66}$$

where $|U_{inj}\sin(\omega_h t) - \Delta u|$ means the fundamental component amplitude of $U_{inj}\sin(\omega_h t) - \Delta u$ and $\hat{L}_{d(q)}$ and $L_{d(q)}$ mean the estimated and theoretical values of *d-q* inductances, respectively.

The voltage error caused by the inverter nonlinearities is different under the different injection conditions. There are inevitable estimation errors in case 1, case 2, and case 3 due to the fundamental voltage error. In case 4, the fundamental voltage error is zero and the DC voltage error does not influence the inductance estimation.

8.4.8 Experimental Results for Signal Processing

The analysis of the inductance identification algorithm has been validated in a 2.2 kW interior IPMSM drive system, as shown in Fig. 3.18. The rated parameters of the IPMSM are listed as follows: 380 V, 5.6 A, 50 Hz, 21 Nm, and 1000 r/min. The intelligent power module FP25R12KT4 is used. The STM32F103 ARM is adopted to execute the whole identification algorithm. The PWM frequency is 6 kHz, and the dead time is 3.2 μs. The frequency of the injected HF voltage signal is 300 Hz. The current reference increases linearly with 0.03 pu/s during the estimation of inverter nonlinearity parameters. A 12-bit absolute encoder is installed to obtain the actual position which is solely used for showing the electrical angle of the rotor during the identification process.

The waveforms of the whole identification process are shown in Fig. 8.43. The a-phase current and the estimated L_d and L_q are given. The rotor position is obtained by initial position identification method, which can realize the identification process at standstill. The supplementary *d*-axis DC current is 0.1pu. The injected HF voltage increases gradually so as to avoid overcurrent and guarantee signal intensity. The *d-q* axis inductances converge to the stable values, 32.4 mH (the theoretical value of L_d is 31.6 mH) and 65.2 mH (the theoretical value of L_q is 62.8 mH), respectively.

Figure 8.44 shows the actual and estimated rotor position waveforms during the inductance identification process with different initial rotor positions. Firstly, the HF voltage signal is injected to obtain the initial rotor position and then a pulse signal is injected to detect the polarity. After acquiring the initial rotor position, the induc-

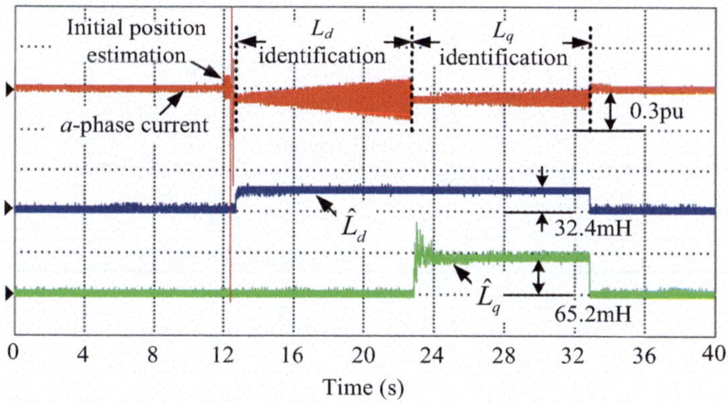

Fig. 8.43 Experimental waveforms of the d-q axis inductance identification

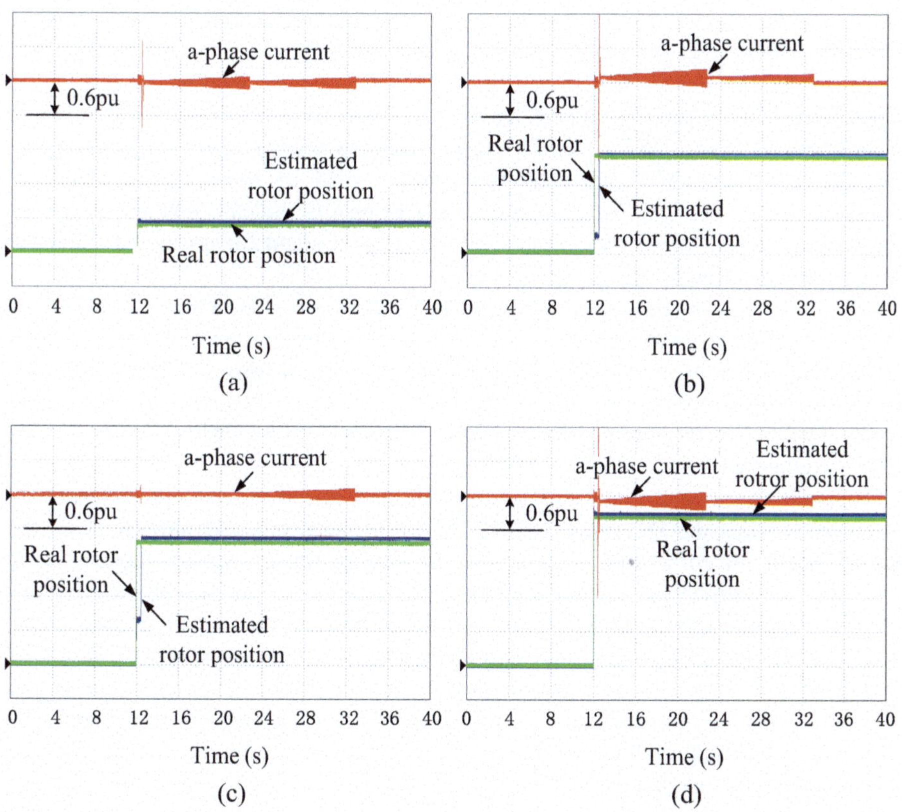

Fig. 8.44 Experimental waveforms of the real and estimated rotor position during the inductance identification process. (**a**) $\theta_e = 46°$, (**b**) $\theta_e = 200°$, (**c**) $\theta_e = 257°$, and (**d**) $\theta_e = 328°$ degree

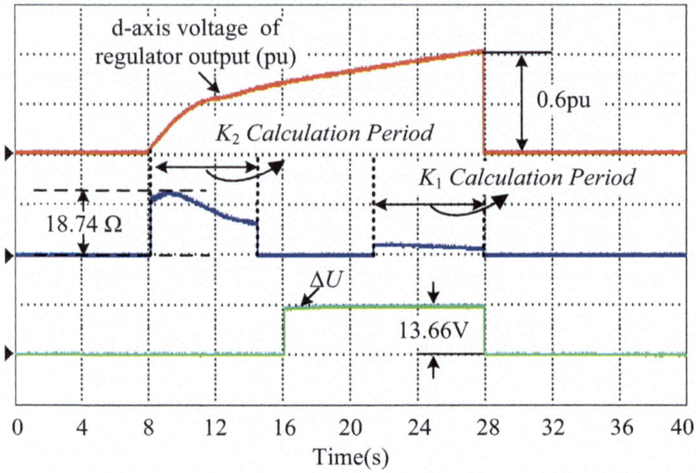

Fig. 8.45 Experimental waveforms of the self-learning process

tance identification process can be started. The actual rotor positions in Fig 8.44a, b, c, and d are 46°, 200°, 257°, and 328°, respectively, while the estimated positions are 50°, 203°, 262°, and 334°, respectively. From Fig. 8.44, the rotor position maintains constant during the identification process. The results indicate that the proposed inductance identification method can be operated at standstill no matter what the initial rotor position is.

Figure 8.45 shows the experimental waveforms of the self-learning process of the inverter nonlinearity parameters. From top to bottom, the *d*-axis voltage reference of the current regulator output, the estimated inverter nonlinearity gradient parameter, and the estimated saturation value of the voltage are given, respectively. According to (8.43), the inverter nonlinearity gradient *K* equals to 15.8 Ω. The saturation value of the voltage error is 13.66 V.

Figure 8.46 shows the harmonic analysis of the induced current with different HF amplitudes when the DC offset is 0.12 pu. From the results, the amplitude of the second harmonic component increases as the HF amplitude becomes larger.

Figure 8.47 shows the harmonic analysis of the induced current with different DC offsets when the HF amplitude is 0.1 pu. It can be seen that the amplitude of the second harmonic component decreases as the DC offset becomes larger.

Figures 8.48 and 8.49 show the theoretical and experimental results of the second and third harmonics of the induced current caused by voltage error with different DC offsets and HF amplitudes, respectively. The experimental results match with the theoretical results closely.

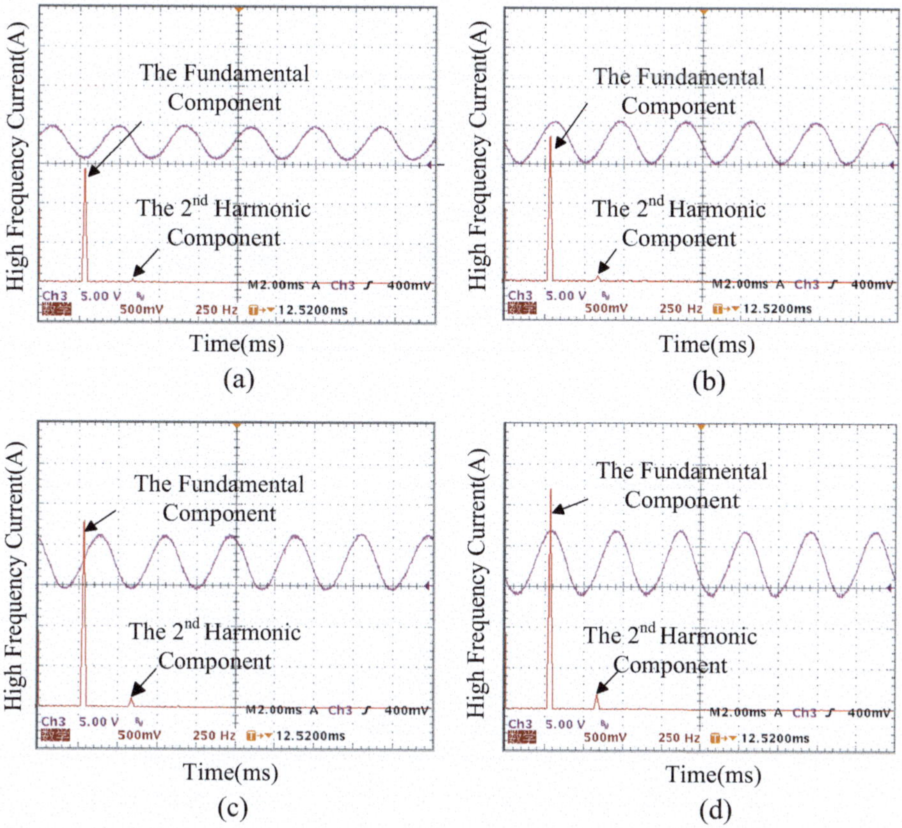

Fig. 8.46 Harmonic analysis of the induced current considering the effects of inverter nonlinearities under different HF amplitudes, (**a**) 0.15 pu, (**b**) 0.2 pu, (**c**) 0.25 pu, and (**d**) 0.3 pu

The experimental result of L_d identification is given to verify the effectiveness of the proposed analysis. Figure 8.50 shows the estimation value of L_d under different injection conditions. It can be seen from the results that with the increasing of DC current component, the estimated value of L_d is more steady and accurate. When the injected DC component and the HF component are as small as case 3 shows, the maximum estimation error reaches 44.9%. Besides, the minimum estimation error is reduced to 4.6% in case 4 when the DC component is high enough.

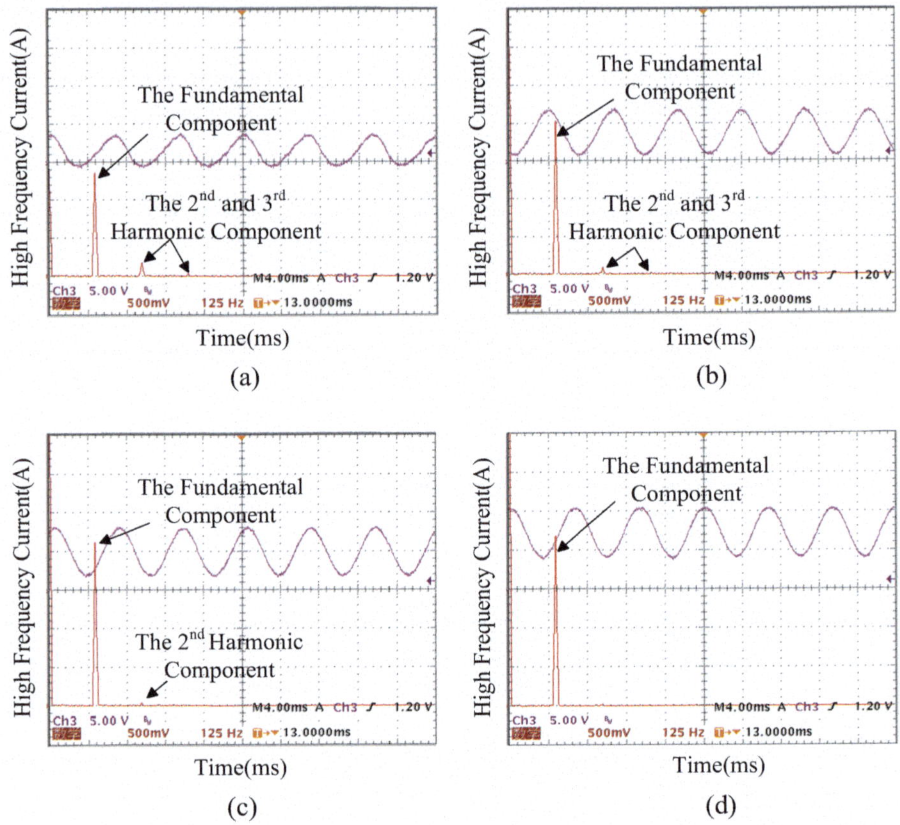

Fig. 8.47 Harmonic analysis of the induced current considering the effects of the inverter nonlinearities, (**a**) DC offset 0.05 pu, (**b**) DC offset 0.15 pu, (c) DC offset 0.2 pu, (**d**) DC offset 0.3 pu

8.5 Summary

A self-commissioning methodology for PMSMs at standstill using a VSI-fed drive system was proposed. The influence of inverter nonlinearities including the parasitic capacitance on the identification at different rotor positions was analyzed. The injected *d*-axis DC current reduces the influence of inverter nonlinearities effectively during the inductance parameter identification. The parameters of inverter nonlinearities can be obtained by self-learning method. Then the current harmonic induced by inverter nonlinearities was analyzed in four different cases with different injection conditions. The nonlinearity model adopted to compensate the identified inductances

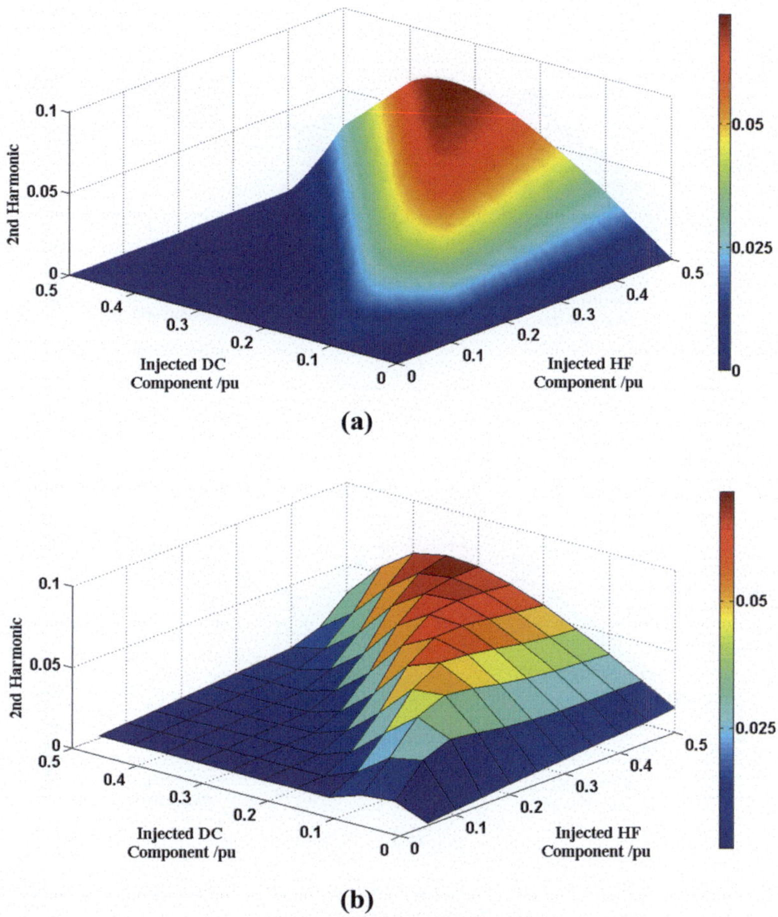

Fig. 8.48 Analysis of the second harmonic caused by voltage error with different DC offset and HF amplitudes. (**a**) The theoretical result, (**b**) the experimental result

enhances the estimation accuracy. During the proposed stator resistance identification, the linearly increasing current is injected to utilize the multipoint information. The linear regression method is a promising and robust way to resist the influence of inverter nonlinearities. In practical applications, the time consumed for the self-commissioning can be adjusted by tuning the change rate of the injected signals. The experimental results show that the proposed self-commissioning method is qualified for general-purpose drive applications.

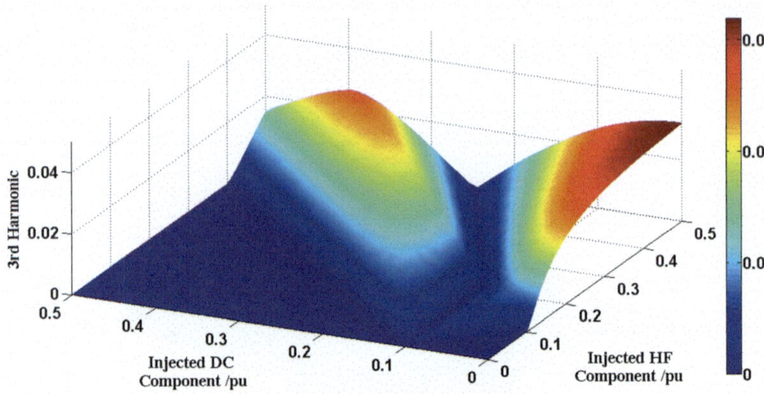

(a)

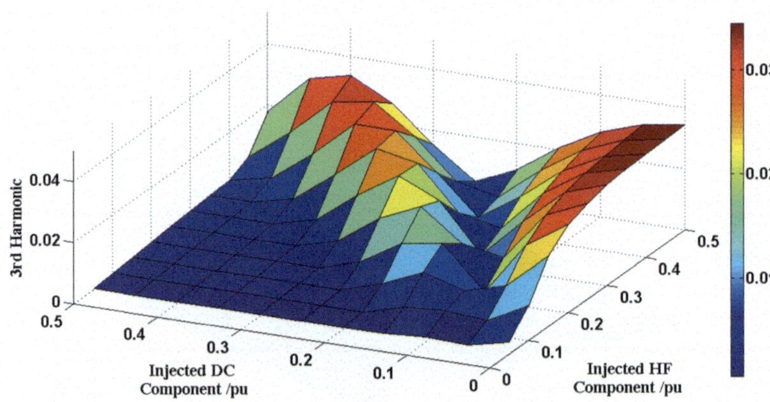

(b)

Fig. 8.49 Analysis of the third harmonic caused by voltage error with different DC offset and HF amplitudes. (**a**) The theoretical result, (**b**) the experimental result

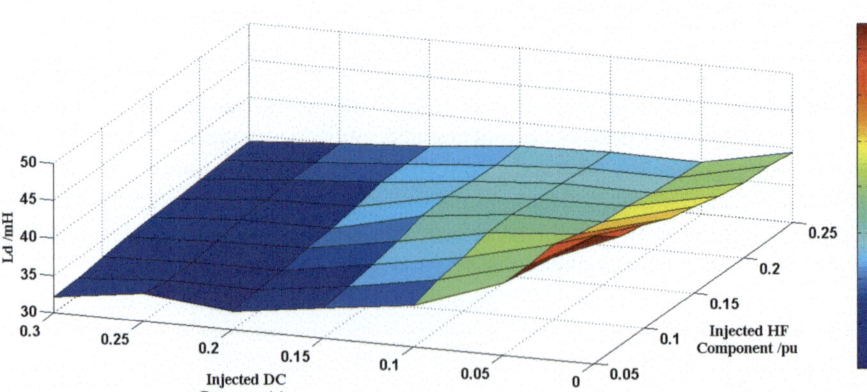

Fig. 8.50 The experimental results of L_d estimation under different injection conditions

References

1. K. Ide, M. Takaki, S. Morimoto, Y. Kawazoe. Saliency-based Sensorless drive of adequate designed IPM Motor for Robot Vehicle Application, *Power Conversion Conference (PCC)*, Nagoya, pp. 1126–1133, 2007
2. S. Yang, S. Yang, J. Hu, Robust initial position estimation of permanent magnet machine with low saliency ratio. IEEE Access **5**, 2685–2695 (2017)
3. R. Antonello, F. Tinazzi, M. Zigliotto, Benefits of direct phase voltage measurement in the rotor initial position detection for permanent-magnet motor drives. IEEE Trans. Ind. Electron. **62**(11), 6719–6726 (2015)
4. G. Wang, G. Zhang, R. Yang, D. Xu, Robust low-cost control scheme of direct-drive gearless traction machine for elevators without a weight transducer. IEEE Trans. Ind. Appl. **48**(3), 996–1005 (May/June 2012)
5. J. Im, R. Kim, Improved saliency-based position Sensorless control of interior permanent-magnet synchronous machines with single DC-link current sensor using current prediction method. IEEE Trans. Ind. Electron. **65**(7), 5335–5343 (2018)
6. G. Wang, L. Qu, H. Zhan, J. Xu, L. Ding, G. Zhang, D. Xu, Self-commissioning of permanent magnet synchronous machine drives at standstill considering inverter nonlinearities. IEEE Trans. Power Electron. **29**(12), 6615–6627, (December 2014)
7. M. A. Herr'an, J. R. Fischer, S. A. Gonz'alez, M. G. Judewicz, D. O. Carrica, Adaptive dead-time compensation for grid-connected PWM inverters of single-stage PV systems. IEEE Trans. Power Electron. **28**(6), 2816–2825 (May 2013)
8. L. M. Gong, Z. Q. Zhu, A novel method for compensating inverter nonlinearity effects in carrier signal injection-based Sensorless control from positive-sequence carrier current distortion. IEEE Trans. Ind. Appl. **47**(3) 557–565 (May/June 2011)
9. J. Holtz, Acquisition of position error and magnet polarity for sensorless control of PM synchronous machines. IEEE Trans. Ind. Appl. **44**(4), 1172–1180 (July/August 2008)
10. Y. Park, S. K. Sul, A novel method utilizing trapezoidal voltage to compensate for inverter nonlinearity. IEEE Trans. Power Electron. **27**(12), 4837–4846 (December 2012)
11. J. M. Guerrero, M. Leetmaa, F. Briz, A. Zamarrón, R. D. Lorenz, Inverter nonlinearity effects in high-frequency signal-injection-based Sensorless control methods. IEEE Trans. Ind. Appl. **41** (2), 618–626 (March/April 2005)
12. G. Wang, Y. Wang, L. Ding, L. Yang, R. Ni, D. Xu, Harmonic analysis of the effects of inverter nonlinearity on the offline inductance identification of PMSMs using high frequency signal injection J. Power Electron. **15**(6), 1567–1576 (November 2015)
13. A. T. d. Almeida, F. J. T. E. Ferreira, J. A. C. Fong, Standards for efficiency of electric motors. IEEE Ind. Appl. Mag. **17**(2), 12–19 (November 2010)

Chapter 9
Practical Issues of Sensorless Control for PMSM Drives

9.1 Efficiency Optimization Control

The MTPA control is a simple and practical method to achieve high-efficiency operation of IPMSM. Recently, many MTPA control methods have been investigated to guarantee the minimum copper loss of IPMSM [1–6]. In most commercial drives, the MTPA control is achieved with the lookup tables or the polynomial fitting. The conventional MTPA control method is parameter dependent, and the effectiveness relies on the knowledge of the accurate motor parameters. The online parameter identification or offline parameter testing is necessary to achieve the robust MTPA operation [4–6]. Recently, the test signals injection method has been proposed, which can allow the MTPA trajectory to be learned online; however the signal injection method looks somewhat complex [1–3]. In most literatures, the online MTPA control methods used in IPMSM drive system should be installed with a position sensor for the control. Some key issues including the current ripple and torque transient should be carefully considered for the sensorless IPMSM drive.

Assuming that the motor parameters are constant, to achieve the MTPA operation, the condition should be

$$\frac{\partial T_e}{\partial \gamma} = \frac{3}{2} P_n I_s \left[\psi_f \cos\gamma + \left(L_d - L_q\right) I_s \cos 2\gamma \right] = 0, \tag{9.1}$$

where γ is the current vector angle.

Therefore, the reference of the current vector angle for the MTPA condition can be obtained from (9.1) and described as

$$\gamma = \arccos\left[\frac{-\psi_f + \sqrt{\psi_f^2 + 8\left(L_d - L_q\right)^2 I_s^2}}{4\left(L_d - L_q\right) I_s} \right]. \tag{9.2}$$

G. Wang et al., *Position Sensorless Control Techniques for Permanent Magnet Synchronous Machine Drives*, https://doi.org/10.1007/978-981-15-0050-3_9

However, L_d and L_q change due to different load, and ψ_f also varies in different temperature. So, the stator flux varies under the different operating condition, which can be expressed as

$$\begin{cases} \psi_d = L_d(i_d)i_d + M_{dq}(i_d, i_q)i_q + \psi_f(t^\circ) \\ \psi_q = L_q(i_q)i_q + M_{qd}(i_d, i_q)i_d, \end{cases} \quad (9.3)$$

where $L_d(i_d)$ and $L_q(i_q)$ are the dq inductances considering the magnetic saturation, $M_{dq}(i_d, i_q)$ and $M_{qd}(i_d, i_q)$ are the mutual inductances, and $\psi_f\,(t^\circ)$ is the permanent flux considering the temperature variation, respectively.

Therefore, the electromagnetic torque considering parameters variation can be obtained.

$$\begin{aligned} T_e = & \frac{3}{2}pI_s\{\psi_f(t^\circ)\sin\gamma + \frac{1}{2}[L_d(I_s,\gamma) - L_q(I_s,\gamma)]I_s\sin 2\gamma \\ & + M_{dq}(I_s,\gamma)I_s\sin^2\gamma - M_{qd}(I_s,\gamma)I_s\cos^2\gamma\}. \end{aligned} \quad (9.4)$$

It can be seen that the MTPA control is difficult to achieve if the motor parameter variations are considered.

Different from the MTPA control schemes that need the motor parameters or the test signals injection, an online auto-tuning MTPA based on searching the minimum current amplitude is proposed [1]. Figure 9.1 shows the MTPA control scheme by tuning the current vector angle γ with a small value $\Delta\gamma$. If γ reaches the optimal angle γ_o, the MTPA operation can be achieved. The d-q axes current references can be decided from γ and I_s^*, where I_s^* is the reference of the current magnitude. I_s^* is outputted from the speed controller. Therefore, the d-q axes current references can be given by

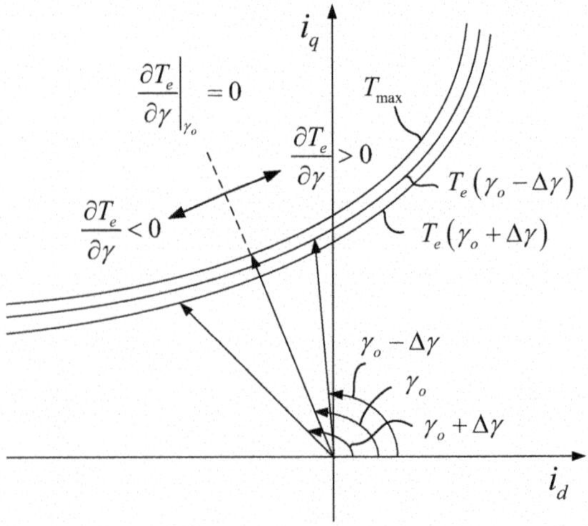

Fig. 9.1 Online MTPA control scheme based on current vector angle regulation

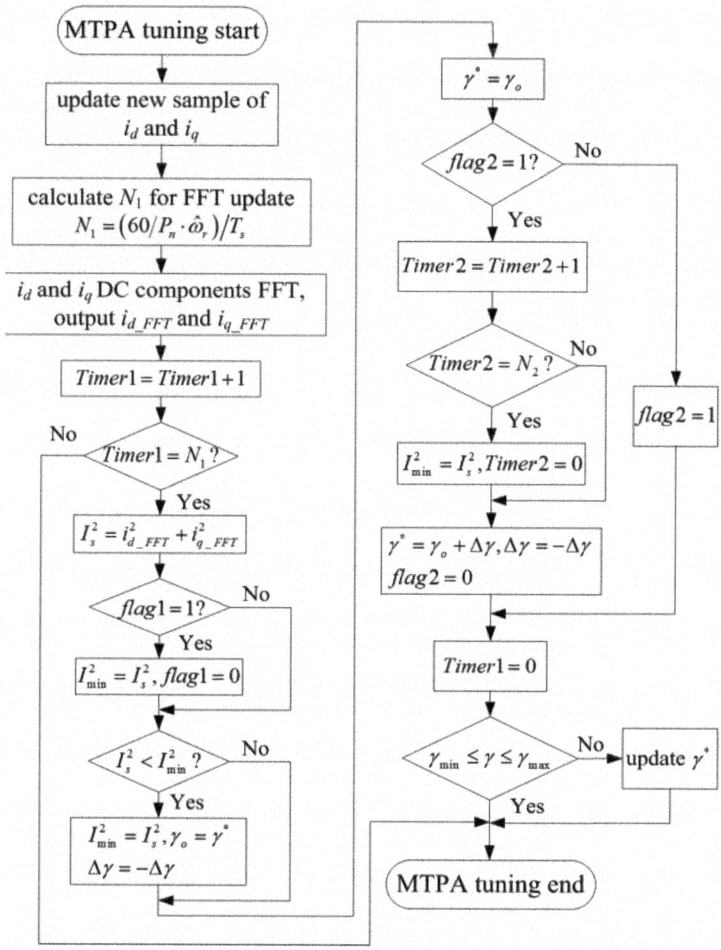

Fig. 9.2 Flow chart of the online MTPA search algorithm in each current control period

$$\begin{cases} i_d^* = I_s^* \cos\gamma \\ i_q^* = I_s^* \sin\gamma \end{cases}. \tag{9.5}$$

The reference of γ can be acquired from the comparison with the magnitude of each current sample value by changing a small angle value $\Delta\gamma$. The flow chart of the MTPA search algorithm is shown in Fig. 9.2. The algorithm is executed in each current control period T_s (the PWM carrier period) after obtaining the new current sample value; and the sample period is equal to T_s. Since the sixth harmonic exists in the d-q axes current components, it is difficult to identify the minimum current operation point directly. Therefore, the d-q current components are inputted to a FFT algorithm to extract the DC components, i_{d_FFT} and i_{q_FFT}. Then i_{d_FFT} and i_{q_FFT}

are used to calculate the fundamental current magnitude I_s. The period of the fundamental current is denoted as T_1, and it can be calculated from the estimated rotor speed by $60/(P_n \cdot \hat{\omega}_r)$, where P_n denotes pole pairs number and the unit of $\hat{\omega}_r$ is r/min. The magnitude of the fundamental current can be updated by calculating the average value in each fundamental period. A timing counter *Timer*1 is used to decide the past sample number needed to update the FFT calculation result. The FFT is calculated at every sample time of the process; the sample time is T_s (T_s is set as 100 μs). Therefore, the number of the past samples used for the FFT update is determined by the rotor speed, and it can be calculated according to the sample time and the estimated speed:

$$N_1 = (60/P_n \cdot \hat{\omega}_r)/T_s, \tag{9.6}$$

where N_1 is the number of the past samples used for the FFT update.

The initial value of flag1 is set as 1. When the MTPA is enabled, assign the first value of I_s as the initial minimum current value $I_{\min}$, and then let flag1 = 0. If the motor operates at steady state, the MTPA search algorithm will auto-tune the reference of the current vector angle γ^* with a small positive value $\Delta\gamma$. When *Timer*1 is equal to N_1, the new current amplitude is compared with the previous one. Both the smaller current amplitude $I_{\min}$ and the corresponding vector angle γ_o are recorded. Then $\gamma_o + \Delta\gamma$ is assigned for the reference angle γ^*. After the output of speed controller is steady, compare the new current $I(\gamma_o + \Delta\gamma)$ with $I_{\min}$. If $I(\gamma_o + \Delta\gamma)$ is smaller than $I_{\min}$, then update $I_{\min}$ to $I(\gamma_o + \Delta\gamma)$, and γ_o is also recorded according to the new vector angle. In the next step, $\gamma_o - \Delta\gamma$ is assigned for the reference angle γ^*. Flag2 is used to select which reference angle be in action, namely, γ_o, $\gamma_o - \Delta\gamma$ or $\gamma_o + \Delta\gamma$. In this way, the disturbance with $\Delta\gamma$ will search the optimal angle γ_o continuously. The iteration number taken to find the minimum current is determined by the incremental angle $\Delta\gamma$ and the practical optimal current vector angle γ_o. The bigger $\Delta\gamma$, the faster it will find the minimum, but the accuracy of MTPA will be decreased. In order to compromise the rapidity and the accuracy, we selected $\Delta\gamma$ as 2 electrical degrees in the algorithm. The optimal current vector angle is determined by the saliency characteristics and the current magnitude as shown in (9.2). So, the iteration number taken to find the minimum current can be calculated through γ_o divided by $\Delta\gamma$.

The advantage of the online tuning method is robust to the parameter variation. The FFT can obtain the fundamental amplitude effectively for the sensorless control applications. However, the search algorithm will take much time to find the optimal current vector angle at low speed, since the FFT processing needs more past samples for calculation. For example, the FFT calculation needs 1 s at 1 Hz. It will take several seconds to find the minimum. So, the MTPA search algorithm is not suitable for low-speed operation. The other important issue is that if the load torque increases rapidly, the stator current will inevitably increase. In this case, I_s will never be smaller than $I_{\min}$ no matter how γ^* is adjusted. To solve the problem, a timing update counter *Timer*2 is added to the search algorithm, as shown in the right side of Fig. 9.2. In every fixed time interval N_2 (N_2 is selected according to the dynamic of the control loop, usually is several times of N_1), the timing auxiliary section will update $I_{\min}$ with the latest I_s. Additionally, the value of γ^* is limited by a saturation function. Thus, the problem of transient state regulation can be solved.

9.2 Active Disturbance Rejection Control for Dynamic Performance Improvement

With the development of sensorless control, the research focus has gradually evolved from the retrieve of rotor position to the performance improvement. Some efforts have been made to improve the demodulation method [7]. Also some researches concentrated on applying additional observer to improve the disturbance rejection capability [8]. The others achieved the performance improvement by directly designing the controller with strong robustness [9–10]. ADRC is effective in tolerating disturbances and achieving the desired dynamic performance. The generalized concept of total disturbances and the disturbance estimation capability of extended state observer (ESO) allow ADRC to require the least amount of plant information but exhibit the relatively excellent robustness. Thus, ADRC shows great prospects in industrial applications.

Nonlinear ADRC (NLADRC) was first proposed by J. Han [11] with the nonlinear function and subsequently parameterized by Z. Gao [12] to shape linear ADRC (LADRC) for more practicability. There is no doubt that the quality of ADRC hinges on the disturbance estimation capability of ESO. Nonlinear ESO (NLESO) employs a nonlinear function that can be regarded as a linear function with variable gains, making it more effective than linear ESO (LESO) potentially. However, it is difficult to perform theoretical analysis. By comparison, LESO makes up for the deficiencies of NLESO. Whereas the fixed gains weaken the efficiency.

9.2.1 Structure of Conventional LADRC and NLADRC

In sensorless PMSM drive, ADRC is employed as speed regulator aimed at improving the robustness to the total disturbances collecting model uncertainties and external disturbances. The schematic diagram of conventional ADRC is shown in Fig. 9.3, which is constituted by three parts: (1) tracking differentiator (TD), (2) ESO, and (3) control law (CL). In the arrangement of the transient process, TD is employed to track the reference input rapidly without overshoot. Meanwhile the differential is given for plants of second-order or above. Instead of concerning about

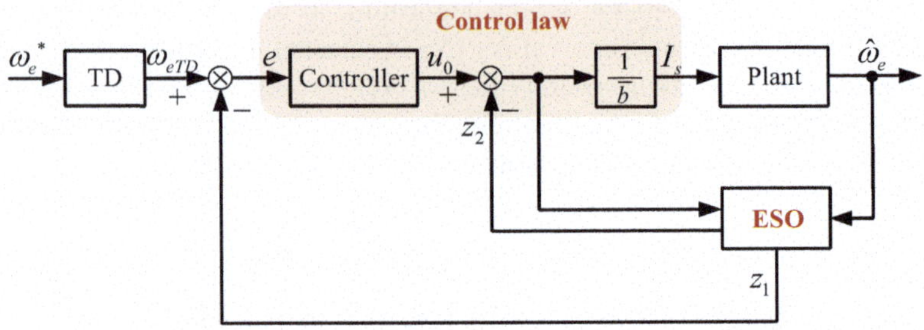

Fig. 9.3 Schematic diagram of conventional ADRC

the detailed disturbance model, ADRC concentrates on estimating the total disturbances via ESO in real time and compensating for these by designing CL. The inputs of ESO are the estimated electrical angular speed and the amplitude of the stator current I_s used to generate the dq-axes current reference.

To facilitate the establishment of ESO, the state equation of plant is first constructed as follows:

$$\begin{cases} \dot{\hat{\omega}}_e = f + \bar{b} I_s \\ \dot{f} = h, \end{cases} \tag{9.7}$$

where

$$b = \frac{3}{2J} {P_n}^2 \psi_f \sin \gamma, \tag{9.8}$$

$$\bar{b} = \frac{3}{2\hat{J}} {P_n}^2 \hat{\psi}_f \sin \hat{\gamma}, \tag{9.9}$$

$$f = \frac{3}{2} \frac{{P_n}^2}{J} (L_d - L_q) {I_s}^2 \sin \gamma \cos \gamma - \frac{P_n}{J} T_L - \frac{B}{J} \omega_e + \left(b - \bar{b}\right) I_s + \left(\dot{\hat{\omega}}_e - \dot{\omega}_e\right). \tag{9.10}$$

$\bar{b}$ is the approximation of the control input gain b. f represents the total disturbances to be estimated and compensated. Meanwhile, assume that the derivative of the total disturbances, h, is bounded. b is related to the moment of inertia J, the number of pole pairs P_n, the permanent magnet flux linkage ψ_f, and the torque angle γ. In (9.10), superscript "∧" denotes the approximation of the actual value. Besides, T_L is the load torque, B is the viscous friction coefficient, and ω_e denotes the actual electrical angular speed.

The quality of ADRC is closely related to the disturbance estimation capability of ESO. In order to track the state variables in (9.7) in real time, the mathematical model of conventional NLESO is established as follows:

$$\begin{cases} e_1 = z_1 - \hat{\omega}_e \\ \dot{z}_1 = z_2 - \beta_{01} e_1 + \bar{b} I_s \\ \dot{z}_2 = -\beta_{02} \mathrm{fal}(e_1, \alpha, \delta), \end{cases} \tag{9.11}$$

where z_1 and z_2 serve to track the estimated speed $\hat{\omega}_e$ and total disturbances f, respectively. β_{01} and β_{02} are observer gains. Since the derivative of the power function near the origin does not exist when the power is in the range of 0–1, it is likely to cause the oscillation of the state reconstruction. Therefore, a linear function is usually used near the origin to form the conventional piecewise nonlinear function fal(e_1, α, δ), whose function expression is

$$\mathrm{fal}(e_1, \alpha, \delta) = \begin{cases} \dfrac{e_1}{\delta^{1-\alpha}}, & |e_1| \leq \delta \\ |e_1|^{\alpha} \operatorname{sgn}(e_1), & |e_1| > \delta, \end{cases} \tag{9.12}$$

with the range of linearity 2δ and the power of nonlinearity α. In the case of $0 < \alpha < 1$, fal(e_1, α, δ) helps the state errors converge rapidly in some conditions. Particularly, NLESO completely becomes a LESO in the case of $\alpha = 1$.

The control law is designed as follows:

$$I_s = \frac{u_0 - (z_2 + s_2)}{\bar{b}}, \tag{9.13}$$

where u_0 is a proportional controller expressed as follows with gain k_p and speed reference $\omega_e{}^*$.

$$u_0 = k_p \cdot (\omega_e{}^* - \hat{\omega}_e). \tag{9.14}$$

According to (9.7) and (9.11), the state errors equation of conventional NLESO is expressed as follows:

$$\begin{cases} \dot{e}_1 = e_2 - \beta_{01} e_1 \\ \dot{e}_2 = -\beta_{02} \mathrm{fal}(e_1, \alpha, \delta) - h' \end{cases} \tag{9.15}$$

where e_1 and e_2 denote speed tracking error $z_1 - \hat{\omega}_e$ and disturbance estimation error $z_2 - f$, respectively. If the conventional nonlinear function fal(·) is substituted by e_1, then (9.16) represents the state errors equation of LESO. When the state variables access steady state, the derivatives of the state errors will approach zero. For the common disturbance in the form of step or ramp, it can be assumed that the derivative of disturbance, h, is a constant in steady state. Subscripts L and N are used to represent the conventional LESO and NLESO, respectively. Then, the steady-state errors of the LESO can be expressed as follows:

$$\begin{cases} e_{L1\infty} = \dfrac{-h}{\beta_{L02}} \\ e_{L2\infty} = \beta_{L01} e_{L1\infty} \end{cases}. \tag{9.16}$$

To facilitate analysis, assume that the observer gains of conventional NLESO are consistent with those of LESO, i.e., $\beta_{N01} = \beta_{L01} = 2\omega_0$, and $\beta_{N02} = \beta_{L02} = {\omega_0}^2$. ω_0 indicates the observer bandwidth, which reflects the compromise between tracking performance and noise-filtering capability. Under these conditions, the steady-state errors of conventional NLESO are described in (9.17) and (9.18) when $e_{N1\infty}$ lies in the linearity and nonlinearity, respectively.

$$\begin{cases} e_{N1\infty} = \dfrac{-h\delta^{1-\alpha}}{\beta_{N02}} = \delta^{1-\alpha} e_{L1\infty} \\ e_{N2\infty} = \beta_{N01} e_{N1\infty} = \delta^{1-\alpha} e_{L2\infty} \end{cases}, \tag{9.17}$$

$$\begin{cases} e_{N1\infty} = \left(\frac{|h|}{\beta_{N02}}\right)^{\frac{1}{\alpha}} \operatorname{sgn}(-h) = e_{L1\infty} \left(\frac{|h|}{{\omega_0}^2}\right)^{\frac{1}{\alpha}-1} \\ e_{N2\infty} = \beta_{N01} e_{N1\infty} = e_{L2\infty} \left(\frac{|h|}{{\omega_0}^2}\right)^{\frac{1}{\alpha}-1} \end{cases}. \tag{9.18}$$

Generally, δ is less than 1. Based on (9.16)–(9.18), the steady-state speed tracking errors of LESO and conventional NLESO are intuitively depicted in Fig. 9.4. It is worth noting that the comparison results of disturbance estimation errors are consistent with those from Fig. 9.4. Then the following conclusions can be drawn.

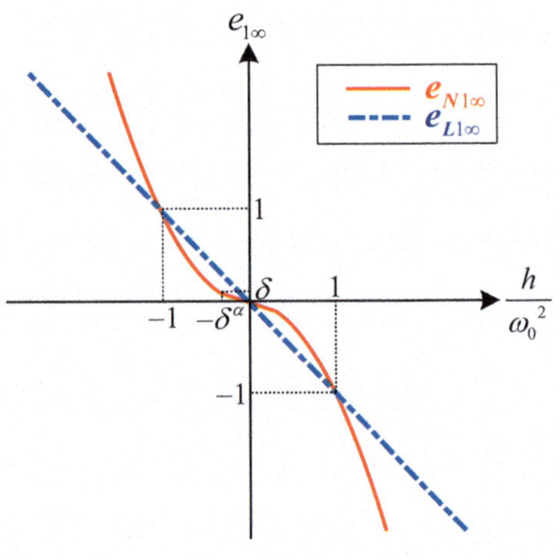

Fig. 9.4 Speed tracking errors comparisons of the conventional LESO and NLESO

1. When $|h| \leq {\omega_0}^2$, whether $e_{N1\infty}$ ultimately locates in the linearity or nonlinearity, the employment of conventional NLESO helps for smaller state errors.
2. When $|h| > {\omega_0}^2$, $e_{N1\infty}$ eventually falls in the nonlinearity. The states tracking accuracy of LESO is much higher, compared with that of conventional NLESO. Further, the greater the absolute value of h, the more prominent the advantages of LESO.

From the perspective of states tracking accuracy, LESO and conventional NLESO have their own optimal range of disturbance to deal with. How to integrate the two and give full play to their respective advantages are worth considering.

9.2.2 Enhanced NLADRC Strategy

In order to facilitate the state errors to converge quickly, a linear function works as a candidate when the performance of power function deteriorates, and an offset is added for continuity and smoothness [10]. Hence, the enhanced nonlinear function efal(e_1, α, δ) can be established as follows:

$$\mathrm{efal}(e_1,\alpha,\delta) = \begin{cases} \dfrac{e_1}{\delta^{1-\alpha}}, & |e_1| \leq \delta \\ |e_1|^{\alpha}\,\mathrm{sgn}\,(e_1), & \delta < |e_1| < \alpha^{\frac{1}{1-\alpha}}\,. \\ e_1 + \left(\alpha^{\frac{\alpha}{1-\alpha}} - \alpha^{\frac{1}{1-\alpha}}\right)\mathrm{sgn}\,(e_1), & |e_1| \geq \alpha^{\frac{1}{1-\alpha}} \end{cases} \tag{9.19}$$

In particular, when α is chosen as 0.5, efal(e_1, 0.5, δ) is embodied as follows:

$$\mathrm{efal}(e_1,0.5,\delta) = \begin{cases} e_1\delta^{-0.5}, & |e_1| \leq \delta \\ \sqrt{|e_1|}\,\mathrm{sgn}\,(e_1), & \delta < |e_1| < 0.25\,. \\ e_1 + 0.25\,\mathrm{sgn}\,(e_1), & |e_1| > 0.25 \end{cases} \tag{9.20}$$

Figure 9.5a shows the contrastive images of linear function e_1, conventional nonlinear function fal(e_1, 0.5, δ), and enhanced nonlinear function efal(e_1, 0.5, δ). Based on state error equation, the speed tracking error in steady state is the solution of $f(e_1) = -h/{\omega_0}^2$. Therefore, the distance between the function and the vertical axis reflects steady-state tracking accuracy. As can be seen from Fig. 9.5a, compared with the other two conventional functions, efal(e_1, 0.5, δ) is closest to the vertical axis. As a result, it improves the tracking accuracy effectively in steady state.

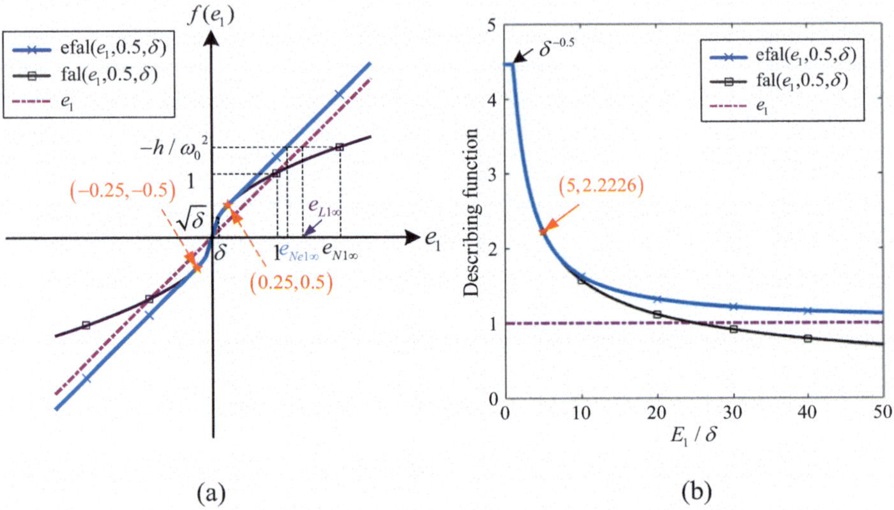

Fig. 9.5 Comparisons of linear function e_1, conventional nonlinear function fal(e_1, 0.5, δ), and enhanced nonlinear function efal(e_1, 0.5, δ). (**a**) Function imagines comparisons. (**b**) Describing functions comparisons

Based on the principle of minimum mean square error, the describing function of the enhanced nonlinear function is first established. Considering the input is sinusoidal as $e_1 = E_1 \sin \omega t$, the fundamental component through the nonlinearity is used to approximate the actual one. Then, the describing function of efal(e_1, 0.5, δ) can be derived. It is worth noting that efal(e_1, 0.5, δ) is an odd function with single value, so the describing function is only the amplitude dependent. The contrastive describing functions of the conventional functions and the enhanced nonlinear function are shown in Fig. 9.5b. As can be seen, the describing function of efal(·) is above that of e_1 or fal(·), which indicates the larger equivalent gain and the better tracking performance in the same case.

To reduce the disturbance estimation burden and further improve the capability of disturbance rejection, the cascaded NLESO is employed for estimating the total disturbances in real time [10]. Based on the enhanced nonlinear function efal(e_1, 0.5, δ) proposed above, NLESO1 is responsible for the preliminary disturbance estimation, while the remaining disturbance is addressed by NLESO2. The schematic diagram of conventional ADRC is shown in Fig. 9.6. The mathematical models of NLESO1 and NLESO2 are established, respectively, as follows:

$$\begin{cases} e_1 = z_1 - \hat{\omega}_e \\ \dot{z}_1 = z_2 - \beta_1 e_1 + \overline{b} I_s \\ \dot{z}_2 = -\beta_2 \mathrm{efal}(e_1, 0.5, \delta) \end{cases}, \tag{9.21}$$

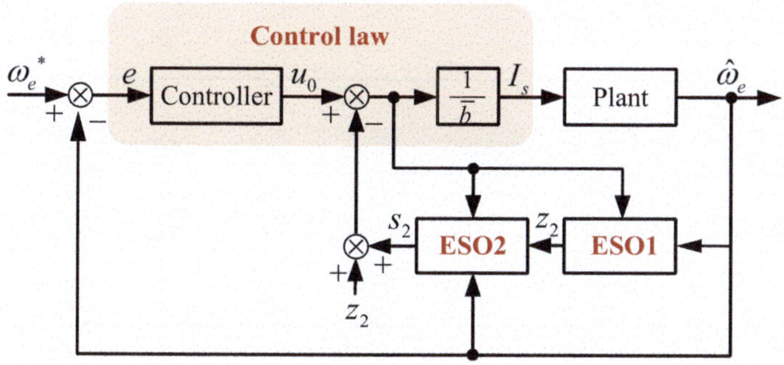

Fig. 9.6 Schematic diagram of enhanced ADRC

$$\begin{cases} e_{s1} = s_1 - \hat{\omega}_e \\ \dot{s}_1 = s_2 - \beta_3 e_{s1} + z_2 + \bar{b} I_s, \\ \dot{s}_2 = -\beta_4 \mathrm{efal}(e_{s1}, 0.5, \delta) \end{cases} \tag{9.22}$$

where z_1 and s_1 are all used for tracking the estimated speed $\hat{\omega}_e$. Accordingly, e_1 and e_{s1} are used to denote the speed tracking error of NLESO1 and NLESO2, respectively. More importantly, z_2 cooperates with s_2 in reconstructing the total disturbances f rapidly and accurately. β_1, β_2, β_3, β_4 are observer gains.

In order to achieve the control objective of the integral system, the disturbance compensation has to be made by the sum of the estimations from the NLESO1 and NLESO2; thus the control law is designed as follows:

$$I_s = \frac{u_0 - (z_2 + s_2)}{\bar{b}}, \tag{9.23}$$

where u_0 is a proportional controller expressed in (9.14).

9.2.3 Experimental Results

The proposed enhanced NLADRC strategy is applied in the HF square-wave signal injection-based sensorless PMSM drive and verified via an experimental platform with a 2.2-kW PMSM, which is consistent with the platform shown in Fig. 3.18.

The disturbance estimation performance of ESO is crucial to the quality of ADRC. Therefore, it is evaluated first with the actual speed and rotor position. When PMSM operates at 100 r/min with a ramp rated-load torque in 0.5 s, the disturbance estimation performance comparisons of conventional ESOs, NLESO with efal(·), and cascaded NLESO are carried out, as shown in Fig. 9.7. Figure 9.7a,

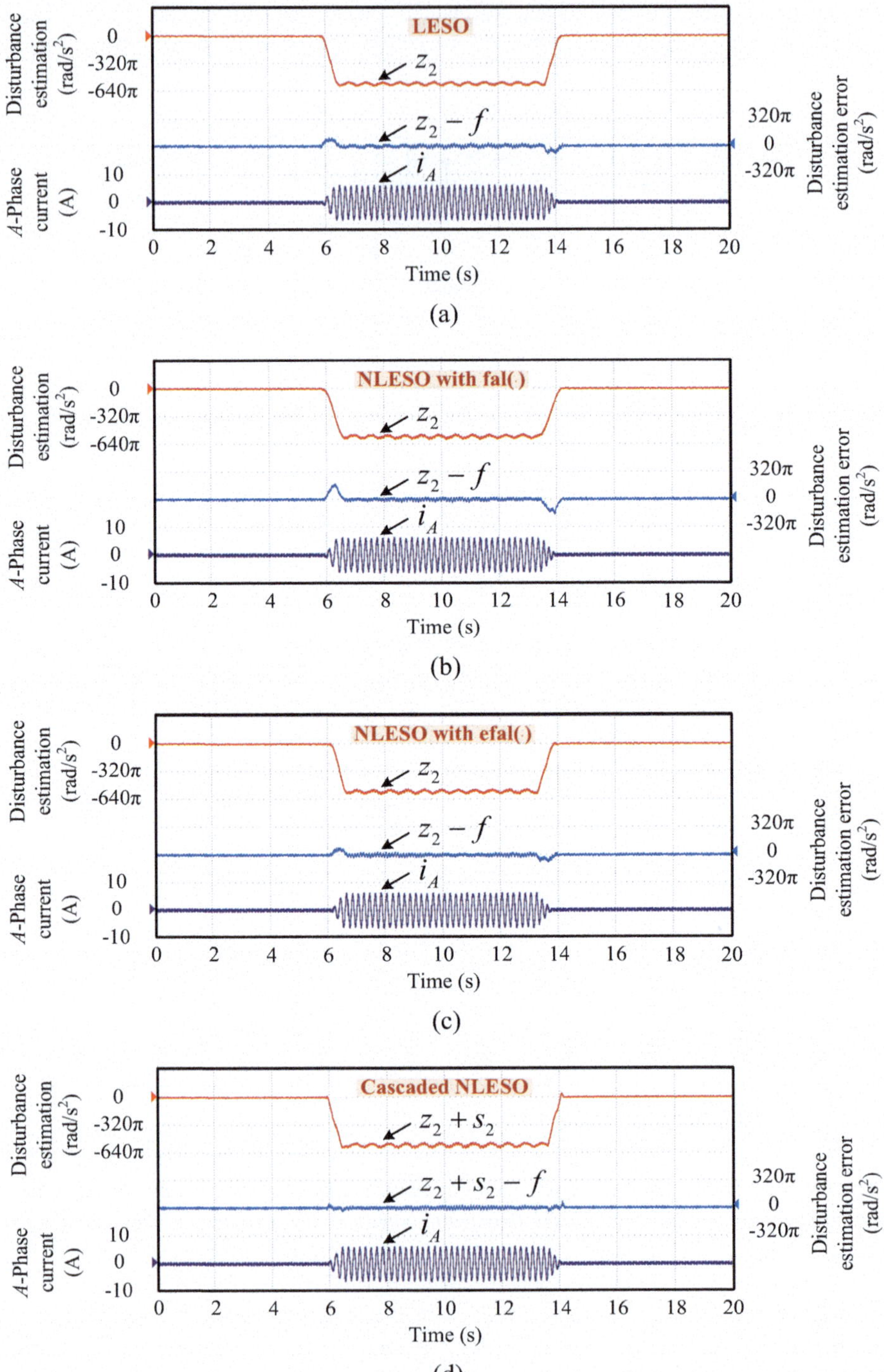

Fig. 9.7 Disturbance estimation performance comparisons with a ramp rated-load torque in 0.5 s at 100 r/min operating condition. (**a**) LESO. (**b**) NLESO with conventional nonlinear function fal(·). (**c**) NLESO with proposed nonlinear function efal(·). (**d**) Cascaded NLESO with proposed nonlinear function efal(·)

and b show estimation and estimation error of disturbance when adopting LESO and conventional NLESO, respectively. It can be seen that the maximum disturbance estimation error of LESO is approximately half that of NLESO when the load torque is applied and removed, which implies the disturbance changes relatively rapidly on this occasion based on the aforementioned theoretical basis. Obviously, LESO is more tolerant of disturbance in such cases. However, the ripple of the disturbance estimation error is quite small with the adoption of conventional NLESO under full load. It is proved that conventional NLESO is better at coping with the disturbance that changes tardily. Moreover, the experimental results employing NLESO with proposed nonlinear function efal(·) are shown in Fig. 9.7c. As can be observed, it combines the advantages of LESO and conventional NLESO and performs well in the whole operation condition. The maximum disturbance estimation error in Fig. 9.7c decreases by 14%, compared with LESO. Figure 9.7d shows the experimental waveforms with the adoption of cascade NLESO. As can be seen, the steady-state error of disturbance estimation is eliminated completely. Further, the rapid and accurate estimation of disturbance is pulled off.

Figure 9.8 shows the robustness comparisons of conventional ADRCs, NLADRC with efal(·), and enhanced NLADRC when the PMSM operates at 100 r/min with a ramp rated-load torque in 0.5 s. With the adoption of conventional LADRC and NLADRC, the actual speed and speed estimation are shown in Fig. 9.8a, and b, respectively. A conclusion consistent with that from Fig. 9.7a, and b can be drawn that LADRC has less speed fluctuation when loading or unloading, which is about half that of NLADRC, while the speed ripple of NLADRC at full load is comparatively unconspicuous. Compared with Fig. 9.8a, and b, it can be seen from Fig. 9.8c that the NLADRC with proposed nonlinear function efal(·) is more tolerant to the disturbance, and the system exhibits stronger robustness in the entire operating state. However, there is a speed offset in the loading and unloading of Fig. 9.8a–c in the presence of steady-state disturbance estimation errors. To solve the problem better, the enhanced NLADRC based on cascaded NLESO is employed, and the experimental results are shown in Fig. 9.8d. As can be observed, the enhanced NLADRC-based system eliminates the speed offset effectively and shows strong robustness to the disturbances. All this can be attributed to the rapid and accurate estimation of the disturbances via the cascaded NLESO.

9.3 Summary

The practical issues for the application of the PMSM drives system are reviewed in this section. First, the MTPA control strategy is introduced to improve the efficiency of the entire system. And then, the ADRC control strategy are further discussed to enhance the dynamic performance of the PMSM sensorless control.

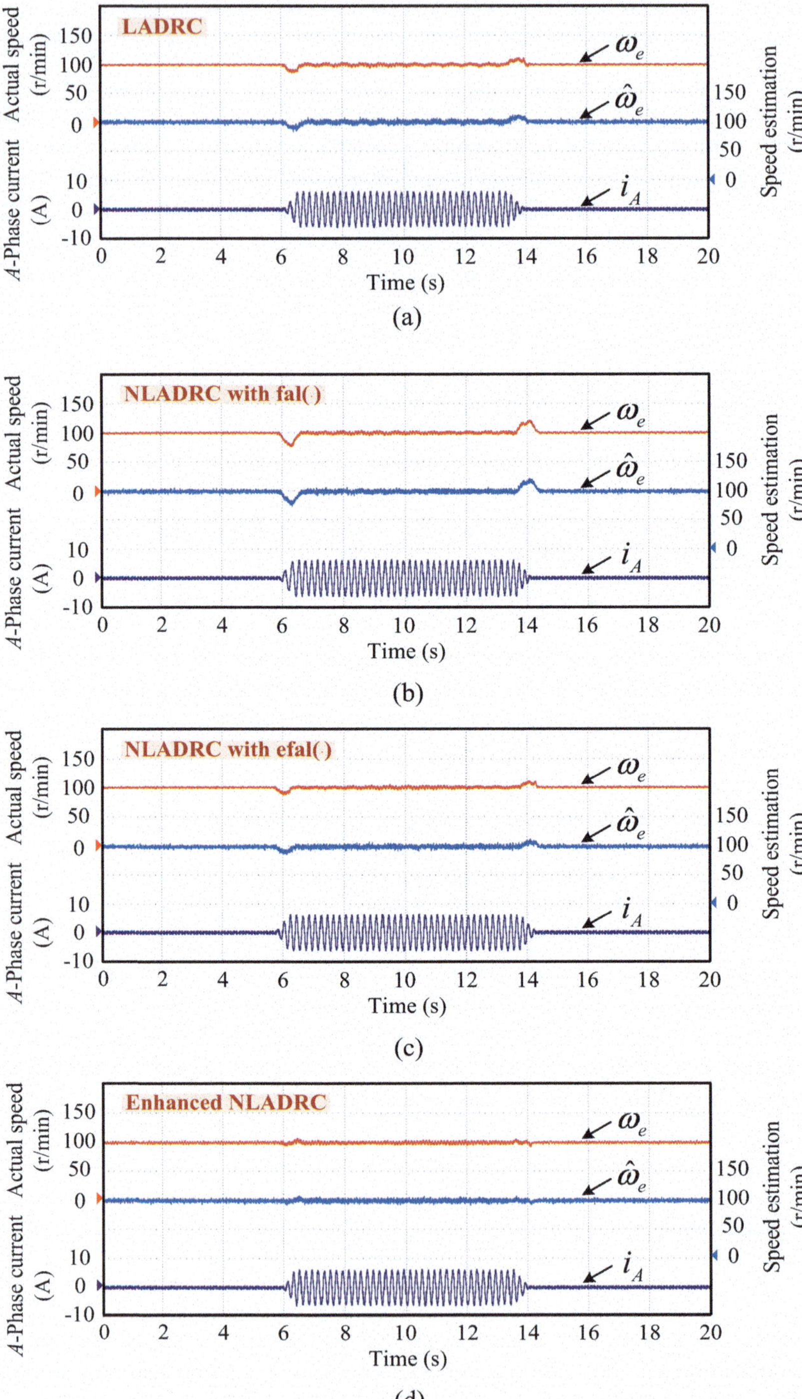

Fig. 9.8 Robustness comparisons with a ramp rated-load torque in 0.5 s at 100 r/min operating condition. (**a**) LADRC. (**b**) NLADRC with conventional nonlinear function fal(·). (**c**) NLADRC with proposed nonlinear function efal(·). (**d**) Enhanced NLADRC based on cascaded NLESO

References

1. G. Wang, Z. Li, G. Zhang, Y. Yu, D. Xu, Quadrature PLL-based high-order sliding-mode observer for IPMSM sensorless control with online MTPA control strategy. IEEE Trans. Energy Convers. **28**(1), 214–224 (March 2013)
2. S. Bolognani, R. Petrella, A. Prearo, L. Sgarbossa, Automatic tracking of MTPA trajectory in IPM motor drives based on AC current injection. IEEE Trans. Ind. Appl. **47**(1), 105–114 (2011)
3. S. Kim, Y. D. Yoon, S. K. Sul, K. Ide, Parameter independent maximum torque per ampere (MTPA) control of IPM machine based on signal injection. In *Applied Power Electronics Conference and Exposition*, 2010, pp. 103–108
4. C.B. Butt, M.A. Hoque, M.A. Rahman, Simplified fuzzy-logic-based MTPA speed control of IPMSM drive. IEEE Trans. Ind. Appl. **40**(6), 1529–1535 (2004)
5. A.R.I. Mohamed, T.K. Lee, Adaptive self-tuning MTPA vector controller for IPMSM drive system. IEEE Trans. Energy Convers **21**(3), 636–644 (2006)
6. C.T. Pan, S.M. Sue, A linear maximum torque per ampere control for IPMSM drives over full-speed range. IEEE Trans. Energy Convers. **20**(2), 359–366 (2005)
7. G. Wang, D. Xiao, G. Zhang, C. Li, X. Zhang, D. Xu, Sensorless control scheme of IPMSMs using high-frequency orthogonal square-wave voltage injection into stationary reference frame. IEEE Trans. Power Electron.. https://doi.org/10.1109/TPEL.2018.2844347
8. S. C. Yang, R. D. Lorenz. Surface permanent-magnet machine self-sensing at zero and low speeds using improved observer for position, velocity, and disturbance torque estimation. IEEE Trans. Ind. Appl. **48**(1), 151–160 (January/February 2012)
9. B. Du, S. Wu, S. Han, S. Cui. Application of linear active disturbance rejection controller for sensorless control of internal permanent-magnet synchronous motor IEEE Trans. Ind. Electron. **63**(5), 3019–3027 (May 2016)
10. G. Wang, R. Liu, N. Zhao, D. Ding, D. Xu, Enhanced linear ADRC strategy for HF pulse voltage signal injection based sensorless IPMSM drives. IEEE Trans. Power Electron.. https://doi.org/10.1109/TPEL.2018.2814056
11. J. Han. From PID to active disturbance rejection control. IEEE Trans. Ind. Electron. **56**(3), 900–906, (March 2009)
12. Z. Gao. Scaling and bandwidth-parameterization based controller tuning. Proc. 2003 Am. Control Conf. **6**, 4989–4996 (June 2003)

Nomenclature

α, β	Stationary reference frame axes
d, q	Rotary reference frame axes
R	Stator resistance
L_d, L_q	d- and q-axis inductances
u_d, u_q	d- and q-axis stator voltages
u_α, u_β	α- and β-axis stator voltages
i_d, i_q	d- and q-axis stator currents
i_α, i_β	α- and β-axis stator currents
$\boldsymbol{u}$	Stator voltage vector
$\boldsymbol{i}$	Stator current vector
e_α, e_β	α- and β-axis extended EMF components
ψ_f	Permanent flux linkage
θ_e	Rotor electrical position
ω_e, ω_r	Rotor electrical and mechanical speeds
T_e	Electromagnetic torque
P_n	Pole pairs number
p	Differential operator
s	Laplace transform variable
k_p, k_i	Proportional and integral gains
δ	Sliding-mode boundary layer
z	Sliding-mode control function
$*$	Reference value
^	Estimated value
~	Estimation error
I_s	Amplitude of the stator current
b	Control input gain
$\overline{b}$	Approximation of b
f	Total disturbances
h	Derivative of the total disturbances

G. Wang et al., *Position Sensorless Control Techniques for Permanent Magnet Synchronous Machine Drives*, https://doi.org/10.1007/978-981-15-0050-3

γ	Torque angle
J	The moment of inertia
T_L	Load torque
B	Viscous friction coefficient
β_{01}, β_{02}	Observer gains
δ	Range of linearity
α	Power of nonlinearity α
u_0	Output of proportional controller
k_p	Proportional gain of P controller
L, N	Conventional LESO and NLESO
ω_0	Observer bandwidth
$e_{1\infty}$, $e_{2\infty}$	Steady-state value
z_1, s_1	Speed tracking of ESO1 and ESO2
e_1, e_{s1}	Speed tracking error of ESO1 and ESO2
e_2	Disturbance estimation error
Superscript "r"	Rotor reference frame
Subscript "h"	High-frequency component
Subscript "R"	Random signal
Subscript "pu"	Normalized signal
$\boldsymbol{R}(\theta)$	Transformation matrix
J_O	Actual inertia
J_I	Measured inertia
$\Delta\theta$	Rotor position estimation errors

The manufacturer's authorised representative in the EU is Springer Nature Customer Service Centre GmbH, Europaplatz 3, 69115 Heidelberg, Germany. If you have any concerns regarding our products, please contact ProductSafety@springernature.com

Printed and bound by CPI Group (UK) Ltd, Croydon, CR0 4YY
30/09/2026
02225247-0003